Mobile Electronic Commerce

Foundations, Development, and Applications

INDUSTRIAL AND SYSTEMS ENGINEERING SERIES

Series Editor
Waldemar Karwowski

PUBLISHED TITLES:

Managing Professional Service Delivery: 9 Rules for Success
Barry Mundt, Francis J. Smith, and Stephen D. Egan Jr.

Laser and Photonic Systems: Design and Integration
Shimon Y. Nof, Andrew M. Weiner, and Gary J. Cheng

Design and Construction of an RFID-enabled Infrastructure: The Next Avatar of the Internet
Nagabhushana Prabhu

Cultural Factors in Systems Design: Decision Making and Action
Robert W. Proctor, Shimon Y. Nof, and Yuehwern Yih

Handbook of Healthcare Delivery Systems
Yuehwern Yih

FORTHCOMING TITLES:

Mobile Electronic Commerce: Foundations, Development, and Applications
June Wei

Supply Chain Management and Logistics: Innovative Strategies and Practical Solutions
Zhe Liang, Wanpracha Art Chaovalitwongse, and Leyuan Shi

Nanomaterials: Advances in Manufacturing Processes and Multi-functional Applications
Ben Wang and Zhiyong Liang

Manufacturing Productivity in China
Li Zheng, Simin Huang, and Zhihai Zhang

Mobile Electronic Commerce

Foundations, Development, and Applications

Edited by June Wei

CRC Press is an imprint of the
Taylor & Francis Group, an **informa** business

CRC Press
Taylor & Francis Group
6000 Broken Sound Parkway NW, Suite 300
Boca Raton, FL 33487-2742

Printed on acid-free paper
Version Date: 20140909

International Standard Book Number-13: 978-1-4665-9090-8 (Hardback)

Visit the Taylor & Francis Web site at
http://www.taylorandfrancis.com

and the CRC Press Web site at
http://www.crcpress.com

To my dear husband, Hongbin, for his everlasting support.

Contents

Preface

Mobile electronic commerce technologies and solutions have already influenced critical aspects of many sectors and the quality of people's lives. This book seeks to answer how industrial and systems engineering as well as information communication technology, such as systems integration, process simulation, mobile-based collaboration, wireless technology, and human–computer interface, among others, can address and accelerate the solutions of mobile electronic commerce challenges in society, economy, organization, and government. This inspires new concepts, theories, models, tools, and methods to further improve quality of people's lives. This book also addresses implications for industry, academia, scientists, engineers, professionals, and students to develop and apply cutting-edge innovative mobile electronic technologies and emerging mobile electronic commerce systems in the near and long term. The impacts of mobile electronic commerce innovations and solutions on the development and applications in the areas of society, economics, culture, organizations, government, industries, and individual's daily life are also addressed.

The role of mobile electronic commerce technologies and systems on innovation in cutting-edge solutions and emerging systems is a cross-disciplinary topic across academia and industry. This book brings together researchers from academia and industry across disciplines to integrate and synthesize the role for professionals, academics, researchers, and managers in the field of mobile electronic commerce. This book also brings together chapters written by a multidisciplinary group of experts in order to understand the ways in which mobile electronic commerce systems and technologies influence mobile electronic commerce innovations and integrated solutions for their development and application. The book aims to stimulate new thinking in the development and application of mobile technology that helps professionals, academic educators, and policy-makers working in mobile electronic commerce field to disseminate cutting-edge knowledge. Throughout the book, particular emphasis in laid on the incorporation of the emerging research and development and application frontiers that provide promising contributions to mobile electronic commerce. These include new theories, state-of-the-art mobile electronic commerce innovations, cutting-edge mobile information and communication technology infrastructures, and emerging mobile electronic systems.

To balance representative authors and researchers from academia and industry, as well as authors globally, the book also intends to overcome cultural barriers to accelerate mobile technological and mobile commerce system changes in the global economy. It also covers the whole spectrum of methods, tools, and guidelines for designing mobile electronic commerce systems and services in different cultures. Finally, it summarizes the highlights and discusses emerging frontiers such as challenges and future research opportunities in mobile electronic commerce technologies, systems, and innovations.

Editor

June Wei, PhD, is a professor in the Department of Management and Management Information Systems, University of West Florida. She earned her PhD from Purdue University, a master's degree from Georgia Institute of Technology, and another master's degree from Zhejiang University. Her research interests cover electronic supply chain management, information security and privacy in mobile electronic commerce, and mobile electronic commerce systems development. She is editor-in-chief of three peer-reviewed leading journals on information systems: *International Journal of Mobile Communications, Electronic Government, an International Journal*, and *International Journal of Electronic Finance*. She is also an associate editor of three peer-reviewed journals: *Journal of Electronic Commerce Research, Journal of Information Privacy and Security*, and *International Journal of E-Entrepreneurship and Innovation*. She teaches courses on business information systems development, systems design and analysis, operations management problems, and electronic business fundamentals. She has published over 120 peer-reviewed papers, 55 peer-reviewed journal articles, and three book chapters. Dr. Wei has won many awards for her research, including the Distinguished Writing Award of the Foundation for Information Technology Education and the Education Special Interest Group of the Association of Information Technology Professionals (2004 and 2005) and the Best Paper Award sponsored by Emerald Publisher in International Conference on Accounting, Business, Leadership, and Information Management (2010). She has been recognized at a Florida Board of Trustees meeting (2011). She has also been recognized by the University of West Florida with the Distinguished Research and Creative Activities Award (2010) and the Distinguished Teaching Award (2012), and the Dyson Award for Excellence in Research from the College of Business (2005, 2007, 2009, 2011, and 2013). She also received educational awards and funding from Marconi Communications, Inc., and Bausch & Lomb. She is currently president of the Southwest Decision Sciences Institute.

Contributors

Preeti Aggarwal
Ansal University
Haryana, India

Shilpa Bahl
Ansal University
Haryana, India

Raffaello Balocco
Department of Management, Economics and Industrial Engineering
Politecnico di Milano
Milan, Italy

Richard Boateng
Operations and Management Information Systems Department
University of Ghana Business School
Accra, Ghana

Inés Küster Boluda
Department of Marketing
University of Valenci
Valencia, Spain

Joseph Budu
Operations and Management Information Systems Department
University of Ghana Business School
Accra, Ghana

Manmohan Chaturvedi
Ansal University
Haryana, India

Te Fu Chen
Department of Business Administration
Lunghwa University of Science and Technology
Guishan Shiang, Taiwan, Republic of China

Edward T. Chen
Operations and Information Systems Department
University of Massachusetts Lowell
Lowell, Massachusetts

Kasun De Zoysa
University of Colombo School of Computing
Colombo, Sri Lanka

Abel E. Ezeoha
Department of Banking and Finance
Ebonyi State University
Abakaliki, Nigeria

Mariana Florea
Department of Decision and Information Science
Stetson University
DeLand, Florida

Christian Damián García
Department of Marketing
University of Valencia
Valencia, Spain

Antonio Ghezzi
Department of Management, Economics and Industrial Engineering
Politecnico di Milano
Milan, Italy

Sam S. Gill
San Francisco State University
San Francisco, California

Carlos F. Gomes
School of Economics
Institute of Systems and Robotics
University of Coimbra
Coimbra, Portugal

Amila Karunanayake
University of Colombo School of Computing
Colombo, Sri Lanka

Mark R. Leipnik
Department of Geography and Geology
Sam Houston State University
Huntsville, Texas

James Scott Magruder
Department of Finance, Real Estate, and Business Law
The University of Southern Mississippi
Hattiesburg, Mississippi

Sathiadev Mahesh
Department of Management
University of New Orleans
New Orleans, Louisiana

Sapna Malik
Ansal University
Haryana, India

Sanjay S. Mehta
Department of Management and Marketing
Sam Houston State University
Huntsville, Texas

Harsh Mishra
Department of Journalism and Creative Writing
Central University of Himachal Pradesh
Himachal Pradesh, India

Pradeep Nair
Department of Mass Communication and Electronic Media
Central University of Himachal Pradesh
Himachal Pradesh, India

Wei Ning
Division of International Business and Technology Studies
Sanchez School of Business
Texas A&M International University
Laredo, Texas

Anselm Nkalemu
Department of Banking and Finance
Ebonyi State University
Abakaliki, Nigeria

Vijayaprabha Rajendran
Department of Geography
and Geology
Sam Houston State University
Huntsville, Texas

Andrea Rangone
Department of Management,
Economics and Industrial
Engineering
Politecnico di Milano
Milan, Italy

Carla Ruiz-Mafe
Department of Marketing
University of Valencia
Valencia, Spain

Electra Safari
Department of Product and
Systems Design Engineering
University of the Aegean
Syros, Greece

Cezar Scarlat
Department of Management
Doctoral School of
Entrepreneurship, Business
Engineering and Management
University "Politehnica" of
Bucharest
Bucharest, Romania

Norazah Mohd Suki
Labuan Faculty of International
Finance
Universiti Malaysia Sabah, Labuan
International Campus, Universiti
Malaysia Sabah
Labuan, Malaysia

Norbayah Mohd Suki
Faculty of Computing and
Informatics
Universiti Malaysia Sabah, Labuan
International Campus, Universiti
Malaysia Sabah
Labuan, Malaysia

Pedro M. Torres
School of Economics
University of Coimbra
Coimbra, Portugal

Rakhi Tripathi
FORE School of Management
New Delhi, India

Nils Walravens
iMinds, Centre for Studies
on Media, Information and
Telecommunication
Vrije Universiteit Brussel
Brussels, Belgium

Haibo Wang
Division of International Business
and Technology Studies
Sanchez Jr. School of Business
Texas A&M International
University
Laredo, Texas

Wei Wang
Division of International Business
and Technology Studies
Sanchez School of Business
Texas A&M International
University
Laredo, Texas

Joseph M. Woodside
Department of Decision and Information Science
Stetson University
DeLand, Florida

Raluca-Andreea Wurster
Doctoral School of Entrepreneurship, Business Engineering and Management
University "Politehnica" of Bucharest
Bucharest, Romania

and

International Management at FOM University of Applied Sciences
Essen, Germany

Mahmoud M. Yasin
Department of Management and Marketing
East Tennessee State University
Johnson City, Tennessee

Dimitrios Zissis
Department of Product and Systems Design Engineering
University of the Aegean
Syros, Greece

section one

Mobile electronic commerce foundations and theories

chapter one

Enhancing the effectiveness of mobile electronic commerce strategy

A customer orientation approach

Mahmoud M. Yasin, Pedro M. Torres, and Carlos F. Gomes

Contents

Learning objectives

Upon completing this chapter, the student should be able to have an understanding of the following:

1. The emerging opportunities in the mobile electronic commerce competitive environment
2. The nature of business organizations as open systems and their reasons for implementing and effectively utilizing mobile electronic commerce
3. The organizational modifications needed toward an effective implementation process of mobile electronic commerce
4. The role of performance management toward improving organizational performance in the mobile electronic commerce environment
5. The importance of the customer orientation in the mobile electronic commerce environment
6. The operational and strategic outcomes of mobile electronic commerce in achieving the customer orientation and an organizational strategic competitive advantage

1.1 Introduction

In recent years, the landscape of the competitive environment has been colored by a host of digital-based technologies, strategies, and competitive methods, which are aimed at enhancing the customer orientation, operational efficiency, and strategic competitiveness. In this context, mobile electronic commerce (m-commerce) and electronic commerce (e-commerce), among others, are becoming very popular as components of the growing and upcoming digital competitive strategy wave.

As a response to the unmistakable and growing digital competitive pressures, today's open system organizations are finding it to be a necessity, rather than a mere luxury, to jump on the digital strategic bandwagon. In this context, many organizations are going digital just to join the crowd. As such, these organizations do not want to be left behind. In recent years, organizations operating in e-business cultures, as well as those operating in more traditional business cultures, have been attempting to enhance their customer orientation through the adoption of some sort of the digital approach. Such approach includes the applications of m-commerce and e-commerce, among other e-technologies aimed at the customer.

While the potential benefits of well-designed digital-based strategic approaches, such as m-commerce and e-commerce, have not been questioned, achieving such benefits has been a subject of concern among practicing managers and scholars alike. In their rush to join the e-crowd, many organizations have utilized a me-too approach to the implementation of m-commerce and e-commerce. These organizations are lacking the systematic and integrated strategy needed to realize the potential benefits of digital technologies strategically. In essence, these organizations do not have a complete, well-integrated digital strategy. In this context, a piecemeal approach, coupled with me-too orientation most likely will result in ineffective implementation of strategic e-solutions, such as m-commerce and e-commerce. This leads to investing in attractive technologies without having well-defined strategy to use such technologies strategically. This approach to the implementation of digital-based technologies, such as m-commerce and e-commerce, among others, is bound to fail, as it lacks the strategic focus needed to support the customer orientation. To avoid wasting resources and efforts, organizations must approach the process of implementation and utilization of these technologies strategically and systematically. This process must be guided by a well-defined, systematic, and integrated strategy. To facilitate such an orientation, organizations must address the following questions and concerns, prior to going digital:

1. Why do we want to implement digital-based strategy, such as m-commerce or e-commerce strategy?
2. What existing systems, processes, cultural elements, and relations with customers and suppliers must be reengineered or modified, prior to the implementation phase?
3. What are the important factors, approaches, and methods that must be considered during the implementation process?
4. How are we going to assess and manage the performance of the implemented m-commerce and e-commerce strategic initiatives? What measures (outcomes) and measurement approaches should our performance management system have in place?
5. How are we going to gauge, monitor, and continuously improve the customer orientation, and other operational and strategic goals resulting from the utilizations of the implemented strategy?

The conceptual framework in Figure 1.1 will be utilized to guide the material covered in this chapter. It is to be noted that the material presented in this chapter is based on the findings of studies, which included e-business cultures environments such as the United States, as well as more traditional business cultures, such as the case of Arabian Gulf States and Iran. Also included are findings from the Portuguese business culture, which could be considered somewhat between the e-business cultures and the traditional business cultures. This chapter attempts to enhance our understanding of the relevant issues pertaining to the theory and practice of effective implementation of digital strategies. The reader is referred to the studies performed by the authors and utilized in this chapter (Czuchry and Yasin 2000, 2001, 2003, 2007; Czuchry et al. 2001, 2002; Sallmann et al. 2004;

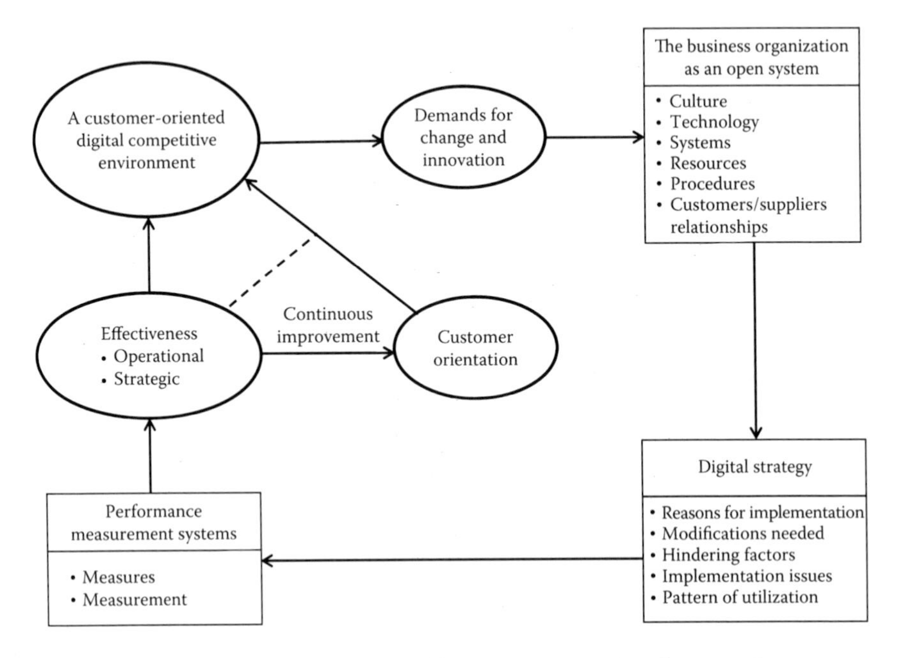

Figure 1.1 A systematic approach to effective e-strategy implementation.

Yasin et al. 2006, 2007, 2010, 2011, 2014; Ashrafi et al. 2007; Yasin and Yavas 2007; Torres et al. 2011). The work of other scholars used in this chapter is also cited and included in the reference list.

1.2 Digital competitive environment

1.2.1 Background

Digital technology, as applied in different organizational cultures, is an evolving and dynamic process. The introduction of the Internet and associated web technologies launched an ever changing applications of such terminologies. While there is a growing jungle of jargon and definitions related to applications and technologies, the potential benefits of these digital technologies are far reaching. This is especially true if these technologies are integrated into a well-designed digital strategy.

Mobile commerce (m-commerce) is often considered a subset of e-commerce, which, in turn, is viewed as a subset of e-business (Jelassi and Enders 2005; Ngai and Gunasekaran 2007). Thus, in this perspective m-commerce is an extension of e-commerce, being similar to the latter, except for the underlying technology, since transactions are conducted wirelessly using a mobile device. Therefore, m-commerce should refer to "any transaction with monetary value that is conducted via a mobile device" (Clarke 2001, p. 133). However, some researchers believe that m-commerce is much more than merely an extension of e-commerce (Chong 2013a), because of the different kinds of interactions with users, usage patterns, and value chains, which make the case for new business models, which are not available on e-commerce, such as location marketing.

1.2.2 *Growth and potential*

Despite the different definitions of m-commerce, one fact is for sure: in recent years, there was a huge increase in the usage of wireless and mobile networks and devices all around the world. The proliferation of Android and iOS devices has led to new mobile usage habits, and the ownership of smartphones has been disseminating and growing over recent years, especially in the young consumers segment. Moreover, there was an explosive increase of online mobile usage in relatively traditional business cultures of Asia and Africa, mostly due to economic reasons, because mobile access responds to a necessity in more efficient way. In fact, mobile devices are nowadays the primary Internet access devices for many individuals, especially for young people. In this context, the convergence of technology trends and the generalization of the broadband Internet access worldwide are changing the marketing game. This represents a business opportunity that most companies should consider in order to improve their competitive position.

Furthermore, the global reach of web technologies enables cost-efficient means of reaching out to new markets, attracting new customers, and delivering products and services (Chatterjee et al. 2002); and business conducted over the Internet changed the way organizations interact with their customers and partners (Jelassi and Enders 2005). Thus, the Internet has accelerated value innovations in service dimensions such as speed, convenience, personalization, and price, thereby changing the customer value proposition (Kalakota and Robinson 2000). The mobile customers have access anywhere and anytime, and they recognize differences between m-commerce and e-commerce. This tends to promote the design of more usable and user-friendly mobile environment (Mahatanankoon et al. 2005).

Moreover, as the digital technologies become more available, customers' power tends to increase because of lower switching costs and higher familiarization with the technology (Porter 2001), and with mobile they gain more flexibility along the time and space dimensions (Balasubramanian et al. 2002).

In this context, considering the opportunities given by mobility and the significance of users' location, in order to determine the drivers for consumer adoption and usage of m-commerce, it is important to understand why and when flexibility is valuable to customers (Mahatanankoon et al. 2005). However, there could be some differences in how convenience is perceived in mobile commerce based on gender. For example, the link between interface design and ease of use may have a special relevance for females (Okazaki and Mendez 2013). Furthermore, it should be noted that users of m-commerce are usually young users and this should be considered when designing applications and strategies (Chong 2013b).

In turn, for corporations, the use of mobile net could give access to enormous customer databases and provide the ability to communicate with individual consumers wherever they might be and whenever they might choose. As a result, corporate managers could enhance their knowledge of consumer preferences and improve their value proposition working on new kinds of products and new ways to market them. For example, retailers have the ability to advertise prices that are time and location sensitive.

Nevertheless, in spite of the possibility of having instantaneous access to one-to-one database marketing, companies should provide true value and utility, and clearly separate digital noise from drawing meaningful attention, in order to succeed in the m-commerce space. Thus, beyond the replication of current e-commerce models to mobile applications, the development of unique m-commerce value propositions rooted in the dimensions of ubiquity, convenience, localization, and personalization could enable companies to achieve a competitive advantage (Clarke 2001).

However, in spite of consumers' growing preference for using apps instead of mobile web, the majority of applications are never used again after being downloaded and some of them are not used more than once. Therefore, considering the rising importance of mobile technologies on business and society, it is important to know which m-commerce user requirements are needed and to explore consumer perception of mobile applications, distinguishing and comparing consumer intention to adopt m-commerce from continual usage intention.

Moreover, although responses to mobile advertising depend on technology-based evaluations (utilitarian considerations) and emotion-based evaluations (hedonic considerations), it is to be noted that mobile advertising is affected by the characteristics of ad communication and by users' voluntary choices of mobile technology (Yang et al. 2013). Furthermore, previous research corroborated that incentive and prior permission influence the responses to mobile advertising, therefore impacting on campaign outcomes (Varnali et al. 2012).

In the final analysis, m-commerce is changing the competitive business landscape and must be considered when formulating business strategies. This is especially true, given its potential and the shifting of consumer and user behavior toward mobile devices utilization, such as smartphones and tablets, when doing business, emailing, searching for locations or information, or entertainment, among other activities.

1.3 *Business organizations as open systems*

In recent times, business organizations moved closer and closer toward the open system of operations and strategy. In the process, these organizations got closer to their customers and suppliers as well as other entities within the competitive environment. In the process, these organizations have undertaken major e-engineering efforts, which focused on processes, procedures, technologies, and strategies, among other organizational aspects. Today's open system organizations are attempting to enhance their customer orientation through the implementation and utilization of digital-based strategies such as m-commerce and e-commerce. In this context, innovative telecommunication technologies have contributed significantly to breaking the barriers between the organizations and their environment.

Despite the ability of the open system organizations to read the competitive environment changes, threats, and opportunities, they must be able to reengineer themselves accordingly. In this context, the open system organizations must have an organic structure, which is conducive to innovation and learning. Such organizational characteristics are required toward the effective adoption of digital-based strategies and applications. In such organizations, cultures, systems, procedures, and linkages with customers and suppliers are relatively easier to modify and reengineer in preparation for new technologies to become an integral component of the business digital strategies. In the final analysis, this is required to transfer technology to effective strategy.

1.4 *Digital strategy*

1.4.1 *Reasons behind mobile electronic commerce implementation*

Research has shown that top management typically tends to champion the initiation of the m-commerce effort. However, other entities such as customers, suppliers, and departments within the organization might also be the initiation agent behind the m-commerce effort. The same entities have also been found to be active in the planning for the m-commerce implementation initiative.

Specifically, in relation to reasons behind the implementation effort, several reasons, related to the industry, competition, suppliers and customers, were mainly cited among the reasons behind the e-commerce effort of the organization. It is to be noted that these reasons tended to be similar, regardless of the prevalent business culture. The motivation behind e-commerce implementation was strategic. It was aimed at improving the customer orientation, which is essential toward achieving and sustaining the competitive advantage. In addition, the reasons behind implementation are operational efficiency, suppliers and customers relationships, employee development, and effective external and internal communication.

1.4.2 *Modifications required for mobile electronic commerce implementation*

Organizational modification efforts prior to the implementation process tended to emphasize reengineering of the different aspects of the organization. These reengineering efforts included systems, procedures, policies, processes related to internal operations and human resource training, as well as relationships and agreements with customers and suppliers. To facilitate this new orientation, these organizations had to make technological-related investments. Again, for the most part, these modifications appeared to be similar, regardless of the business culture involved. One exception stands out. Organizations operating in traditional business cultures tended to invest more in employee training, upgrading and integrating, and/or even replacing existing technologies. Perhaps this reflects the difference in e-readiness among organizations operating in different technological and business cultures. On the other hand, US organizations paid more attention to the integration of different technologies and establishing stronger linkages with customers and suppliers. In this context, US organizations were emphasizing more integrated systems and solutions, while their counterparts in traditional-based cultures were approaching the e-commerce effort incrementally and discreetly.

Recent studies suggested that, although business intelligence (BI) and information systems (IS) applications could make m-commerce more powerful, there is clearly a noticeable gap between users' expectations with regard to the integration of BI and IS into m-commerce implementation and the actual situation (Xu and Yang 2012).

1.4.3 *Hindering factors of mobile electronic commerce implementation*

The factors that tended to hinder the e-commerce implementation process appeared to be both technical and nontechnical in nature. In most cases, these factors were related to the organization itself. They included factors such as lack of qualified staff, rushing the implementation process, and the lack of technological resources. In addition, the lack of cooperation from suppliers and customers was a major hindering factor of the implementation effort. Moreover, the lack of planning and top management support both appeared to be important hindering agents of the effective e-commerce implementation process. The lack of qualified employees and technological resources tended to be more vivid as hindering factors in the case of organizations of traditional business cultures relative to their US counterparts.

1.4.4 Implementation issues

Once the digital strategy is formulated, it must be implemented. Some organizations find a top-down approach as more effective, while others utilize a bottom-up approach to implementation more effectively. Either way, cross-functional teams representing the different departments must be involved in the actual implementation. In some cases, organizations resort to consultants to manage the implementation process. Project management tools are often utilized to ensure an on-time implementation. In addition, the organization must decide to determine which change strategy to follow as it changes over from the old to the new system. In this context, some organizations use a cutcover strategy, while others choose a phased approach strategy. It is to be noted that while top management support is essential in championing the case for digital strategy, top management should not get into the details of specialized implementation. That should be left for the experts.

1.4.5 Patterns of utilization of mobile electronic commerce

The actual organizational utilization of e-commerce after implementation was found to be associated with several factors. These factors tended to stress the organizational use of e-commerce to strengthen internal and external communication. In addition, the use of e-commerce contributed to helping the organization reach new customers and markets. In the process, it facilitated the management of the purchasing activities. The improvements in internal and external communication appeared to be a significant by-product of the e-commerce implementation. This increased organizational connectivity, both internally and externally. The enhanced connectivity facilitated strategic gains in terms of new markets and customers. In addition, e-commerce often is used to strengthen strategic alliances with suppliers. Also, applications of e-commerce tended to improve the customer orientation, through improving customer service and satisfaction.

The benefits or outcomes due to e-commerce implementation appeared to be both strategic and operational in nature. In general, organizations reported improved customer/supplier and employee relations due to the implementation of e-commerce. Overall, it appeared that the e-commerce philosophy and related technologies tended to present similar opportunities and challenges to organizations operating in e-business cultures as well as traditional business cultures. Therefore, organizations operating in traditional business cultures stand to benefit from organized and systematic benchmarking efforts of their counterparts in e-business cultures before launching the e-commerce business model. The lack of the appropriate technologies and qualified human resources in organizations operating in traditional business cultures appear to present serious challenges. In this context, employee training should help. However, the lack of the appropriate technological hardware and software represents a long-term problem. This solution of such problem requires immediate and long-term investments.

Benchmarking the e-commerce practices of developing countries might reduce the learning cycle for organizations operating in more traditional business cultures. Such benchmarking effort should eliminate the waste associated with the piecemeal and the trial and error approaches to e-commerce implementation and utilization.

Our findings, coupled with several recent studies published in reputable journals, suggested several factors that could improve the effectiveness of e-commerce strategy. Further qualitative and quantitative research is necessary to validate and refine the art of the effective implementations of m-commerce strategy. For example, in the case of Iran, information technology (IT) readiness, or lack of IT, is a very important hindering factor for Iranian small and medium enterprises (SMEs) (Ghobakhloo et al. 2011). Also, a systematic approach to technical innovation that simultaneously develops both the e-commerce business model and IT is lacking. The technological innovation–related investments could be used as motivating agents toward more progress in relation to e-commerce utilization (Bagherinejad 2006). However, without an implementation framework that includes factors that influence customer online buying behaviors (Ghasemaghaei et al. 2009), business benefits will be difficult to obtain. This current lack of understanding of e-commerce benefits is evident in traditional business cultures, such as in the case of Iran due to the lack of appropriate education of the customers specifically in rural areas (Jalali et al. 2011).

The educational component could have significant benefits when deploying an improved e-commerce strategy in traditional business cultures. The educational approach combined with technology innovations tends to create *intangible value*. Huarng and Yu (2011) asserted that such an approach could promote collaboration and cooperation throughout the value chain and persuade more SMEs to pursue e-commerce. Finally, the e-commerce strategy should stress removing some of the infrastructure barriers, which contributes to lackluster implementation of e-commerce strategies among the SMEs (Sarlak and Hastiani 2008).

Additional effort is necessary to finalize a framework for improved e-commerce deployment. Such a framework will promote systematic benchmarking in the road toward effective e-commerce implementation. The frameworks presented in this chapter are a modest step in that direction. Others, including Fathian et al. (2008), Bagherinejad (2006), and Amiri and Salarzehi (2010), attempted to offer their perspectives of such frameworks. The two conceptual frameworks offered in this chapter are attractive as they are logical, simple, and practical in nature.

1.5 *Performance measurement systems*

A performance measurement system (PMS) should be designed to manage the multifaceted aspects of the organizational performance. In this context, it should be integrated and balanced. Such a system must be able to measure, track, monitor, and improve specific operational measures and outcomes. In general, two types of organizational performance evaluation platforms are needed in order to have an effective and dynamic PMS, which has a broader organizational perspective on performance, which include and integrate all the specific nature of key performance areas in the organization (Gomes et al. 2004, 2007, 2011; Gomes and Yasin 2013).

Platform A is designed to gauge the organization's competitive efforts in response to market tendencies. On the other hand, platform B is more closely tied to the organizational structure in order to support and maintain an effective operational culture (Gomes and Yasin 2011).

The first evaluation, platform A, has a more global, corporate management orientation. As such, this platform focuses mainly on a few performance measures that reflect critical organizational performance dimensions. The emphasis of this platform is on the effective flow of products/services to markets. The measures used in this platform must be directly related to the strategic objectives of the organization. This platform should incorporate and support both organizational effectiveness measurement and competitive external benchmarking efforts.

The second evaluation, platform B, maintains a measure-specific perspective. This platform defines the relationship between specific measures and the organizational unit responsible for such measure. In this context, individual performance measures could be used to evaluate the efficiency, reliability, and quality components of operations pertaining to a specific unit or function. To accomplish this, diverse measures should be utilized individually and/or in small groups. These measures are critical to detecting and dealing with specific efficiency-related problems. The key to performance improvement under this measure-specific platform is the effective training and development of employees in order to promote responsibility and accountability.

The performance measurement of the mobile electronic commerce should be an integral part of the digital strategic effort. As such, it should be considered as a subsystem of the corporate performance measurement system. In this context, it should be integrated, flexible, dynamic, and practical. If it is designed carefully, the operational, strategic, and customer orientation aspects of the mobile electronic commerce organizations would be readily managed toward achieving the competitive advantage.

When analyzing the literature related to online user interaction, factors such as website design, fulfillment/reliability, privacy/security and customer service, and merchandising have been suggested to be predictive of e-business performance (Szymanski and Hise 2000; Wolfinbarger and Gilly 2003; Trabold et al. 2006). In some other markets, convenience and site design emerged as the most important drivers of e-performance (Evanschitzky et al. 2004).

In this context, business organizations could be better off by making their websites more useful and enjoyable, rather than spending money increasing store familiarity and store style (van der Heijden and Verhagen 2004). Thus, another important e-business performance dimension is website usability (Agarwal and Venkatesh 2002). This performance dimension was found to be one of the determinants of overall service quality (Yang et al. 2005).

Customer service has been identified as an important competitive e-business factor (Huang et al. 2009). However, due to the pace of e-market competition, order winning e-service features seemingly become order qualifiers overnight (Trabold et al. 2006).

Although several studies have attempted to empirically shed some light on the different aspects of performance measurement in e-business, considering the evolution of virtual market in recent years and the lack of solid and integrated measurements of e-business, further research is still very much needed. In fact, the development of theoretical models that satisfy the generic requirements of e-commerce applications constitutes a research challenge (Fink 2006). The importance of determining measures and metrics for e-supply chains is also stressed by literature as another worthy challenge (Sambasivan et al. 2009). Therefore, considering that, even to assess the success of e-supply initiatives, managers should, at least, capture information on user utilization and transactions statistics (Sammon and Hanley 2007), the development of a measurement profile that matches customers' most important requirements regarding the online experience could be a good starting point.

According to more recent literature, the following dimensions emerged as important components of e-commerce performance. These dimensions included attractiveness and education, completeness and practicality, customer orientation, customization and responsiveness, uniqueness, advertising form, customer service, customer accessibility, customer confidence and use, pricing information, customer interest, overall quality, and service convenience and innovation (Torres et al. 2013).

1.6 Conclusion

This chapter is based on the studies of different aspects of the mobile electronic commerce strategy. The studies were conducted in the United States, Portugal, Arab Gulf States, and Iran. The chapter emphasizes the integrated and systematic approach to the implementation of mobile electronic commerce (m-commerce).

The traditional business cultures studies, in general, lacked the training and know-how needed for effective digital strategy. In this context, benchmarking the modifications, planning, and implementation efforts of their e-business culture counterparts should prove useful.

The competitive environment of today's open system organization has been shaped by dynamic, innovational, operational, and strategic technologies, such as e-communication technologies. The growth of these technologies, coupled with an increasing demand from customers, is presenting the open system organizations with some serious threats and many opportunities.

To meet such threats and capitalize on promising opportunities, today's open system organizations are attempting to adopt e-solutions, through the deployment of operational and telecommunication technologies in order to become more efficient, effective, and customer oriented. With the growth of the Internet applications, digital initiatives such as m-commerce are shaping the competitive strategic landscape of today's organizations. In recent years, it appears that the road to the competitive strategic advantage is paved with digital options.

This chapter explores the effectiveness of digital applications such as m-commerce from an operational, strategic, and customer orientation perspective. In this context, the digital strategy must be systematically integrated with a well-defined strategic customer focus. In the process, reasons for implementation, required modifications, and implementation issues must be approached strategically. In this context, piecemeal and me-too approaches will leave the organization without the needed strategic control of its m-commerce strategy. The role of the performance management system is critical in ensuring that the organization is measuring and continuously improving all of the critical elements of the mobile electronic commerce performance.

In their effort to approach the e-initiatives strategically, managers are left with many questions and few answers. In the next section, a conceptually based roadmap is offered to facilitate the goal of achieving an effective total digital strategy, which is consistent with the demands faced by today's open system organizations.

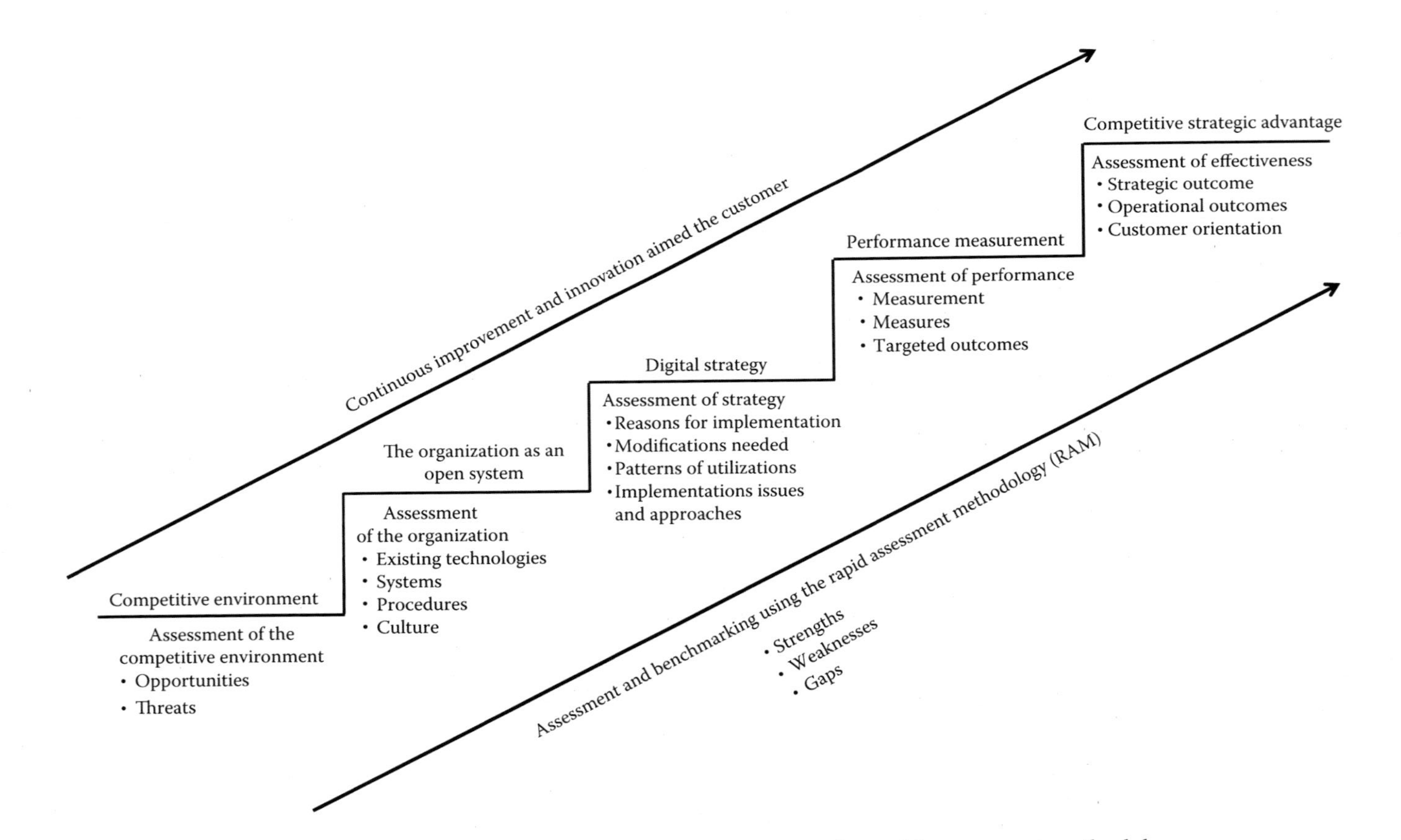

Figure 1.2 The road toward effective digital strategy implementations using the rapid assessment methodology.

1.7 Implications

In today's tight resources economy, managers are being asked to do more with less. In this context, the complete strategic planning process is time consuming and costly. The rapid assessment methodology (RAM) approach has been used in different operational contexts as a quick and less costly approach to assess the weakness, strengths, and potential gaps of organizations in different situations (Yasin et al. 1999; Czuchry and Yasin 2000, 2001, 2003). The RAM could be utilized to assess the strategic readiness of organizations to implement a digital strategy effectively. This methodology attempts to capture the options of panel of executives and/or managers. In this context, the RAM uses a scale between 1 and 5. Each participant answers the question (item) using this scale. For example, why do we want to adopt digital mobile e-strategy? Options might include improving customer orientation, improving operational efficiency, or improving market share, among other options. This process uncovers the reason for implementation, which could be classified into operational, supply-related, customer-related, efficiency, innovation, and so on. Other questions (items) might include the following: What are the current weakness and strengths we have? Can we close the gaps? This process is repeated for the other facets of the effective m-commerce implementation cycle.

The staircase diagram in Figure 1.2 attempts to illustrate the RAM, in association with the assessment needed for an effective implementation of digital strategy. In this context, it could serve as a starting point, in an otherwise all out costly strategic planning process. The advocated process is, somewhat, a step between let us jump into it and do it, and let us wait until we do a complete and expensive detailed strategic plan. The role of top management support in the RAM approach as well as championing the initiation of the mobile electronic commerce strategy is very critical.

References

Agarwal, R. and V. Venkatesh. 2002. Assessing a firm's web presence: A heuristic evaluation procedure for the measurement of usability. *Information Systems Research* 13(2): 168–186.

Amiri, Y. and H. Salarzehi. 2010. Fuzzy sketch for implementation of e-business plan in Iran SMEs (Case Study: Yazd Industrial Town-Iran). *International Business Research* 3(4): 172–180.

Ashrafi, R., M.M. Yasin, A. Czuchry, and Y. Al-Hinai. 2007. E-commerce practices in the Arabian Gulf GCC Business Culture: Utilization and outcomes patterns. *International Journal of Business Information Systems* 2(4): 351–371.

Bagherinejad, J. 2006. Cultivating technological innovations in middle eastern countries: Factors affecting firms' technological innovation behaviour in Iran. *Cross Cultural Management: An International Journal* 13(4): 361–380.

Balasubramanian, S., R.A. Peterson, and S.L. Jarvenpaa. 2002. Exploring the implications of m-commerce for markets and marketing. *Journal of the Academy of Marketing Science* 30(4): 348–361.

Chatterjee, D., R. Grewal, and V. Sambamurthy. 2002. Shaping up for e-commerce: Institutional enablers of the organizational assimilation of web technologies. *MIS Quarterly* 26(2): 65–89.

Chong, A.Y.-L. 2013a. A two-staged SEM-neural network approach for understanding and predicting the determinants of m-commerce adoption. *Expert Systems with Applications* 40(4): 1240–1247.

Chong, A.Y.-L. 2013b. Mobile commerce usage activities: The roles of demographic and motivation variables. *Technological Forecasting and Social Change* 80(7): 1350–1359.

Clarke, I. 2001. Emerging value propositions for m-commerce. *Journal of Business Strategies* 18(2): 133–148.

Czuchry, A. and M.M. Yasin. 2000. Achieving global competitiveness: A rapid assessment methodology approach. *Journal of Global Competitiveness* 3(1): 21–27.

Czuchry, A. and M.M. Yasin. 2001. Enhancing global competitiveness of small and mid-sized firms: A rapid assessment methodology approach. *Advances in Competitiveness Research* 9(1): 87–99.

Czuchry, A. and M.M. Yasin. 2003. Improving e-business with a Baldrige-based methodology. *Information Systems Management* 20(3): 29–38.

Czuchry, A. and M.M. Yasin. 2007. Effective marketing of technical innovation. *International Journal of Business Innovation and Research* 1(4): 448–463.

Czuchry, A., M.M. Yasin, and P. Bayes. 2001. Are you in control of your e-commerce strategy? *Information Strategy* 17(3): 6–11.

Czuchry, A., M.M. Yasin, and L. Robertson. 2002. Entrepreneurial e-commerce: Strategy and tactics. *Information Strategy* 18(2): 42–48.

Evanschitzky, H., G.R. Iyer, J. Hesse, and D. Ahlert. 2004. E-satisfaction: A re-examination. *Journal of Retailing* 80(3): 239–247.

Fathian, M., P. Akhavan, and M. Hoorali. 2008. E-readiness assessment of non-profit ICT SMEs in a developing country: The case of Iran. *Technovation* 28(9): 578–590.

Fink, D. 2006. Value decomposition of e-commerce performance. *Benchmarking: An International Journal* 13(1): 81–92.

Ghasemaghaei, M., S.A. Monadjemi, and B. Ranjbarian. 2009. Effective factors on Iranian consumers behavior in internet shopping: A soft computing approach. *Journal of Computer Science* 5(3): 172–176.

Ghobakhloo, M., D. Arias-Aranda, and J. Benitez-Amado. 2011. Adoption of e-commerce applications in SMEs. *Industrial Management & Data Systems* 111(8): 1238–1269.

Gomes, C.F. and M.M. Yasin. 2011. A systematic benchmarking perspective on performance management of global small to medium-sized organizations: An implementation-based approach. *Benchmarking: An International Journal* 18(4): 5.

Gomes, C.F. and M.M. Yasin. 2013. An assessment of performance-related practices in service operational settings: Measures and utilization patterns. *The Service Industries Journal* 33(1): 73–97.

Gomes, C.F., M.M. Yasin, and J.V. Lisboa. 2004. An examination of manufacturing organizations' performance evaluation: Analysis, implications and a framework for future research. *Journal of Operations & Production Management* 24(5): 488–513.

Gomes, C.F., M.M. Yasin, and J.V. Lisboa. 2007. An empirical investigation of manufacturing performance measures utilization: The perspectives of executives and financial analysts. *International Journal of Productivity and Performance Management* 56(3): 187–204.

Gomes, C.F., M.M. Yasin, and J.V. Lisboa. 2011. Performance measurement practices in manufacturing firms revisited. *International Journal of Operations & Production Management* 31(1): 5–30.

Huang, J., X. Jiang, and Q. Tang. 2009. An e-commerce performance assessment model: Its development and an initial test on e-commerce applications in the retail sector of China. *Information & Management* 46(2): 100–108.

Huarng, K. and T. Yu. 2011. Entrepreneurship, process innovation and value creation by a non-profit SME. *Management Decision* 49(22): 284–296.

Jalali, A., M.R. Okhovvat, and M. Okhovvat. 2011. A new applicable model of Iran rural e-commerce development. *Procedia Computer Science* 3: 1157–1163.

Jelassi, T. and A. Enders. 2005. *Strategies for e-Business: Creating Value through Electronic and Mobile Commerce*. Prentice Hall, Essex, England.

Kalakota, R. and M. Robinson. 2000. *e-Business 2.0: Roadmap for Success*. Addison-Wesley, Upper Saddle River, NJ.

Mahatanankoon, P., H. Joseph Wen, and B. Lim. 2005. Consumer-based m-commerce: Exploring consumer perception of mobile applications. *Computer Standards & Interfaces* 27(4): 347–357.

Ngai, E.W.T. and A. Gunasekaran. 2007. A review for mobile commerce research and applications. *Decision Support Systems* 43(1): 3–15.

Okazaki, S. and F. Mendez. 2013. Exploring convenience in mobile commerce: Moderating effects of gender. *Computers in Human Behavior* 29(3): 1234–1242.

Porter, M.E. 2001. Strategy and the internet. *Harvard Business Review* 79(3): 63–78.

Sallmann, F., A. Czuchry, and M.M. Yasin. 2004. An applied e-business approach for reinsurance services. *Marketing Intelligence and Planning* 22(7): 716–731.

Sambasivan, M., Z.A. Mohamed, and T. Nandan. 2009. Performance measures and metrics for e-supply chains. *Journal of Enterprise Information Management* 22(3): 346–360.

Sammon, D. and P. Hanley. 2007. Becoming a 100 per cent e-corporation: Benefits of pursuing an e-supply chain strategy. *Supply Chain Management* 12(4): 297–303.

Sarlak, M.A. and A. Hastiani. 2008. E-business barriers in Iran's free trade zones. *Journal of Social Sciences* 4(4): 329–333.

Szymanski, D.M. and R.T. Hise. 2000. e-Satisfaction: An initial examination. *Journal of Retailing* 76(3): 309–322.

Torres, P., J. Lisboa, and M. Yasin. 2011. Strategies employed by e-commerce firms in Portugal: An empirical investigation. *Portuguese Journal of Management Studies* 16(1): 43–56.

Torres, P.M., C.F. Gomes, and M.M. Yasin. 2013. Determinants of e-satisfaction: An empirical research. In *Proceedings of the 22nd International Business Research Conference*, pp. 1–11. Madrid, Spain.

Trabold, L.M., G.R. Heim, and J.M. Field. 2006. Comparing e-service performance across industry sectors: Drivers of overall satisfaction in online retailing. *International Journal of Retail & Distribution Management* 34(4): 240–257.

van der Heijden, H. and T. Verhagen. 2004. Online store image: Conceptual foundations and empirical measurement. *Information & Management* 41(5): 609–617.

Varnali, K., C. Yilmaz, and A. Toker. 2012. Predictors of attitudinal and behavioral outcomes in mobile advertising: A field experiment. *Electronic Commerce Research and Applications* 11(6): 570–581.

Wolfinbarger, M. and M.C. Gilly. 2003. eTailQ: Dimensionalizing, measuring and predicting retail quality. *Journal of Retailing* 79(3): 183–198.

Xu, H. and J. Yang. 2012. Do M-commerce user's expectations reflect reality? *IJEBM* 10(4): 322–331.

Yang, B., Y. Kim, and C. Yoo. 2013. The integrated mobile advertising model: The effects of technology- and emotion-based evaluations. *Journal of Business Research* 66(9): 1345–1352.

Yang, Z., S. Cai, Z. Zhou, and N. Zhou. 2005. Development and validation of an instrument to measure user perceived service quality of information presenting web portals. *Information & Management* 42(4): 575–589.

Yasin, M.M., J. Alavi, A. Czuchry, and R. Shafieyoun. 2014. An exploratory investigation of factors shaping electronic commerce practices in Iran: Benchmarking the role of technology and culture. *Benchmarking: An International Journal* 21(6).

Yasin, M.M., M. Augusto, A. Czuchry, and J. Lisboa. 2010. Effect of implementation reasons, implementation plans and system modifications on e-commerce outcomes: A structural equation approach. *International Journal of Management* 27(2): 303–319.

Yasin, M.M., M. Augusto, J. Lisboa, and A. Czuchry. 2007. Effective e-commerce implementation: A systematic approach. *Journal of Global Business and Technology* 18(35): 39–53.

Yasin, M.M., M. Augusto, J. Lisboa, and P. Miller. 2011. Assessing the competitive effectiveness of hospitals: The role of quality improvement initiatives. *Total Quality Management and Business Excellence* 22(4): 433–442.

Yasin, M.M., A. Czuchry, M. Gonzales, and P. Bayes. 2006. E-commerce implementation challenges: Small to medium-sized versus large organisations. *International Journal of Business Information Systems* 1(3): 256–275.

Yasin, M.M., A.J. Czuchry, D. Jennings-Jarvis, and C. York. 1999. Managing the quality effort in health care setting: An application. *Health Care Management Review* 24(1): 45–56.

Yasin, M.M. and U. Yavas. 2007. An analysis of e-business practices in the Arab culture: Current inhibitors and future strategies. *Ross Cultural Management: An International Journal* 14(1): 68–73.

chapter two

Dominant issues and conceptual approaches in mobile business research from 2005 to 2013

Joseph Budu and Richard Boateng

Contents

2.1 Introduction: Backgrounds and rationale

The ever-increasing rate of adoption and use of mobiles and related services has numerous effects on individuals and businesses. These effects have fairly spurred a number of valuable studies seeking to study and understand the phenomenon and its interrelationships with business (e.g., Schierholz et al., 2007; Lee and Park, 2008; Frempong, 2009; Gonçalves and Ballon, 2011; Ghezzi, 2012, Jaramillo and Harting, 2013). This chapter undertakes a review and classification of mobile business (m-business) research to indicate the current state and direction of research issues in the field. Reviews are needed to facilitate the advancement of knowledge, theory development, close saturated research areas, and uncover new areas for research (Webster and Watson, 2002). Generally, mobile research is endowed with several such reviews (e.g., Fouskas et al., 2005; Scornavacca et al., 2006; Ngai and Gunasekaran, 2007; Ladd et al., 2010). Scornavacca et al. (2006) categorized and analyzed 235 journal and conference articles between 2000 and 2004 to suggest future research areas (1) about business and organizational applications, (2) about empirical research, and (3) toward theory development. Similarly, Fouskas et al. (2005) derived a research roadmap after an in-depth review of m-business dimensions. The roadmap proposes short-, medium-, and long-term research directions including but not limited to the study of organizational capacity to integrate mobile applications (m-applications) and mobile services (m-services) within work culture and business processes, content pricing, and interoperability. In addition, a more recent review seems to have narrowed down on a specific m-application area, that is, mobile commerce (m-commerce) research and applications (Ngai and Gunasekaran, 2007). In addition, there exists an overarching review of over 800 articles that states that within the broader mobile computing research area "the technology itself was initially of interest, followed by business applications, user concerns, and finally research related to commerce applications of the technology" (Ladd et al., 2010).

Ladd et al.'s review somewhat suggests that despite these existing valuable reviews, none of them seems to satisfy the need to understand the theoretical and conceptual approaches to m-business research. Furthermore, considering the very dynamic nature of the m-business area, there is room to contribute with a review of the current state of the area. The purpose of this chapter is thus to understand the dominant issues and conceptual approaches to m-business research from 2005 to 2013 to identify gaps and to suggest future research areas. This chapter is structured into seven sections. Section 2.1 provides a background to, and the purpose of, this chapter; Section 2.2 discusses m-business definitions toward proposing a more comprehensive conceptualization; and Section 2.3 describes the methodology of the review. Section 2.4 presents the results of the review, based on m-business research themes. A discussion on the theoretical and conceptual frameworks of the previous research is presented in Section 2.5. Section 2.6 concludes with the gaps in the issues and conceptual approaches to provide some future research directions.

2.2 Framing mobile business research

Balasubramanian et al. (2002) observe that the term "m-business" lacks a formal conceptualization. This void has led to seemingly ad hoc definitions of the term in literature. For instance, Fouskas et al. (2005) defined it as "...communication, information exchange and transactions conducted over mobile or wireless networks," while Giaglis (2006) defined it as "...the ways in which mobile communication technologies can be applied to address the requirements of mobile users that need to access a varied range of applications and services through wireless access devices." These definitions, as listed in Table 2.1, conceptualize m-business as either a service (e.g., Muller-Veerse, 1999; Clarke, 2001 and Barnes, 2002 as cited in Woolfall, 2006), technology (e.g., Fouskas et al., 2005; Giaglis, 2006), or as a product (e.g., Lehner, 2003 and Zobel, 2001 as cited in Schierholz et al., 2007). Woolfall (2006) provides a definition that depicts an interaction of technology, product, and service by defining m-business as "...the means by which multiple actors conduct discrete or relational exchanges of economic or social value via a wireless network." Based on the foregoing definitions (Fouskas et al., 2005; Giaglis, 2006; Woolfall, 2006; Schierholz et al., 2007) and aiming to advance and develop m-business theory, we conceptualize m-business as the application of mobile devices to execute a firm's functions internally and in dealing with its external stakeholders (i.e., partners, suppliers, and customers). This definition attempts to combine the definitions of Giaglis (using mobiles to address user requirements), Woolfall (multiple actors conducting exchange), and Fouskas et al. (information exchange over mobile networks). Here, the users or multiple actors could be either an individual or a firm; the exchange could be done among or between individuals and firms, but nevertheless enabled by

Table 2.1 Some Definitions of Mobile Business (Authors' Construct)

Author(s)	Definition/conceptualization
Muller-Veerse (1999), Clarke (2001), and Barnes (2002) as cited in Woolfall (2006, p. 266)	Any monetary transaction conducted directly or indirectly via a mobile telecommunications network
Lehner (2003) and Zobel (2001) as cited in Schierholz et al. (2007)	The application of mobile technologies to improve or extend business processes and open new market segments
Fouskas et al. (2005)	Communication, information exchange, and transactions conducted over mobile or wireless networks
Giaglis (2006)	The ways in which mobile communication technologies can be applied to address the requirements of mobile users that need to access a varied range of applications and services through wireless access devices
Woolfall (2006)	The means by which multiple actors conduct discrete or relational exchanges of economic or social value via a wireless network

mobile or wireless technologies (e.g., mobile phones and tablet computers). This definition not only captures customer issues, but firm-level issues too, creating a broader basis for m-business discussions.

2.3 Methodology for the review

Only peer-reviewed journal articles are included in this review because they contain the core arguments raised in working papers and conference proceedings (Webster and Watson, 2002). Articles used were downloaded from Emerald FullText, JSTOR, Science Direct, and Wiley Online Library. The search descriptors used in all four databases were *mobile, mobile business,* and *m-business*. The results from each database were sorted according to relevance. The final papers included in the review were those published between 2005 and 2013 inclusive and related to m-business subcategories found in Ladd et al. (2010). Also, the coauthors adopted the following method to write the review; the first author searched for, identified, and downloaded all the papers used in this review. Based on the m-business themes discussed in Section 2.4, these papers were placed in, respectively, labeled folders. The first author read and discussed the issues, evidence, and conceptual approaches for the mobile business applications/services and economics, strategy, and business models themes, while the second author did the same for the consumer acceptance/adoption theme. We then merged the descriptions and discussions of conceptual approaches in extant m-business research into a comprehensive piece. We discussed and recorded possible future research directions based on previous researchers' calls for studies and the identification of arguably less-researched m-business themes from the issues presented in the evidence.

Furthermore, the authors presented the initial resultant write-up at the Association of Information Systems-affiliated 12th International Conference on Mobile Business held in June 2013 (see Budu and Boateng, 2013). The comments and suggestions from conference track chairs and participants served as useful input in enhancing the preparation of the final write-up for this chapter.

2.4 Mobile business research: Issues and evidence

The broad nature of research on mobile technology has culminated in the identification of several themes posited in existing reviews as shown in Table 2.2.

Examining the coverage of the four reviews listed in Table 2.2 suggests that the review of Ladd et al. (2010) captures all the themes posited in the previous reviews. Hence, we adopt R4, that is, Ladd et al. (2010) as the basis for categorizing the papers discussed in the ensuing sections.

Table 2.2 Themes from Previous Mobile-Related Reviews (Authors' Construct)

Review no.	References	Themes/categories from review
R1	Fouskas et al. (2005)	Technology, value, service, enablers
R2	Scornavacca et al. (2006)	Consumer, business, technology, general, industry
R3	Ngai and Gunasekaran (2007)	Cases and applications, wireless network infrastructure, mobile middleware, wireless user infrastructure, mobile commerce theory and research
R4	Ladd et al. (2010)	Mobile theory and research, mobile user, mobile business, mobile cases and applications, mobile technology

The discussions border on the dominant issues in previous m-business research. Owing to length constraints, we focused on the mobile business category that explores the mobile computing phenomenon from the business or industry perspective. This category has seven subcategories including (1) mobile business applications/services, (2) economics, strategy, and business models, (3) consumer acceptance/adoption, (4) macroeconomic cycles, (5) government–industry interaction, (6) interorganizational networks, and (7) infrastructure. Specifically, this chapter focuses only on subcategories (1), (2), and (3). The rest are not included because (1) there is limited discussion on those topics and (2) the authors wanted to attain some fidelity in the discussion by examining the chosen subcategories (Ladd et al., 2010).

2.4.1 *Mobile business applications/services*

The mobile business applications/services subcategory is related to the specific m-applications that firms adopt and use in their processes. Research under this subcategory has focused on firm-level adoption of m-business applications from two perspectives: mandatory adoption—where firms are required by some business association or government regulation to use a mobile technology (see Lee and Park, 2008) and issues surrounding voluntary adoption (see Liang et al., 2007; Balocco et al., 2009). Either form of adoption is expected to yield some benefits. The benefits include reduction in transportation cost, flexibility, organizational efficiency, individual productivity and effectiveness, transparency within business processes (Schierholz et al., 2007), removal of overlapping functions, reduced document handling at the back-office, reduction in typing errors and amount of clerical work needed, and single document flow (Rossi et al., 2007). Despite these benefits, there are some challenges observed with the use of mobile technologies. These challenges include but not limited to high subscription cost, customer insistence on face-to-face transactions, and the cost of on-net and off-net calls (Chiware and Dick, 2008; Frempong, 2009). To meet these challenges, a framework has been proposed to serve as guidelines for the implementation of mobile initiatives by firms (see Sinisalo et al., 2007). The dominant applications discussed include mobile library services (Cummings et al., 2010; Canuel and Crichton, 2011; Johnstone, 2011; Ballard and Blaine, 2013; Lo et al., 2013), mobile customer relationship management (Schierholz et al., 2007), and mobile supply chain systems (Pan et al., 2013). One trend worth noting about mobile business applications is the increasing possibility of using m-applications as a service, hence *Mobile Apps as a Service* (*MAaas*) (Jaramillo and Harting, 2013). Therefore, business may no longer need to be developing or acquiring their m-application servers; they can outsource development and/or use already built m-application platforms.

2.4.2 Economics, strategy, and business models

This subcategory is related to the production and distribution of mobile-related products and services as well as firms' strategy and business models. There seems to be a fair distribution of research along all three dimensions of this subcategory. Some studies provide frameworks to obtain predictions for costs and benefits of a mobile solution (Gruhn et al., 2007), while others dispute the economic impact of mobiles (Rohman, 2012). On the other hand, numerous studies show evidence of economic, social, and developmental impacts of mobiles (Cassidy, 2006; Gani and Clemes, 2006; Horst, 2006; Ishii, 2006; Mutula, 2008; Dunn, 2009; Best et al., 2010; Ndung'u and Waema, 2011; Hamade, 2012; Ilahiane and Sherry, 2012). Au contraire, some of the negative outcomes of mobile technology which research have discovered include phone use by criminals, infringement of privacy rights, and the creation of a dishonest society, which have the potential to affect the economics of adopting mobiles (Cassidy, 2006; Ndung'u and Waema, 2011).

Furthermore, some studies focus on the strategic implications of adopting mobiles, for example, to improve work processes, increase internal communication and knowledge sharing, and enhance sales and marketing effectiveness (Sheng et al., 2005). Other studies provide guidelines for firms within the m-business ecosystem to overcome competition, for instance, by cooperating with network providers, releasing new products quickly (Chang et al., 2009), and having market orientation and customer focus (Kristensson et al., 2008; Bose and

Chen, 2010; Isoherranen and Kess, 2011; Rahman and Azhar, 2011; Jeng and Bailey, 2012; Kuriyan et al., 2012). Some studies also capture how firms interpret and react to changes in the technological landscape, for example, impact of wireless local area network (W-LAN) on the activities of mobile network operators (MNOs) (Madjdi and Husig, 2011). The issues to be addressed in designing new strategy in the mobile landscape are also delineated by Peppard and Rylander (2006).

Due to the potential for customers to affect firm strategy, some studies are dedicated to studying how firms can respond to customer needs (Aydin et al., 2005; Jayawardhena et al., 2009; Yeh and Li, 2009; Santouridis and Trivellas, 2010; Aleke et al., 2011; Srinuan et al., 2011; Zhou, 2011; Awwad, 2012; Hung et al., 2012; Tobbin, 2012; Wang and Lin, 2012). Concerning business models, issues discussed include the creation of customer value for m-services (Methlie and Pederson, 2007); viability, reconfiguration, and sustainability of business models (de Reuver and Haaker, 2009; Ghezzi, 2012; Johansson et al., 2012); search for additional revenue sources (Gonçalves and Ballon, 2011); and how to capture low-income customers (Anderson and Kupp, 2008).

2.4.3 Consumer acceptance/adoption

The importance of consumers reflects in arguments supporting the inclusion of clients in the delivery of m-services (see Martin et al., 2012). This argument could thus explain the dominance of consumer research in m-business. Research in this subcategory primarily concerns factors influencing consumers' adoption of mobile devices and m-services (e.g., Shim et al., 2006; Laukkanen, 2007; Laukkanen et al., 2007; Bouwman et al., 2008; Kumar and Lim, 2008; Mannukka, 2008; Salmi and Sharafutdinova, 2008; Yang and Jolly, 2008; Deng et al., 2010; Karim et al., 2010; Petruzzellis, 2010; Riquelme and Rios, 2010; Wessels and Drennan, 2010; Gomez-Barroso et al., 2012; Mokhlis and Yaakop, 2012; Tobbin, 2012; Hsiao, 2013; Parreño et al., 2013). While others provide customer categorization schemes in terms of what influences their mobile adoption behavior (Kimiloglu et al., 2010), others study the factors that inhibit customers' adoption (Dunn, 2009; Lu and Su, 2009; Koenig-Lewis et al., 2010; Püschel et al., 2010; Dey et al., 2011). From the perspective of the actual consumers, others study the actual consumer use of mobiles (Donner, 2008; de Angoitia and Ramirez, 2009; Sey, 2009; Dey et al., 2013), switching behavior (Shin and Kim, 2007; Lee et al., 2011; Srinuan et al., 2011; Nikbin et al., 2012), and satisfaction (Chakraborty and Sengupta, 2013).

As noted earlier about previous mobiles-related reviews, there is still a void in terms of knowing the dominant conceptual approaches to m-business research. Knowing this could contribute to future research extending existing knowledge by proposing relatively underused conceptual approaches and some possible applications. The next section achieves this aim.

2.5 Mobile business research: Conceptual approaches

This section discusses theoretical- and framework-based approaches to m-business research to suggest conceptual gaps for future research. M-business research guided by theoretically based approaches tends to be dominated by consumer acceptance/adoption theme. The theories employed include the technology acceptance model (TAM) in either its original form (Amin, 2008; Yang and Jolly, 2008; Zhou, 2011; Tobbin, 2012) or extended form by adding on certain constructs (Okazaki et al., 2008; Wessels and Drennan, 2010; Akturan and Tezcan, 2012; Lule et al., 2012). TAM is sometimes combined with the theory of reasoned action (e.g., Lee et al., 2007; Kim et al., 2009; Lu and Su, 2009; Koenig-Lewis et al., 2010; Liu and Li, 2010; Riquelme and Rios, 2010) and the task technology fit model (Pagani, 2006) to form a theoretical basis. On the other hand, some studies simply modify original theories to explore their research objectives. For instance, TAM whose arguments are based on volition, seems deficient for use in the measure of mandatory technology adoption. Hence, TAM has been modified to create the technology satisfaction model (Lee and Park, 2008). Other evident clear-cut theories used include unified theory of acceptance and use of technology (Yang, 2010), game theory (Woolfall, 2006), theory of disruptive competition and innovation (Gillwald and Mureithi, 2011), economic theory (Au and Kauffman, 2008), the expectation confirmation model (Hung et al., 2012), and media richness theory (Lee et al., 2007).

The seeming limitations in technology adoption models also reflect in their inability to explain continuous use and effective use of technology, hence the addition of macroenvironmental factors and individual skills, knowledge, and experience to study mobile technology appropriation in various socio-economic contexts (Dey et al., 2013).

Studies with framework-based approaches include the following: providing directions for an organization to choose a suitable mobile solution (Gruhn et al., 2007), analyzing the business side of mobile/wireless technologies (Kang et al., 2011), developing viable business models (de Reuver and Haaker, 2009), and ensuring collaboration among m-business ecosystem players (Jing and Xiong-Jian, 2011). Other (popular) frameworks employed include the strategy orientation framework (Isoherranen and Kess, 2011) and the development of a fit-viability model for the adoption of mobile technology in business (Liang et al., 2007). Table 2.3 illustrates m-business subcategories, research issues, and their related conceptual approaches.

Table 2.3 Conceptual Approaches to Mobile Business Research (Authors' Construct)

M-business subcategory	Research issue	Classification of conceptual approach	Article references
Mobile business application services	Modeling and analysis of mobile business processes	Framework	Gruhn et al. (2007)
	Framework for the application of mobiles to customer relationship management	Framework	Schierholz et al. (2007) Sinisalo et al. (2007)
	Integrating perceived loss of control with user satisfaction and technology acceptance model (TAM) to explain B2B market performance	Theoretical	Lee and Park (2008)
	A conceptual framework to examine the effect of relationship conflict and cooperation on the business processes of a partnership	Theoretical	Woolfall (2006)
	Security risks associated with the use of mobile social media	N/A	He (2013)
	Adoption model of mobiles in supply chain management	Theoretical	Pan et al. (2013)
Economics, strategy, and business models	Business model techniques and analysis	Framework	Johansson et al. (2012) Methlie and Pederson (2007) Ballon (2007a) de Reuver and Haaker (2009) Braet and Ballon (2008) Anderson and Kupp (2008) Ghezzi (2012) Hawkins and Ballon (2007)

(*Continued*)

Table 2.3 (Continued) Conceptual Approaches to Mobile Business Research (Authors' Construct)

M-business subcategory	Research issue	Classification of conceptual approach	Article references
Consumer acceptance/ adoption	Determining of factors accounting for firm's perceived performance of mobile commerce using technology–organization–environment framework	Theoretical	Martin et al. (2012)
	Strategic analysis of mobile phone industry using from a value chain perspective	Framework	Chang et al. (2009)
	Examination of initial trust in mobile banking adoption	Theoretical	Zhou (2011)
	Three elements of TAM used to identify differences in adoption of mobile data services for two age cohorts	Theoretical	Yang and Jolly (2008)
	Adoption of mobile technology for fashion goods	Theoretical	Kim et al. (2009)
	Quality factors that influence customer trust	Theoretical	Yeh and Li (2009)
	Determinants and motivations of consumer mobile technology adoption and/or use	Theoretical	Yang (2010) Püschel et al. (2010) Liu and Li (2010) Akturan and Tezcan (2012) Tobbin (2012) Amin (2008) Hung et al. (2012) Okazaki et al. (2008) Wessels and Drennan (2010) Jayawardhena et al. (2009)

(*Continued*)

Table 2.3 (Continued) Conceptual Approaches to Mobile Business Research (Authors' Construct)

M-business subcategory	Research issue	Classification of conceptual approach	Article references
	Service quality's impact on customer satisfaction and customer loyalty	Framework	Santouridis and Trivellas (2010) Chakraborty and Sengupta (2013)
	Impact of telecommunications development on the quality of life	Theoretical	Rohman (2012)

2.6 Discussions

As observed by Scornavacca et al. (2006), and evidenced herein, m-business research is dominated by consumer issues. It is, however, encouraging that some research exists in the business side too. This review's contributions lie in the attempt to provide a holistic definition of m-business and to highlight the dominant themes and conceptual approaches to m-business research from 2005 to 2013. The evidence presented in this review implies that first, apart from gaps potentially overlooked in the *infrastructure, macroeconomic cycles, government–industry interaction,* and *interorganizational networks* subcategories, there is a need for ongoing research into the *economics, strategy, and business models* subcategory because of high dynamics in the mobile industry (Ladd et al., 2010). Second, there is room to contribute to the general *mobile theory and research* category by adopting and/or adapting more theoretically grounded approaches in m-business research to fill the paucity especially in the subcategories discussed under Sections 2.4.1 and 2.4.2. In addition, the following sections discuss the evidence and suggest further research areas.

2.6.1 Discussion of issues and evidence

First, concerning mobile business applications/services, firms face numerous uncertainties and challenges in the adoption of mobile technology for business activities. Studying possible means of overcoming such challenges would be both helpful to firms and as a way of m-business transformation (Tsai and Gururajan, 2007). Detailed case studies documenting how firms overcome such challenges would provide useful insight and a roadmap for others seeking to implement mobile technologies. Another area that could be studied is the impact of mobile technology on a firm's managerial functions. Overall, these pointers and discussions in previous reviews suggest that the ongoing and future research still focus on the adoption of mobile applications by firms that are noncore members of the m-business ecosystem (see Liang et al., 2007; Chiware and Dick, 2008; Lee and Park, 2008; Balocco et al., 2009; Frempong, 2009). Such studies naturally study the related impact of mobiles adoption. Meanwhile, Heeks (2006) suggests four stages within the Informatics Life cycle: development, adoption, use, and impact. With current research seemingly focused on adoption, use, and impact, there is room for research into the development phase—how m-business applications are created. In addition, current trends in mobile technology usage suggest that firms are encouraging employees to bring mobile devices to the workplace (Miller et al., 2012). With such trends come benefits that have just been mentioned, but not studied and tested empirically. The need to study such issues arises because of the concerns raised about the potential for the use of mobile technology, for instance, to access social media and affect an organization's security (He, 2013). Since mobile technology has and continues to create mobile versions of existing hitherto personal computer-based applications, there is a need to study the issues arising from such *mobilizations*.

Second, in economics, strategy, and business models, previous studies focus on economic and noneconomic impact of adoption, strategic implications of mobile adoption, overcoming competition, how to respond to technological changes, and responding to customer needs. However, within mobile business organizations (MBOs), there is the need to understand their ability to combine internal and external resources to create new resource combinations (Koruna, 2004). The seeming paucity of empirical studies about how firms combine resources is evidenced in an m-business special issue by *The Journal of Policy, Regulation and Strategy for Telecommunications* (Ballon, 2007a). There is the need to go beyond the current frontier to study managerial processes within MBOs, especially due to the many unresolved business challenges pertaining to the deployment and management of value-added services (Giaglis, 2006).

Third, research under the consumer acceptance/adoption subcategory needs extension in the face of advocacy for the inclusion of customers in the delivery of m-services (Martin et al., 2012). Further research could thus explore how MBOs harness customer perceptions as part of their value creation process. Also, to attract low-income consumers, companies have been advised to reconfigure their value chain according to a strategic pattern depending on their business model type (Ghezzi, 2012). This raises the need for empirical research into how such reconfiguration is done and what best practices could be followed. Such a study would help provide knowledge as to the existence or otherwise of any company that has succeeded or failed in a reconfiguration effort.

Generally, it seems that firm-level m-business research is dominated by studies on MNOs (see Ballon, 2007b; Anderson and Kupp, 2008;

Chang et al., 2009; Gonçalves and Ballon, 2011; Srinuan et al., 2011). To provide a holistic insight into m-business, other firms within the ecosystem need to be studied. This need is supported by the observation of focus shifting to value addition and content creation and hence from mobile network providers to other players in the mobile ecosystem (Peppard and Rylander, 2006). Furthermore, since the usage of third-party applications has a bright future (Methlie and Pederson, 2007; Verkasalo and Hämmäinen, 2007), firms like content providers are also important members of the mobile ecosystem that need to be studied. Other such firms include service providers, service creators, mobile operators, and handset manufacturers (Smura et al., 2009). Value creation in MBOs would be a promising area for research especially into the effects of business model choices on performance, and into the actual process of creating value.

2.6.2 *Discussion of conceptual approaches*

The focus of the issues as discussed in Section 2.4 mirrors a concentration of conceptual approaches to studying consumer applications and user perceptions toward them. The availability and the applicability of mobile technology in many areas of human endeavor result in various applications some of which are yet to be researched. Nevertheless, a good starting point in keeping track of such applications, and extending existing research into such individual m-applications, could be the creation of a framework that provides a holistic classification of m-business applications/services.

Currently, one important and ever-growing application of mobiles for business activities is m-commerce. Due to the different nature of firms and their activities, there is a possibility of implementing m-commerce at different rates or for varying purposes across firms. To provide knowledge of firms succeeding in m-commerce implementation, research could study the use of mobiles in commercial activities. Yet, because of different levels of firm-level factors (such as resources, capabilities, and readiness) and other external factors, this m-application does not guarantee the same level of benefits for all firms, and even similar firm types. For instance, the existence of different levels of e-commerce capabilities makes us argue for the possible existence of different levels of achieving m-commerce and the broader form of m-business (Boateng et al., 2008). Future research could adapt the e-commerce capabilities framework toward understanding the levels of m-business in firms.

In addition, as pointed in the gaps in Section 2.6.1, there is the need to understand the development and deployment of firm resources to create products (m-business applications) and achieve benefits (for the firm). None of the conceptual approaches identified in this review provides the pathway to attaining this knowledge, hence the need to introduce one that captures this gap succinctly. A conceptual approach of this nature may not exist specifically for the m-business area. However, deriving one from the broader strategic management circle could inform and provide guidelines in the creation of an m-business-specific framework. This review thus suggests the use of an established framework for studying how firms configure resources to create value and achieve firm benefits, for example, resource-based theory, and its later extension—the dynamic capability framework (Teece et al., 1997).

Similarly, while we have some evidence about companies involving customers creating value (Kristensson et al., 2008), we should consider asking whether customers are (external) resources and to what extent they influence the value creation process. Obviously, customer feedback would have to be captured to inform the value creation process, but from the m-business perspective, what are the mobile technologies specifically for this purpose—call it mobile knowledge management (KM). What KM frameworks can help understand this? Similarly, scientific research needs to evaluate the applicability and fit of value artifacts created with customer input. In this regard, design science theory provides some useful guidelines.

In addition, there is the need for frameworks that capture mobile use and benefits obtained by firms in various industries. So far, attempts have been made to understand this dimension from the perspective of individual users (e.g., Boateng, 2011). There is the need for similar extensions to understand firm-level impacts of mobile adoption. According to the evidence presented, research so far has borrowed from and extended the benefits of using information and communication technologies (ICTs) in commerce or trade (Amit and Zott, 2001; Boateng et al., 2008) to its mobile equivalent. For instance, the mobile for development perspective posits three effects that mobiles have on adopters, that is, incremental, transformational, and production (Batchelor and Scott, 2004; Heeks and Jagun, 2007). Incremental effects are those benefits from using mobile phones to improve what a firm already does, for example, communicating with customers and partners. Transformational effects are benefits from using the mobile phone in creating new things or accessing new services, for example, mobile banking. Production effects are benefits from trading in or selling mobile phones and related services. Current benefits captured by research relate to incremental and transformational effects; hence, there is room for research into the production effects of mobiles, for example, how mobile applications are created and sold to customers—both individual and corporate.

The evidence provided in the previous sections suggests that technology acceptance frameworks and theories are replete in m-business research. Yet, since m-business focuses on the organizational or industrial aspects of mobile technology, such theories have been seen to be inadequate in addressing interorganizational factors that have the potential to affect firms' adoption intention and behavior in terms of mobile technology (Pan et al., 2013). Thus, in the move toward more organization-oriented studies, researchers should go beyond the boundaries of such theories, by either operationalizing their constructs in specific mobile contexts or developing new frameworks best suited for studying the particular context (see examples in Section 2.5).

2.7 Conclusion and future research directions

With a categorization and discussion of salient existing m-business literature, we attempted to provide a general picture of the characteristics of past and current research into the m-business subcategories of mobile business applications/services; economics, strategy, and business models; and consumer acceptance/adoption. The main motivations for the review are that first, m-business is a dynamic area that needs continuous ongoing research (Ladd et al., 2010). Reviews of this nature are therefore important to identify and close over-researched areas, while uncovering new areas. Second, even though past m-business research had been criticized to be more of intuition-based reasoning and conceptual analysis than empirical (Scornavacca et al., 2006), the past reviews did not analyze the conceptual approaches nor the methodological approaches used in those previous studies. This creates a gap in knowing which theories are replete and which need attention. Third, this review explores the assertion that m-business research was focused on consumer applications (Anckar et al., 2003 as cited in Scornavacca et al., 2006).

The findings in this review provide evidence that initial perceptions are quite true. A transformation of the m-business research field as was predicted by Scornavacca et al. (2006) can, however, not be guaranteed. Nonetheless, a clear focus for future researchers would help grow the area better. Especially, two main focal points informed by the issues, evidence, and past theoretical approaches discussed here seem fertile for future m-business research. The first is research into business dimension of m-business: current research is skewed toward adoption issues—among both individual and corporate consumers. There is the need to understand the process of developing firm resources to realize the consumer value (as proposed by the business model argument). The internal and external conditions and factors/resources that facilitate the process could also be explored to a holistic understanding of the value creation process. The second is theory development. Even though the IS discipline has key theories, for example, TAM, there is the need for the development of key theories specifically for m-business in order to grow the area. This may be achieved by adapting and/or testing such IS theories with the m-business or m-computing subdiscipline.

References

Akturan, U. and Tezcan, N. (2012). Mobile banking adoption of the youth market: Perceptions and intentions. *Marketing Intelligence and Planning*, 30(4), 444–459.
Aleke, B., Ojiako, O., and Wainwright, D. W. (2011). ICT adoption in developing countries: Perspectives from small-scale agribusinesses. *Journal of Enterprise Information Management*, 24(1), 68–84.
Amin, H. (2008). Factors affecting the intentions of customers in Malaysia to use mobile phone credit cards. *Management Research News*, 31(7), 493–503.
Amit, R. and Zott, C. (2001). Value creation in e-business. *Strategic Management Journal*, 22, 493–520.
Anckar, B., Carlsson, C., and Walden, P. (2003). Factors affecting consumer adoption decisions and intents in mobile commerce: Empirical insights. *Proceedings of the 16th Bled eCommerce Conference*, Bled, Slovenia.
Anderson, J. and Kupp, M. (2008). Serving the poor: Drivers of business model innovation in mobile. *Info*, 10(1), 5–12.
Au, Y. A. and Kauffman, R. J. (2008). The economics of mobile payments: Understanding stakeholder issues for an emerging financial technology application. *Electronic Commerce Research and Applications*, 7, 141–164.
Awwad, M. S. (2012). An application of the American Customer Satisfaction Index (ACSI) in the Jordanian mobile phone sector. *The TQM Journal*, 24(6), 529–541.
Aydin, S., Özer, G., and Arasil, Ö. (2005). Customer loyalty and the effect of switching costs as a moderator variable: A case in the Turkish mobile phone market. *Marketing Intelligence and Planning*, 23(1), 89–103.
Balasubramanian, S., Peterson, R. A., and Jarvenpaa, S. L. (2002). Exploring the implications of m-commerce for markets and marketing. *Journal of the Academy of Marketing Science*, 30(4), 348–361.
Ballard, T. L. and Blaine, A. (2013). A library in the palm of your hand. *New Library World*, 114(5/6), 251–258.
Ballon, P. (2007a). The redesign of mobile businesses. *Info*, 9(5), 6–9.
Ballon, P. (2007b). Changing business models for Europe's mobile telecommunications industry: The impact of alternative wireless technologies. *Telematics and Informatics*, 24, 192–205.
Balocco, R., Mogre, R., and Toletti, G. (2009). Mobile internet and SMEs: A focus on the adoption. *Industrial Management and Data Systems*, 109(2), 245–261.
Barnes, S. (2002). The mobile commerce value chain: analysis and future developments. *International Journal of Information Management*, 22(2), 91–108.
Batchelor, S. and Scott, N. (May 18, 2004). The role of ICTs in the development of sustainable livelihoods, a set of tables. Retrieved July 30, 2012, from Gamos Ltd. website: http://gamos.org.uk/sustainbleicts/livelihoods.htm.
Best, M. L., Smyth, T. N., Etherton, J., and Edem, W. (2010). Uses of mobile phones in post-conflict Liberia. *Information Technologies and International Development*, 6(2), 91–108.
Boateng, R. (2011). Mobile phones and micro-trading activities—Conceptualising the link. *Info*, 13(5), 48–62.
Boateng, R., Heeks, R., Molla, A., and Hinson, R. (2008). E-commerce and socio-economic development: Conceptualising the Link. *Internet Research*, 18(5), 35–46.
Bose, I. and Chen, X. (2010). Exploring business opportunities from mobile services data on customers: An inter-cluster analysis approach. *Electronic Commerce Research and Applications*, 9, 197–208.

Bouwman, H., Carlsson, C., Walden, P., and Molina-Castillo, F. J. (2008). Trends in mobile services in Finland 2004–2006: From ringtones to mobile internet. *Info*, 10(2), 75–93.

Braet, O. and Ballon, P. (2008). Cooperation models for mobile television in Europe. *Telematics and Informatics*, 25, 215–236.

Budu, J. and Boateng, R. (2013). Dominant issues and conceptual approaches to mobile business research from 2005–2012, Conference Paper—*2013 International Conference on Mobile Business, Paper 4*. Available at http://aisel.aisnet.org/icmb2013/4.

Canuel, R. and Crichton, C. (2011). Canadian academic libraries and the mobile web. *New Library World*, 112(3), 107–120.

Cassidy, S. (2006). Using social identity to explore the link between a decline in adolescent smoking and an increase in mobile phone use. *Health Education*, 106(3), 238–250.

Chakraborty, S. and Sengupta, K. (2013). An exploratory study on determinants of customer satisfaction of leading network providers—Case of Kolkata, India. *Journal of Advances in Management Research*, 10(2), 279–298.

Chang, C.-Y., Wang, F., and Fu, H.-P. (2009). A strategic analysis of the mobile telephone industry in Mainland China. *Journal of Manufacturing Technology Management*, 20(4), 489–499.

Chiware, E. R. and Dick, A. L. (2008). The use of ICTs in Namibia's SME sector to access business information services. *The Electronic Library*, 26(2), 145–157.

Clarke, I. (2001). Emerging value propositions for m-commerce. *Journal of Business Strategies*, 18(2), 133–148.

Cummings, J., Merrill, A., and Borrelli, S. (2010). The use of handheld mobile devices: Their impact and implications for library services. *Library Hi Tech*, 28(1), 22–40.

de Angoitia, R. and Ramirez, F. (2009). Strategic use of mobile telephony at the bottom of the pyramid: The case of Mexico. *Information Technologies and International Development*, 5(3), 35–53.

de Reuver, M. and Haaker, T. (2009). Designing viable business models for context-aware mobile services. *Telematics and Informatics*, 26, 240–248.

Deng, Z., Lu, Y., Wei, K. K., and Zhang, J. (2010). Understanding customer satisfaction and loyalty: An empirical study of mobile instant messages in China. *International Journal of Information Management*, 30, 289–300.

Dey, B., Newman, D., and Prendergast, R. (2011). Analysing appropriation and usability in social lives: An investigation of Bangladeshi farmers' use of mobile telephony. *Information Technology and People*, 24(1), 46–63.

Dey, B. L., Binsardi, B., Prendergast, R., and Saren, M. (2013). A qualitative enquiry into the appropriation of mobile telephony at the bottom of the pyramid. *International Marketing Review*, 30(4), 297–322.

Donner, J. (2008). The rules of beeping: Exchanging messages via intentional "missed calls" on mobile phones. *Journal of Computer-Mediated Communication*, 13, 1–22.

Dunn, H. S. (2009). From voice ubiquity to mobile broadband: Challenges of technology transition among low-income Jaimaicans. *Info*, 11(2), 95–111.

Fouskas, K. G., Giaglis, G. M., Kourouthanassis, P. E., Karnouskos, S., Pitsillides, A., and Stylianou, M. (2005). A roadmap for research in mobile business. *International Journal of Mobile Communications*, 3(4), 350–373.

Frempong, G. (2009). Mobile telephone opportunities: The case of micro- and small enterprises in Ghana. *Info*, 11(2), 79–94.

Gani, A. and Clemes, M. D. (2006). Information and communications technology: A non-income influence on economic well being. *International Journal of Social Economics*, 33(9), 649–663.

Ghezzi, A. (2012). Emerging business models and strategies for mobile platform providers: A reference framework. *Info*, 14(5), 36–56.

Giaglis, G. M. (2006). Mobile business research: Recent advances and future prospects. *Business Process Management Journal*, 12(3), 261–264.

Gillwald, A. and Mureithi, M. (2011). Regulatory intervention or disruptive competition? Lessons from East Africa on the end of international mobile roaming charges. *Info*, 13(3), 32–46.

Gomez-Barroso, J. L., Bacigalupo, M., Nikolov, S. G., Campano, R., and Feijoo, C. (2012). Factors required for mobile search going mainstream. *Online Information Review*, 36(6), 846–857.

Gonçalves, V. and Ballon, P. (2011). Adding value to the network: Mobile operators' experiments with Software-as-a-Service and Platform-as-a-Service models. *Telematics and Informatics*, 28, 12–21.

Gruhn, V., Kohler, A., and Klawes, R. (2007). Modeling and analysis of mobile business processes. *Journal of Enterprise Information Management*, 20(6), 657–676.

Hamade, S. (2012). The impact of mobile technology on low-income communities in Lebanon. *Digest of Mobile East Studies*, 21(1), 24–38.

Hawkins, R. and Ballon, P. (2007). When standards become business models: Reinterpreting "failure" in the standardisation paradigm. *Info*, 9(5), 20–30.

He, W. (2013). A survey of security risks of mobile social media through blog mining and an extensive literature search. *Information Management and Computer Security*, 21(5), 381–400.

Heeks, R. (2006). Theorizing ICT4D research. *Information Technology and International Development*, 3(4), 1–4.

Heeks, R. and Jagun, A. (2007). Mobile phones and development: The future in new hands? *Id21 Insights*, 69, 1–6.

Horst, H. A. (2006). The blessings and burdens of communication: Cell phones in Jamaican transnational social fields. *Global Networks*, 6(2), 143–159.

Hsiao, K. (2013). Android smartphone adoption and intention to pay for mobile internet. *Library Hi Tech*, 31(2), 216–235.

Hung, M.-C., Yang, S.-T., and Hsieh, T.-C. (2012). An examination of the determinants of mobile shopping continuance. *International Journal of Electronic Business Management*, 10(1), 29–37.

Ilahiane, H. and Sherry, J. W. (2012). The problematics of the "Bottom of the Pyramid" approach to international development: The case of micro-enterpreneurs' use of mobile phones in Morocco. *Information Technologies & International Development*, 8(1), 13–26.

Ishii, K. (2006). Implications of mobility: The uses of personal communication media in everyday life. *Journal of Communication*, 56, 346–365.

Isoherranen, V. and Kess, P. (2011). Analysis of strategy focus vs. market share in the mobile phone case business. *Technology and Investment*, 2, 134–141.

Jaramillo, S. and Harting, D. C. (2013). The utility of mobile apps as a service (MAaas): A case study of BlueBridge digital. *Journal of Technology Management in China*, 8(1), 34–43.

Jayawardhena, C., Kuckertz, A., Karjaluoto, H., and Kautonen, T. (2009). Antecedents to permission based mobile marketing: An initial examination. *European Journal of Marketing*, 43(3), 473–499.

Jeng, D. J.-F. and Bailey, T. (2012). Assessing customer retention strategies in mobile telecommunications: Hybrid MCDM approach. *Management Decision*, 50(9), 1570–1595.

Jing, Z. and Xiong-Jian, L. (2011). Business ecosystem strategies of mobile network operators in the 3G era: The case of China Mobile. *Telecommunications Policy*, 35, 165–171.

Johansson, J., Malmstrom, M., Chroneer, D., Styven, M. E., Engstrom, A., and Bergvall-Kareborn, B. (2012). Business models at work in the mobile service sector. *iBusiness*, 4, 84–92.

Johnstone, B. T. (2011). Boopsie and librarians: Connecting mobile learners and the library. *Library Hi Tech News*, 28(4), 18–21.

Kang, J. -S., Lee, H. -Y., and Tsai, J. (2011). An analysis of interdependencies in mobile communications technology: The case of WiMAX and the development of a market assessment model. *Technology in Society*, 33, 284–293.

Karim, N. S., Oyebisi, I. O., and Mahmud, M. (2010). Mobile phone appropriation of students and staff at an institution of higher learning. *Campus-Wide Information Systems*, 27(4), 263–276.

Kim, J., Ma, Y. J., and Park, J. (2009). Are US consumers ready to adopt mobile technology for fashion goods? An integrated approach. *Journal of Fashion Marketing and Management*, 13(2), 215–230.

Kimiloglu, H., Nasir, V. A., and Nasir, S. (2010). Discovering behavioral segments in the mobile phone market. *Journal of Consumer Marketing*, 27(5), 401–413.

Koenig-Lewis, N., Palmer, A., and Moll, A. (2010). Predicting young consumers' take up of mobile banking services. *International Journal of Bank Marketing*, 28(5), 410–432.

Koruna, S. (2004). Leveraging knowledge assets: Combinative capabilities—Theory and practice. *R&D Management*, 34(5), 506–516.

Kristensson, P., Matthing, J., and Johansson, N. (2008). Key strategies for the successful involvement of customers in the co-creation of new technology-based services. *International Journal of Service Industry Management*, 19(4), 474–491.

Kumar, A. and Lim, H. (2008). Age differences in mobile service perceptions: Comparison of Generation Y and baby boomers. *Journal of Services Marketing*, 22(7), 568–577.

Kuriyan, R., Nafus, D., and Mainwaring, S. (2012). Consumption, technology, and development: The "poor" as "consumer". *Information Technologies and International Development*, 8(1), 1–12.

Ladd, D. A., Datta, A., Sarker, S., and Yu, Y. (2010). Trends in mobile computing within the is discipline: A ten-year retrospective. *Communications of the Association for Information Systems*, 27(Article 17), 285–306.

Laukkanen, T. (2007). Internet vs. mobile banking: Comparing customer value perceptions. *Business Process Management Journal*, 13(6), 768–797.

Laukkanen, T., Sinkkonen, S., Kivijärvi, M., and Laukkanen, P. (2007). Innovation resistance among mature consumers. *Journal of Consumer Marketing*, 24(7), 419–427.

Lee, M. K., Cheung, C. M., and Chen, Z. (2007). Understanding user acceptance of multimedia messaging services: An empirical study. *Journal of the American Society for Information Science and Technology*, 58(13), 2066–2077.

Lee, S.-G., Yu, M., Yang, C., and Kim, C. (2011). A model for analysing churn effect in saturated markets. *Industrial Management and Data Systems*, 111(7), 1024–1038.

Lee, T. M. and Park, C. (2008). Mobile technology usage and B2B market performance under mandatory adoption. *Industrial Marketing Management*, 37, 833–840.

Lehner, F. (2003). Mobile und *drahtlose Informationssysteme: Technologien, Anwendungen, Märkte*, Berlin, Germany: Springer.

Liang, T.-P., Huang, C.-W., Yeh, Y.-H., and Lin, B. (2007). Adoption of mobile technology in business: A fit-viability model. *Industrial Management and Data Systems*, 107(8), 1154–1169.

Liu, Y. and Li, H. (2010). Mobile internet diffusion in China: An empirical study. *Industrial Management and Data Systems*, 110(3), 309–324.

Lo, L., Coleman, J., and Theiss, D. (2013). Putting QR codes to the test. *New Library World*, 114(11/12), 459–477.

Lu, H.-P. and Su, P. Y.-J. (2009). Factors affecting purchase intention on mobile shopping websites. *Internet Research*, 19(4), 442–458.

Lule, I., Omwansa, T. K., and Waema, T. M. (2012). Application of technology acceptance model (TAM) in m-banking adoption in Kenya. *International Journal of Computing and ICT Research*, 6(1), 31–43.

Madjdi, F. and Husig, S. (2011). The response strategies of incumbent mobile network operators on the disruptive potential of public W-LAN in Germany. *Telecommunications Policy*, 35, 555–567.

Mannukka, J. (2008). Customers' purchase intentions as a reflection of price perception. *Journal of Product and Brand Management*, 17(3), 188–196.

Martin, S. S., Lopez-Catalan, B., and Ramon-Jeronimo, M. A. (2012). Factors determining firm's perceived performance of mobile commerce. *Industrial Management and Data Systems*, 112(6), 946–963.

Methlie, L. B. and Pederson, P. E. (2007). Business model choices for value creation of mobile services. *Info*, 9(5), 70–85.

Miller, K. W., Voas, J., and Hurlburt, G. F. (2012). BYOD: Security and privacy considerations. *IT Professional*, 14(5), 53–55.

Mokhlis, S. and Yaakop, A. Y. (2012). Consumer choice criteria in mobile phone selection: An investigation of Malaysian university students. *International Review of Social Sciences and Humanities*, 2(2), 203–212.

Müller-Veerse, F. (1999). Mobile commerce Report, Durlacher Research Limited, London, U.K.

Mutula, S. M. (2008). Digital divide and economic development: Case study of sub-Saharan Africa. *The Electronic Library*, 26(4), 468–489.

Ndung'u, M. N. and Waema, T. M. (2011). Development outcomes of internet and mobile phones use in Kenya: The households' perspectives. *Info*, 13(3), 110–124.

Ngai, E. and Gunasekaran, A. (2007). A review of mobile commerce research and applications. *Decision Support Systems*, 43, 3–15.

Nikbin, D., Ismail, I., Marimuthu, M., and Armesh, H. (2012). perceived justice in service recovery and switching intention. *Management Research Review*, 35(3), 309–325.

Okazaki, S., Skapa, R., and Grande, I. (2008). Capturing global youth: Mobile gaming in the US, Spain, and the Czech Republic. *Journal of Computer-Mediated Communication*, 13, 827–855.

Pagani, M. (2006). Determinants of high speed data services in the business market: Evidence for a combined technology acceptance model with task technology fit model. *Information and Management*, 43, 847–860.

Pan, Y., Nam, T., Ogara, S., and Lee, S. (2013). Adoption model of mobile-enabled systems in supply chain. *Industrial Management and Data Systems*, 113(2), 171–189.
Parreño, J. M., Sanz-Blas, S., Ruiz-Mafe, C., and Aldas-Manzano, J. (2013). Key factors of teenagers' mobile advertising acceptance. *Industrial Management and Data Systems*, 113(5), 732–749.
Peppard, J. and Rylander, A. (2006). From value chain to value network: Insights for mobile operators. *European Management Journal*, 24(2–3), 128–141.
Petruzzellis, L. (2010). Mobile phone choice: Technology versus marketing. The brand effect in the Italian market. *European Journal of Marketing*, 44(5), 610–634.
Püschel, J., Mazzon, J. A., and Hernandez, J. M. (2010). Mobile banking: Proposition of an integrated adoption intention framework. *International Journal of Bank Marketing*, 28(5), 389–409.
Rahman, S. and Azhar, S. (2011). Xpressions of generation Y: Perceptions of the mobile phone service industry in Pakistan. *Asia Pacific Journal of Marketing and Logistics*, 23(1), 91–107.
Riquelme, H. E. and Rios, R. E. (2010). The moderating effect of gender in the adoption of mobile banking. *International Journal of Bank Marketing*, 28(5), 328–341.
Rohman, I. K. (2012). Will telecommunications development improve the quality of life in African countries? *Info*, 14(4), 36–51.
Rossi, M., Tuunainen, V. K., and Pesonen, M. (2007). Mobile technology in field customer service: Big improvements with small changes. *Business Process Management Journal*, 13(6), 853–865.
Salmi, A. and Sharafutdinova, E. (2008). Culture and design in emerging markets: The case of mobile phones in Russia. *Journal of Business and Industrial Marketing*, 23(6), 384–394.
Santouridis, I. and Trivellas, P. (2010). Investigating the impact of service quality and customer satisfaction on customer loyalty in mobile telephony in Greece. *The TQM Journal*, 22(3), 330–343.
Schierholz, R., Kolbe, L. M., and Brenner, W. (2007). Mobilizing customer relationship management: A journey from strategy to system design. *Business Process Management Journal*, 13(6), 830–852.
Scornavacca, E., Barnes, S. J., and Huff, S. L. (2006). Mobile business research published in 2000–2004: Emergence, current status, and future opportunities. *Communications of the Association for Information Systems*, 17(1), 119.
Sey, A. (2009). Exploring mobile phone-sharing practices in Ghana. *Info*, 11(2), 66–78.
Sheng, H., Nah, F. F.-H., and Siau, K. (2005). Strategic implications of mobile technology: A case study using value-focused thinking. *Journal of Strategic Information Systems*, 14, 269–290.
Shim, J. P., Ahn, K., and Shim, J. M. (2006). Empirical findings on the perceived use of digital multimedia broadcasting mobile phone services. *Industrial Management and Data Systems*, 106(2), 155–171.
Shin, D. H. and Kim, W. Y. (2007). Mobile number portability on customer switching behavior: In the case of the Korean mobile market. *Info*, 9(4), 38–54.
Sinisalo, J., Salo, J., Karjaluoto, H., and Leppaniemi, M. (2007). Mobile customer relationship management: Underlying issues and challenges. *Business Process Management Journal*, 13(6), 771–787.
Smura, T., Kivi, A., and Töyli, J. (2009). A framework for analysing the usage of mobile services. *Info*, 11(4), 53–67.

Srinuan, P., Annafari, M. T., and Bohlin, E. (2011). An analysis of switching behaviour in the Thai cellular market. *Info*, 13(4), 61–74.

Teece, D. J., Pisano, G., and Shuen, A. (1997). Dynamic capabilities and strategic management. *Strategic Management Journal*, 18(7), 509–533.

Tobbin, P. (2012). Towards a model of adoption in mobile banking by the unbanked: A qualitative study. *Info*, 14(5), 74–88.

Tsai, H.-S. and Gururajan, R. (2007). Motivations and challenges for m-business transformation: A multiple-case study. *Journal of Theoretical and Applied Electronic Commerce Research*, 2(2), 19–33.

Verkasalo, H. and Hämmäinen, H. (2007). A handset-based platform for measuring mobile service usage. *Info*, 9(1), 80–96.

Wang, K. and Lin, C.-L. (2012). The adoption of mobile value-added services: Investigating the influence of IS quality and perceived playfulness. *Managing Service Quality*, 22(2), 184–208.

Webster, J. and Watson, R. T. (2002). Analysing the past to prepare for the future: Writing a literature review. *MIS Quarterly*, 26(2), 13–23.

Wessels, L. and Drennan, J. (2010). An investigation of consumer acceptance of M-banking. *Journal of Bank Marketing*, 28(7), 547–568.

Woolfall, D. (2006). Game and network perspective on m-business partnerships. *Business Process Management Journal*, 13(3), 265–280.

Yang, K. (2010). Determinants of US consumer mobile shopping services adoption: Implications for designing mobile shopping services. *Journal of Consumer Marketing*, 27(3), 262–270.

Yang, K. and Jolly, L. D. (2008). Age cohort analysis in adoption of mobile data services: Gen Xers versus baby boomers. *Journal of Consumer Marketing*, 25(5), 272–280.

Yeh, Y. S. and Li, Y.-M. (2009). Building trust in m-commerce: Contributions from quality and satisfaction. *Online Information Review*, 33(6), 1066–1086.

Zhou, T. (2011). An empirical examination of initial trust in mobile banking. *Internet Research*, 21(5), 527–540.

Zobel, J. (2001). *Mobile Business and M-Commerce - Die Markte der Zukunfterobern*. Munchen, Germany: Hanser.

section two

Mobile electronic commerce technologies

chapter three

Critical infrastructure management for mobile electronic commerce

Security and reliability issues on mobile ad hoc network

Haibo Wang, Wei Ning, and Wei Wang

Contents

3.1 Introduction

The fifth generation of WiFi and new standards of telecommunication protocols make possible the massive implementation of mobile ad hoc network (MANET) in retail business at a lower cost. While the retail industry has started embracing this new technology and mostly by installing a WiFi hotspot in their stores and shopping carts, the issues of security and reliability have become serious concerns to both public and private organizations.

There is a large volume of literature on critical infrastructure of telecommunication system; however, the topic on mobile e-commerce, especially on MANET, still offers researchers many great opportunities for new ideas and designs for the transmission of data. This chapter will present systematic review of the literature on MANET with the objective of addressing security and reliability by studying the state of the art in MANET research and applications. The authors try to identify the gap between academia and industry and to point out the challenge and future research directions. Eighty articles are collected, and the outputs of literature review are categorized into four distinct categories: infrastructure, security, reliability, and academic research and industry solutions. We hope that the findings of this study will serve as a good source for those interested in MANET. The chapter also presents some real-world challenges as well as future research directions.

3.2 M-commerce and infrastructure

3.2.1 Development of MANET and its characteristics

New technology in telecommunication network infrastructure enables its capability to be spread sparsely over large territories at a global scale to provide exceptional coverage and to impact the lives of large numbers of people. The new development of MANET communication caught the attention of investigators in both academia and industry. MANET is becoming an industry standard, and some retailers are promoting new customer services powered by MANET technology. The potential applications of MANET will bring a landslide change to the customer relationship management and supply chain management. Thus, the security and reliability of MANET are receiving immediate attentions from both public and private organizations. The advantage and potential applications of MANET in retailing industry include the following:

1. Sharing product information with and bringing online to off-line (O2O) experiences to customers
2. Monitoring the movement of customer in the stores to understand the customer's behavior
3. Enabling customer to share shopping experience instantly

In an environment where other telecommunication networks are unable to perform, the self-organization capability of MANET is the key, in that it can provide the critical communication infrastructure needed to effectively mobilize resources and share information. However, MANET is also vulnerable to destruction and malicious attacks due to its self-organization capability. Each device (node) in the MANET can act as a router that forwards data packets to other devices (nodes) and forms a very dynamic and unpredictable network topology. The connectivity of MANET depends on many internal and external factors, such as the node's density, the node's count, and the lifetime of each node. The routing protocols of the data packet in different layers of MANET are always subject to destruction and attacks. When considering security and reliability of MANET, researchers often raise questions on two important issues: disruption and influence. For example, in the MANET, the attacker can disrupt routing of data packets by short-circuiting and taking over the control of multiple nodes using a *wormhole* attack. To prevent such disruption attack, one can identify the critical components of the network by utilizing traffic analysis and building a better protocol for data packet routing. In the influence situation, if the objective is to improve the security and reliability of MANET, one can deploy relay nodes to improve the stability of the network and to reduce energy consumption, which should result in better data packet routing protocols and improved loading schemes for the critical components. However, the two problems often interact and coexist in many situations. For example, to prevent the *black hole* attack on MANET, where the attacker pretends to be an authorized node (when authorized nodes are temporarily down due to the power outage or battery life) and responds to misrouted data packets to disrupt the network, in this situation the task is to identify the critical components to promote high availability and minimum downtime for the authorized nodes (lower energy consumption for longer lifetime and less downtime) and to reduce the possibility of compromised networks by creating useful redundancies (relay nodes) to improve the self-healing structure in the MANET.

In addition, unlike other forms of telecommunication network, MANET is extremely mobilized and characterized by complicated configuration. Other issues such as node hopping, node trustiness, and node forward incentive are also needed to be addressed in order to build a reliable MANET for e-commerce. The systematic approach to addressing these issues is to study the critical components in the infrastructure in the real-world setting and to propose solutions to protect the critical components in a real-time manner. All these tasks require sophisticated algorithms and better network topology. The modeling and analysis of the dependencies and interdependencies among the critical components within the same critical infrastructure system are highly complicated, and their behavior is highly nonlinear. Currently, great effort in critical infrastructure management has focused on developing models to simulate the behavior of critical infrastructure systems and to discover the interdependencies and vulnerabilities among the critical infrastructure system.

3.2.2 *Rising popularity of MANET in mobile commerce*

The concept of mobile commerce was developed in the early stage of mobile technology, and many business models were proposed to provide products and services to customer without the limit of the time and location. Tsalgatidou and Pitoura (2001) discussed the business models using handheld terminals in mobile electronic commerce and transaction issues for these business models defined by the special characteristics of mobile terminals and wireless networks. However, these business models put too much emphasis on the voice and text data, thus bringing limited experience to the customer with high transaction cost due to the bottleneck of bandwidth.

The breakthrough of third- and later-generation network technology broke the bandwidth limitation on developing mobile commerce and allowed business to promote complex e-commerce interaction with lower transaction cost and higher transmission rate. The rich media and multidimensional interaction between the customers and retailers have added tremendous value to the mobile e-commerce. The development of social networks on the mobile platform also prompts retailers to chase innovated business model. Some researchers (Kuo and Yu 2006, Hwang et al. 2007) investigated the core business models of 3G and 4G service providers and their roles of developing mobile e-commerce. Recent survey by eDigitalResearch showed that the fast connection on 4G network can accelerate the mobile commerce and enable customer to locate a cheaper product from both online and off-line stores by scanning bar codes (eDigitalResearch 2013). Cloud-based 4G technology can provide customer and retailer a comprehensive platform for interaction and information sharing. In addition, the outdated revenue generation models in the voice service and the declining revenue caused by the intensive competition from service providers created opportunities for new value-added services built on the next-generation network (NGN) communication platform. The demand of content-rich and multimedia experience from the customer keeps the service providers on enhancing the network bandwidth and transmission rate in order to improve the communication efficiency of mobile commerce. These value-added services not only create new revenue sources for service providers, but also help retailers develop new business strategy. For example, the standard components in smartphones such as camera and GPS help retailers to deliver products and services to their customers anytime and anywhere; the sharing of information among customers also demonstrates that social group affects individual decisions on purchasing. The availability of MANET infrastructure is a key to deliver these services and to form a social group in the real time.

3.2.3 Value creation through mobile network

To understand the connection between new value-added services and associated business strategy, Chen and Cheng (2010) reported a study that collected feedback from 35 industry and research institution experts and scholars, using an analytic network process method to evaluate the strategy of mobile service providers on the NGN. The authors presented an evaluation framework for business strategy and discussed how the results of their study can be used as guides for NGN service providers to evaluate, reposition, and improve their service and strategy. In the meantime, mobile networks provide opportunities for technology-based self-services including e-government services and customer relationship management. Kuo and Chen (2006) presented an analytical tool using the fuzzy synthetic evaluation method based on analytic hierarchy process to select and evaluate the best mobile value-added service providers with the most customer satisfaction. The results of this study not only assist consumers to choose the appropriate service providers from the other consumers' opinions, but also help the provider to catch the potential market trend of mobile value-added services based on the prevailing consumer patterns. A recent study by Venkatesh et al. (2012) on a web-based survey of 2465 citizens showed four key attributes that influenced the adoption and use of the e-government services. One of the key attributes is the security provision.

One of the most attractive value-added services is mobile banking in e-commerce. The mobile banking breaks the boundary of time and space limitation of commerce activity and lowers the transaction costs compared to the conventional banking. Many retailers have adopted the mobile banking technology and enhanced their customer services. For example, customers in the Starbuck stores can use their mobile phone to pay for their cups of Java. In many countries, customers can simply use their mobile phones to pay for products and services from soft drinks in the vending machines to movie theater tickets. The mobile cash and advanced authorization methods can improve the security of online payment, thus bring convenience to the customer. The new fingerprint and face recognition technology in the mobile phone can lower the risk associated with using mobile banking and eliminate the complication of credit card purchase in some business environments. The replacement of lost and misplaced credit card by the customers and the stolen credit card data by the retailers cost businesses millions of dollars every year. In some African countries, conventional Internet banking service is not available due to the lack of infrastructure. Mobile banking can now provide low cost and efficient banking service to the customer and business under such conditions. The project using whitespace spectrum initiated by Google creates opportunities to the people living in an isolated community by improving their quality of life; the technology lays the foundation for mobile network infrastructure (Welch 2013). Even in the countries with advanced infrastructure for wired Internet service such as India and China, the conventional money order service is inefficient and inconvenient to customers. Customers spent lots of time in the long waiting lines and shared the high transaction cost. Both countries have large technologically disadvantaged migrant workers population, and customers have to visit banks or post offices to send money to their family due to the lack of computer and home Internet service. Thus, the demand of mobile money transfer creates opportunity for banks and mobile service providers. Singh (2012) reported a study to examine the feasible business model for mobile banking to integrate Indian Post with banking sectors on the mobile communication platform in order to provide efficient and lower cost money transfer services to migrant worker. The authors identified the factors for justifying the demand of mobile money order and analyzed the issues of slow adoption of mobile banking. In fact, in many countries, the lack of consumer confidence in security and reliability of mobile platform is the key reason for resistance of slow adoption of mobile banking as a main banking service. The slow implementation of security and reliability strategy by banking sectors cannot keep pace with the development of mobile commerce. In addition, the mobile commerce growth is always related to the development of mobile payment service markets that is still under transition due to the potential new technology innovation for mobile network infrastructure, such as MANET. Dahlberg et al. (2008) reported a literature review study on the current state of the mobile payment services market. They proposed a framework using a classic model of four contingency and five competitive force factors after analyzing a variety of factors that influence the markets of mobile payment services. They pointed out that the social and cultural factors on mobile payments, as well as comparisons between mobile and traditional payment services, are entirely uninvestigated issues. The directions for future research in this continuing emerging field are discussed.

Another widely developed value-added service is location-based mobile service. The potential of location-based mobile applications and intensive computing demand have attracted the attention from both academy and industry. The hardware and software developers have proposed different mechanisms to process the location-dependent queries. The quality of location-based mobile service depends on the efficiency of the queries and the precision of the computing results. Unlike static coordinating online services on the wired Internet platform, the location-based mobile services required intensive computing and continuous processing due to the changes of coordinates of customer. For example, to collect the data of customer purchasing behavior inside a retail store, the location-based service needs to pinpoint the movement of customers with high precision and record the time spent by the customer on particular items or shelves. The location data in this application will be collected in a fine granularity and precision. To deliver the information on products or services to the customer, based on their locations in such environment, location data need to be refreshed automatically as customers move around the store. It is highly challenging to the service provider on delivering high-quality services without draining the battery power of the mobile devices from the customers. MANET can provide a good performance distributed system to process the queries without expensive computing processes. It not only increases the flexibility of network communication and reduces power consumption of the location queries considerably but also performs the location-based service efficiently. Veijalainen et al. (2006) present a requirement analysis for the location-based e-commerce transactions using a graphic model based on a transaction manager (TM) architecture to protect e-commerce workflows against communication links, applications, or crashes. This system can be applied to monitor constraints related to the security and reliability of critical components in the network.

To provide value-added services to customers, several innovated mobile business models have been proposed. Figge (2004) introduces a situation-dependent service model based on the spatial dimension, personal dimension, and temporal dimension of the customer and service provider, which is considered as a situation provider based on the situation description. The author also discussed issues related to organization, technology, and security of situation dependency in mobile e-commerce. Huang et al. (2009) proposed a hierarchical mobile agent framework for handling key management and access control problems between mobile agent and host. The proposed method can improve the security of key management and mobile control accesses by managing the accessing relationship between the mobile agents and the host and operate storage space efficiently in a distributed environment, which is important to nonspecific network such as MANET. Jiang et al. (2010) proposed a collaborative business model with optimal profit among the portal access service provider (PASP), the product service provider (PSP), and the mobile service provider (MSP) as a multi agent system for mobile business. To achieve the optimal profit, the authors presented an agent evolution algorithm for computation and used a simulation experiment to show how the decision makers choose the profit-oriented strategies. The results of this study indicated that the collaborative mechanism can help players to achieve a better performance at a low level of risk. Esparza et al. (2006) discussed the use of mobile agent technology in brokerage systems to accelerate the development of a massive use of mobile e-commerce application and pointed out the security issues that hinder its use. In the next section, we will discuss the security issues related to MANET infrastructure in mobile e-commerce.

3.3 Security issues on MANET

3.3.1 Urgent need for better MANET security solutions

The self-organized nature of MANET infrastructure leads to the vulnerability of security threats and makes it much easier to suffer from numerous attacks than the conventional wired network. In addition, nodes within MANET have unreliable links due to the energy supply and constantly changing topology. All these features make MANET extremely vulnerable to security attacks that exploit its inherent weakness, and this also leads to the proliferation of attacks specifically designed for MANET. Thus, the security concerns for the MANET become prominent given the rising popularity of its application. Park et al. (2011) provided a brief overview of the state-of-the-art research on the security issues on MANET. Datta and Marchang (2012) discussed a wide range of attacks that are of special research importance. In their study, these attacks are divided into two subgroups: active and passive, determined by the behavioral pattern of such attacks, that is, whether they bring disturbance to the function of the network or just intercept key information from this system. They subsequently presented several basic techniques used for protecting MANET from such attacks, most of which are of research importance.

3.3.2 Attacks exploiting MANET's inherent weakness

The data packets moving among the nodes are controlled by the routing protocol without much security features included. MANET is often characterized by the lack of secured boundaries, Byzantine failure of comprised nodes, absence of centralized management facility, restricted power supply, and poor scalability. For example, the restricted power supply can lead to a denial-of-service (DoS) attack. Researchers classified the attacks on MANET infrastructure as internal and external based on the source and as active and passive based on the behavior. In this section, we discuss the various security issues in MANET infrastructure based on the literature.

First of all, DoS is the common external attack to both wireless and wired network. For example, Ack-storm DoS attacks pose a renewed threat to the network by taking advantage of a flawed design in Transmission Control Protocol (TCP) specifications. Theoretically, such attack could be amplified by unlimited times and they can be easily initiated in a large scale to fight against access network or to block web server even by a very weak man-in-the-middle attacker (MitM). Abramov and Herzberg (2013) found two simple ways to prevent Ack-storm DoS attacks—one through a simple fix in a client or server and the other through a packet-filtering firewall. To prevent DoS attack, Agah et al. (2006) presented a study on the design of security enforcement mechanism through the lens of auction theory. In the new protocol, nodes compete with each other in "forwarding incoming packets and gaining reputation in the network" on the basis of auction theory. Maximum bids are based on the node's utility value; thus, untruthfully bidding nodes will be identified and therefore isolated in order to prevent DoS attack; this novel protocol is called secure auction–based routing. In addition, the authors discussed the defense strategy in light of game theory. They found that eventually in a "noncooperative, two-player, non-zero-sum game between an attacker and a wireless sensor network (WSN)," the game moves toward Nash equilibrium, which finally leads to two new strategies. One synthesizes the overall utility of all the en-route nodes in data packet, "where utility is the difference between gain and cost for each node," and the other utilizes a *rating system* where wireless nodes receive rating from its neighboring nodes. The proposed *game theory–based framework* is supported by simulation results in terms of improving defense strategy for the WSN which is a type of MANET. A more recent attempt to prevent MANET from DoS attacks is done by Jia et al. (2013), in which CapMan, a security mechanism based on participating nodes' capability, is presented. By its design, CapMan is able to monitor both the capability limit of and the real-time traffic flow through a node. Each node's traffic situation is summarized and exchanged within the whole network. Because of this, nodes are able to make informed decision and better regulate the traffic. CapMan is implemented through simulations extensively, and the results show that CapMan can effectively thwart even sophisticated DoS attacks.

3.3.3 *Improvements made on routing protocols*

In MANET, mobile devices are connected via wireless links. In such a network, nodes change locations very often, thus posing special difficulty for routing algorithm to address the changing topology. Given that there are various protocols available and every protocol is best for a particular setting, Joshi (2011) discusses the security issues and what can be done to address those security issues on network layer. Security design is a key function for any type of network. For MANET, which relies on the mobile nodes to accomplish connection and is characterized by dynamic topology, the security design is of special importance. Komninos et al. (2006) discussed the major security issues regarding MANET *at the data link and network layers* and paid attention to the security requirement in security design for these two layers. Komninos et al. (2007) further investigated the design requirement for MANET in light of multiple authentication protocols and reported a study focusing on the authentication process *in a layered approach*. They also performed simulation to test the several such protocols and suggest *multiple lines of defense* to improve the security of MANET infrastructure. In addition, Shim and Lee (2005) identified the security flaws related to the protocols for authentication and key establishment. Their study revealed a new kind of attack: known key-share attack when the communication protocols in MANET failed to provide authentication. In addition, Karyotis and Papavassiliou (2007) studied an attack strategy based on a probabilistic model to maximize the disruption caused by those attacks. Considering the dynamic topology of MANET, this proposed model contains a topology control algorithm which is able to evaluate the capability, resources, and characteristics of a target network. In so doing, those attacks are expected to cause greater impact on target MANET. The test results show that the proposed algorithm is more effective than *any other flat and threshold-based approaches*. Adnane et al. (2013) conducted a comprehensive analysis on the trust-based routing protocol Optimized Link State Routing (OLSR), presented a detailed review of the reasoning process utilized by the node to form a trust value under OLSR protocol, and provided their own modification on the prevailing OLSR protocol. Garcia-Morchon et al. (2013) also examined the mechanism of such routing protocols that exercise trust management. Based on their distinct perspective, a modified trust-based protocol is proposed in this study, which emphasizes the two voting procedures, one for node admission and the other for node elimination. More commonly, building on established and well-performing protocols to make incremental improvement to these protocols is also a practically meaningful approach. von Mulert et al. (2012) presented their Secure Ad hoc On-Demand Distance Vector (AODV) protocol, which is built on the established AODV protocol. Their newly proposed protocol incorporates cryptographic mechanism so as to keep the message transmitted within the network from being *faked* or *altered*.

Given the unstable nature of the nodes in the mobile network, there are also protocols designed to address such security issues of MANET. Malavenda et al. (2012) presented a novel protocol that is delay tolerant and energy saving after a thorough examination of state-of-the-art protocols of such type. Their proposed protocol has been approved as good and even outperformed the most prevailing mechanism given the experimental results.

Since there are many routing protocols available for security improvement, it is equally important to measure the effectiveness of these proposed protocols so that protocols with better safety parameters could stand out. There are some standards that specify the requirements which include

X.800 and X.805. Almomani et al. (2013) based their analysis on these standards and proposed their logic-based security architecture, which can systematically analyze the security requirement a protocol could achieve and give mathematical proof. Based on their proposed architecture, they presented their analysis results for many prevailing protocols.

MANET is considered as a pervasive and ubiquitous network which is vulnerable to internal attack such as the Byzantine failure caused by the compromised nodes. The traditional access control and authentication become inadequate when it comes to pervasive and ubiquitous environment, especially after new technology emerged. To build a reliable pervasive network, Boukerche and Ren (2008) proposed a "reputation-based trust system which assigns credentials to nodes, updates private keys, managing trust value and makes decision." This system is tested to show that it can effectively identify malicious nodes and take proper action to protect the pervasive network. For the pervasive network, flooding is an effective method when it comes to *secure routing application*. But it also has flaws such as *extreme redundancy*, which often causes broadcast storm and even system failure because of the number of collision. Gossip is *a probabilistic algorithm* used to reduce the number of retransmission by calculating the retransmission probability. Burmester et al. (2007) reported a study to reduce the problem of retransmission while retaining as much security as possible by presenting "several new gossip protocols which exploit local connectivity to adaptively correct propagation failures and protect against Byzantine attacks." Because there is no infrastructure in MANET, the communication among nodes is conducted by the nodes themselves. Thus, the cooperation between nodes becomes critical in order to establish a route for communication. For various reasons such as capacities and batteries, some nodes may turn to misbehave and such nodes usually complicate the routing process. Gopalakrishnan and Rhymend Uthariaraj (2012) proposed a solution to mitigate the effect of those misbehaving nodes by introducing a *collaborative polling–based routing security scheme*, which is effective in detecting and isolating misbehaving nodes. This approach is supposed to reduce false detection. The test results show reduced packet drop ratio and malicious drop. As we move to a ubiquitous computing environment, the risk management becomes more and more critical because of the complexity of the connectivity. Hayat et al. (2007) try to find out what effect the traditional information security triangle is going to experience and how the information security requirement is going to be influenced. Special attention has been paid to the need for risk management and *context-based access control* as well as *proactive threat assessment techniques*. Clustering in MANET refers to the method that groups nodes in different clusters. Usually, there is a managing node in each cluster, also called clusterhead. In the current body of knowledge, little concern has been given to the *trust level* of head nodes in the selection process of *clusterhead* for internal attack. Elhdhili et al. (2008) proposed a clustering algorithm for security in MANET called CASAN, which focuses on the trustworthiness, stability, and energy of a potential node. The authors also tested this algorithm through simulation as well as comparison with other clustering techniques, and all these tests show positive evidence in support of CASAN. This approach can be used to identify the critical infrastructure in MANET.

3.3.4 Concerns and strategies for information assurance

In order to safeguard sensitive information and networks, usually we solve such security issues at the cost of the quality of service (QoS), especially *for networks with strong constraints such as MANET.* Aiache et al. (2008) proposed a solution for MANET to more efficiently combat viral attacks without compromising much QoS, and evidence of its effectiveness is provided. This is more effective when the approach is used as an automated process in the route discovery phase because it uses model checking approach to evaluate protocol abstraction. It will evaluate the security properties of MANET when an attacker is attempting to corrupt the process of discovering routes. Andel et al. (2011) developed a new topology reduction technique to lower the computational requirements. The study by Ben Othman and Mokdad (2010) focused on a multipath to improve the confidentiality in transmission. Specifically, in this proposed approach, original message will be *split into* different parts, then encrypted and *transmitted through different disjointed paths* to receivers. The security can be improved in that it is unlikely for attackers to intercept all the parts, increasing the confidentiality of data transmission. The authors also compared this approach with other multipath solutions and provided performance evaluation.

Password authentication is among the most widely used methods in ensuring network security. Its application to the newly emerged network system such as MANET does not provide the same security because the availability and strong protection *against off-line guessing attacks* are not considered conventional password authentication designs. Chai et al. (2007) proposed a *threshold password authentication scheme*, which can solve the problems presented earlier. In this proposed scheme, it only takes two rounds of message exchange to complete a jointly authentication process by several server nodes. This new scheme is of potential popularity because no password table is to be stored in server nodes. Other researchers presented a scheme to improve the security of MANET by mimicking the military structure which personnel at different levels have access to different resources and create a multilevel structure in MANET to restrict the access of critical information to the nodes at different levels (Changda and Shiguang 2008).

3.3.5 Wireless sensor network and its security issues

With more wireless sensors or sensing devices deploying to MANET, the security issues of a WSN become challenging to the business. Since the sensor systems are always used in high-risk environment and the security of sensors is always a concern, Bialas (2010) reported a study to look at a widely used method in maintaining sensor's assurance, which is Common Criteria (ISP/IEC 15408). The result from this study contributes to the existing knowledge by providing *common criteria* (CC)*–related security design patterns and to improve the effectiveness of the sensor development process.* Among security challenges raised by mobile WSNs, clone attack is particularly dreadful since it makes an adversary able to subvert the behavior of a network by just leveraging a few replicas of some previously compromised sensors. Conti et al. (2013) introduced two novel realistic adversary models, the vanishing and the persistent adversary, characterized by different compromising capabilities. The authors then proposed two distributed, efficient, and cooperative protocols to detect replicas: History Information–Exchange Protocol (HIP) and its optimized version (HOP). Both HIP and HOP leverage just local (one-hop) communications and node mobility, and differ for the amount of computation required. Both analysis and simulation results showed that their solutions are effective and efficient to provide high detection rate with the limited overhead. Based on these encouraging results, the authors proposed two distributed protocols to detect replica attacks in mobile WSNs, which provide fast detection even with just a single clone in the network. Actuation poses a new threat on the security of sensor network when it is used in a physical environment. Czarlinska and Kundur (2007) examined a new kind of actuation attacks that focus on sensor fidelity and reliability of the sensor network; they furthermore proposed a strategy to protect sensor network from such actuation attacks based on a *controlled level of random mobility.* They further provided evidence showing this proposed new scheme is effective in *reducing affected nodes over time* as well as improvements in the *resilience and tolerance for fault in network.* To address different requirements on data transmission and network constraints on a special type of wireless sensor and actuator networks (WSANs), Hu et al. (2007) proposed a ripple-zone (RZ)-based routing scheme that characterizes by scalability and energy efficiency to achieve better information security. In this new scheme, a two-level rekeying/rerouting scheme is also proposed; this new scheme is able to *adapt to dynamic topology* as well as to *securely update keys for each data transmission session.* This proposed scheme has been tested through extensive simulations and experiments and is proved effective in information security for WSANs.

3.3.6 *Integrative security framework*

So far, we have discussed the different techniques that one can utilize to strengthen the security of MANET. Since these techniques often have varying focus, grounded in different theoretic basis, and can be applied in different scenarios, thus, many researchers are working on building an integrative framework that incorporate elements from different security strategies to make a comprehensive defense system. Such effort can be illustrated as the framework presented in Lacey et al. (2012). In this study, the authors propose a reputation-based Internet protocol security (RIPsec) framework under which links and nodes are encrypted, nodes' behavior is monitored and graded under a reputation system, and message exchange is secured by certificate protection. This framework incorporates existing techniques and forms a comprehensive security solution; it is also proved to be more effective against existing attacks based on simulation results. Bankovic et al. (2011) put together a network routing protocol and an intrusion detection system (IDS) to form a new security solution. This new solution utilizes a reputation-based scheme to neutralize its internal threats and leverage an unsupervised algorithm to analyze abnormal behavior, and most importantly, identify attacks that may be unknown beforehand. There are also some frameworks and solution designed specifically for unique situation. The security solution presented in Cionca et al. (2012) provides integrative protection to sensor networks that are mostly used in hostile environment.

3.4 *Reliability issues*

3.4.1 *Difficulties to maintain a reliable mobile network*

The vulnerability of critical mobile communication infrastructures in highly mobilized community to extreme events is one of the great challenges in today's society. MANET consists of a set of independent mobile devices (nodes) that communicate with each other in a decentralized manner. It is often highly dynamic in topology since node mobility changes the existence and quality of links. The existence and quality of links, however, impose remarkable impact on the reliability of networks. Given the importance of the quality of links, it is necessary to incorporate path lifetime into routing strategy. The path lifetime, however, is not analytically available, thus, in Yuste et al. (2013), a fuzzy logic–based algorithm is suggested in an attempt to estimate the reliability and quality of a link with higher accuracy after taking path lifetime into account. A successful incorporation of the algorithm is shown to lead to improved routing efficacy and energy consumption.

The highly dynamic nature of MANET makes its reliability analysis hard. In a recent paper of Zhao et al. (2013), a new reliability analysis model has been proposed that takes node mobility effect and node reliability into consideration. The results reveal that node mobility has a greater effect on terminal reliability than the quality of link.

The robustness of MANET configuration poses difficulty to conventional reliability methods (e.g., reliability block diagram), because there is no standard structure or configuration to define all types of MANET. Therefore, many new methods are designed to improve the reliability of MANET by taking into consideration the differential configurations or mobility models. The MANET with better design and configuration have improved stability and reliability, as is shown in Wang (2014), in which two hierarchies' cluster-based MANET configuration is designed. The first hierarchy consists of head nodes and associate nodes, and the second hierarchy has only one head and multiple member nodes. The cluster head is the one that has the most member nodes; the communication between head nodes is conducted through only one associate node; algorithms for merging clusters are also provided. This proposed configuration demonstrates remarkable improvements in terms of key performance parameters. Building on this configuration, a following study, Xiaonan and Huanyan (2013), was conducted to show that the IP address can be better configured under such structure because each cluster can perform this process separately. On the same issue, Wang and Qian (2013) suggested a tree-based configuration and an algorithm for address assignment. In this proposed configuration, IP resource is distributed by proxy nodes and the communication among these proxy nodes is within one hop, which minimizes the cost of communication and the address is automatically retrievable, further reducing resource waste.

3.4.2 *State-of-the-art research on reliability issue of MANET*

Most of the existing works on the reliability of MANET have been considering binary link model w.r.t. transmission range and distance between the transmitter and the receiver. It is obvious that such model lacks practical consideration. In reality, the success of a transmission undergoes a probability, which is determined by the signal strength, fading, interferences, etc. The resiliency of failure during extreme events for the MANET can be defined as the MANET's ability to maintain network communication in the event of external failures. It can be measured by the time to restore service during the external failures. It is critical to ensure the MANET resiliency when the specific critical components fail without warnings caused by external events. Whitson and Ramirez-Marquez (2009) discussed the methods for safeguarding critical infrastructure components of MANET and for optimizing the use of redundancy as a strategic approach to improve MANET resiliency in the future. Despite the availability and easy-to-access of telecommunication technology such as MANET, the major challenge in adopting MANET is the lack of trust among the customers using MANET because of their lack of experience on using the technology. The spread of misleading or inaccurate information within MANET can also lead to unfavorable consequences in the practical situation. Thus, in order to facilitate collaboration on using MANET, the customers have to possess the ability of assessing the trustworthiness of the propagated information by other customers. Some decentralized trust models are proposed to enhance reliable information dissemination (Qureshi et al. 2010, 2012, Zyba et al. 2011). These models contain the distributed recommendation scheme. The scheme is built into decentralized trust model based on an existing membership service for MANET. Additionally, the mechanism for the trust-based information propagation is based on a nature-inspired spreading model that can establish an immediate and robust of trust among the nodes with the feature of prevention of flooding information from the multi-hop routes in MANET. Wang et al. (2011) proposed a trusted route that *considers communication reliability and path length for a reliable and feasible packet delivery in a MANET.* This study considered security as an integrated part of routing protocol in the MANET routing scheme, instead of applying additional layer to the basic routing layer. Nodes in MANET use a set of attributes to calculate the trust levels of others nodes based on the similarity between attributes and adopt a new forwarding rule for information propagation. This new trusted routing protocol generated the better results on preventing *black hole* attacks than the Dynamic Source Routing (DSR) protocol based on the routing performance parameters such as node's account, range, and speed of transmission. Another update made on DSR is the modified dynamic source routing (MDSR) protocol presented in Mohanapriya and Krishnamurthi (2013). Under MDSR, IDS is not activated until a malicious node is detected and this would significantly reduce the energy consumption. This is achieved by configuring an IDS system that detects abnormal data packet forwarding activities. Once anomaly is detected, the nearby IDS is turned into active mode and broadcast blocking messages to nodes that have contact with the malicious nodes hence to isolate such nodes. This modified protocol is found to have a 64% reduction in data packet loss and lower energy usage compared to DSR. In addition to the mechanisms updated through traditional means, research in effective counter-intrusion schemes also draws upon different perspectives. A novel IDS system, which is called biological-based artificial intrusion detection system (BAIDS), is suggested by Sundararajan et al. (2014). Together with hybrid negative selection algorithm, BAIDS is able to differentiate normal behaving nodes and malicious nodes with improved accuracy and it is also designed to combat wormhole attack in MANET.

As MANET is often deployed in malicious attacks environments, it is important to implement intrusion detection techniques to prevent with both internal and external attacks and improve the system reliability. Jin-Hee et al. (2010) reported a study on analyzing the performance of IDS techniques on the reliability of MANET. Unlike the traditional wired network IDS, the IDS on MANET should be performed to maximize the mean time to failure of the system at the optimal rate based on the operational conditions and attacker behaviors. Padmavathy and Chaturvedi (2013) proposed to use a propagation-based link reliability model with nodes failure under a known distribution to study the reliability issue via Monte Carlo simulation. They illustrated their method with an application and gave some imperative conclusions. Cook and Ramirez-Marquez (2008) presented applicable reliability metrics for MANET based on networking resources, such as bandwidth and energy. Chen et al. (2012) developed a random waypoint (RWP) mobility model to investigate the two-terminal reliability of the MANET and used the model to describe the mobility pattern of mobile users in MANET. The authors implemented the analytical method to present the expressions for one- and two-hop connectivity to measure the two-terminal reliability. The central tendency in the RWP model leads to its capability on assessing the effect of node's density and network topology on MANET's reliability where the model is constructed as a source node location function due to the asymptotic spatial distribution of nodes. Zhao et al. (2013) reported a study using Monte Carlo method for two-terminal reliability and modeled and analyzed the node's reliability and mobility effect.

The routing protocol possesses a critical influence on MANET reliability. Xie et al. (2008) proposed a hybrid routing protocol known as the link reliability–based hybrid routing (LRHR). LRHR can establish multiple paths between a pair of source–destination nodes under the *multiple paths* routing strategy that is different from traditional single-path routing and multipath routing strategies. Due to some circumstances such as resource or availability, the traditional multipath routing can be downgraded into a single path and lose the advantages of multipath mechanism. In the new LRHR routing strategy, the change rate of network topology can provide a routing strategy for switching dynamically between table-driven and on-demand manner. In order to improve the source–destination path reliability, the authors presented *an adaptive multipath routing mechanism with path segment, in which multipath can be established as part of the end-to-end path*. Thus, the reliability of the MANET can be improved by dividing the end-to-end path into several short segments. Zhen et al. (2010) presented a parallel forwarding mechanism to guarantee the QoS of multimedia services over MANET using the same strategy. In addition, other routing protocols are proposed for MANET to improve bandwidth and energy utilization. For example, Chang et al. (2009) introduced the Multicast Ad hoc

On-Demand Distance Vector (MAODV) routing protocol for achieving multicast in MANET. It is known that data packets transmission through a multicast tree in the MANET often passes through the unreliable links due to the mobility or lack of energy supply, which may significantly degrade the performance of MANET. MAODV uses a broadcast-type local repair mechanism to locate the alternative route of the multicast tree when lost links occur. The proposed unicast-type multi hop local repair protocol for multicast in MANET can recover the lost links quickly to improve the network reliability. In addition, it can improve the packet delivery rate, reduce the number of control messages, and repair delay. The construction of multicast trees is built upon the optimal number of hops in the multihop neighborhood based on the critical components in the infrastructure. As opposed to the multicasting approach used in MAODV, Liang et al. (2012) introduced a cross-layer unicasting local repair mechanism in trying to avoid such unfavorable outcomes caused by MAODV as extensive message overload and large energy consumption when broken links occur in a network. Their newly proposed mechanism is proved to be able to alleviate the congestion caused by broken links by informing the senders of such situation; it is also able to identify optimal alternative path.

Other broadcast methods such as gossip-based broadcasting are promising solution for efficient information dissemination in large-scale applications under distributed environment of MANET. For better reliability against node failures, Tsuchiya et al. (2006) developed a counter-based optimization for gossip-based broadcasting. The key design is that node rebroadcasts a message when it has received the copy of the message a few times. This way, a node can confirm the status of message dissemination and copy it to the others at appropriate time.

Self-healing routing protocols are indispensable to handle topological changes due to node mobility in dynamic networks such as MANETs. Furthermore, many emerged time-critical MANET applications, e.g., disaster response, rescue missions, and battlefield operations, should be able to support real-time, reliable data streaming with efficient energy consumption. However, most of the current energy-efficient routing protocols depend on configuration parameters which must be obtained prior to the deployment phase. Mukherjee et al. (2009) designed a self-managing, energy-efficient multicast routing protocol suite based on a self-stabilization paradigm. The suite includes

> (i) WECM, a Waste Energy Cost Metric designed for energy-efficient route selection, (ii) SS-SPST-E, a Self-Stabilizing, Shortest-Path Spanning Tree protocol for Energy efficiency based on WECM to maintain an energy-efficient, self-healing routing structure, (iii) SS-SPST-E fc, an enhanced SS-SPST-E with

fault containment to decrease stabilization latency, (iv) AMO, an Analytical Model for Optimization framework to reduce the route maintenance mechanism's energy overhead, and (v) self-configuration mechanisms observing, estimating, and disseminating the optimization parameters.

The proposed suite can address the issues such as wasted overhead energy, change rate of route and link selection under stabilization latency, and the intensity of application data transmission and delivery requirement.

Robust routing protocol should also consider the battery exhaustion. The study by Takeuchi et al. (2013) is the best example in this regard. The authors provided an updated version of their routing protocol named Route-Split Routing (RSR), built on their previously proposed protocol which falls short when node's battery exhausts.

3.5 Academic research and industry solutions

In MANET, each node can perform the function of router to form dynamic links when it moves randomly within the network. Implementing security and improving reliability in such dynamically changing environments are extremely challenging. Both academic and industrial investigators should consider the MANET characteristics in designing efficient solutions to address both issues. Recently, some academic research and industry solutions have been proposed to address security and reliability in MANET in terms of nodes authentication and availability, information confidentiality, routing security, and detecting malicious intrusion. First, there are multiple paths between the nodes in MANET infrastructures. Data transmission is controlled by the multipath routing protocol. The original data can be secured by splitting into packets with encryption and composition mechanisms. Then, a variety of disjointed paths between source and destination are used for data packets transmission. Even if an attacker can obtain one or more transmitted packets successfully, the attacker will have a very low probability to reconstitute the original message. Ben Othman and Mokdad (2010) proposed a securing data–based multipath routing protocol for MANET to prevent eavesdropping. Second, the basic routing protocol between the nodes in MANET is based on the trust levels. Each node will be calculating the trust values and association statuses for all of its neighboring nodes. Bhalaji and Shanmugam (2012) developed a trust-based DSR protocol to identify and isolate the *grayhole* nodes. Hadjichristofi et al. (2011) introduced a new routing protocol for MANET that can utilize connectivity, quality of routes and links, or any other information to improve the routing decision. The new protocol utilizes multiple mechanisms combining routing, security, and monitoring to support the concurrent use of IPv6 and IPv4 addresses and design network topologies with less energy consumption and less interference based on the connectivity information from the routing. It can build trust decisions among the nodes in MANET using the integrated trust component to support IPsec deployment and establish secure paths. Lacey et al. (2012) presented a Reputation-Based Internet Protocol Security (RIPsec) framework that provides security for MANET in terms of encryption, IPsec transport mode, behavior grading, and multipath routing at multiple levels. It protects MANET from external threats using encrypted links and encryption-wrapped nodes and from internal threats using behavior grading to define the node's reputation based on their demonstrated participation in the routing process. Pai and Wu (2011) designed an encryption-based method to nullify wormhole attacks that can seriously impair the routing protocols in MANET used by e-commerce. This method can detect the attack coming from both outside unauthorized intruders and inside authorized attackers. Third, strong authentication schemes are implemented to meet the requirements of MANET due to its dynamic topology and vulnerable structure for attack. In MANET, some server nodes can join together to provide mutual authentication based on the message exchange mechanism. Some service providers are implementing this solution on the mobile devices, which do not require password table stored in the server nodes. This approach can prevent the off-line guessing attacks in the case of compromised node. Others use the public key certificates management procedure between mobile agent and broker in the e-commerce transition. For example, Esparza et al. (2006) described how to use watermarking techniques to detect manipulation attacks and use brokers to punish the malicious hosts. Liebeherr and Dong (2007) proposed a key management and encryption scheme for application layer of MANET in which each node can share secrets with the authenticated neighbor nodes to avoid the global rekeying operation. Workman et al. (2008) developed a socio-biologically inspired approach on enforcing security among mobile agents based on structuration theory. Feng et al. (2009) discussed a security bootstrap model of key presharing mechanisms in MANET using one-way hash functions to prevent the key from being exposed. This mechanism is effective in detecting and blocking the DoS attack and other malicious attacks.

To prevent various attacks on MANET, IDS is extremely effective to both internal and external attacks. However, MANET offers a challenging environment for conventional IDS. It is important to execute the mission at an optimal rate for energy conservation. Jin-Hee et al. (2010) proposed models on Stochastic Petri nets to figure out the optimal rates, which are determined by *operational conditions, system failure definitions, attacker behaviors, and IDS techniques used*. Chizari et al. (2012) discussed their newly developed flooding algorithm, which is called energy efficient multi-point relay (EF-MPR), devised for effective message broadcasting in a network. Based on their extensive simulations, EF-MPR is proved to be effective in reducing energy consumption, minimizing the number of packet collision and the propagation time. In particular, MANET has a dynamic topology with issues of unreliable connectivity, limited node energy supply, and possibly high retention cost of nodes. To overcome these challenges, researchers have developed a variety of agent-based IDS. Stafrace and Antonopoulos (2010) designed a framework combining a military command structure and an agent behavioral model for IDS on MANET. The framework can improve the effectiveness of the IDS without consuming too much resource using a risk-based approach being directly proportional to the risk factor of the route. The effectiveness of this approach is measured through a simulation to detect and recover from a Sinkhole attack in MANET, using the AODV as the routing protocol. The results are reported in terms of the detection precision, data loss, and the communication overheads. von Mulert et al. (2012) presented the Secure AODV SAODV protocol using encryption mechanisms to protect the routing control messages from the attacks such as *medium access control layer misbehavior, resources depletion, black holes, wormholes, jellyfish and rushing attacks*; SAODV provides a number of security schemes such as *multipath routing, incentive schemes, directional antennae, packet leashes, randomized route requests, localized self-healing communities, and a reactive intrusion detection node blacklisting.* Venkatraman and Agrawal (2003) reported a study of AODV on the performance of the network with and without our secured routing scheme based on the behavioral patterns of the nodes. They proposed authentication techniques to isolate compromised nodes from MANET.

Researchers and application developers have presented a variety of solutions to improve the security and reliability on different layers of network communication. For example, Itani and Kayssi (2004) presented a security solution for MANET on the application layer using pure Java component independent from protocols or infrastructure, and applied the solution to a mobile banking application. More recently, Bar et al. (2013) presented a novel protocol in order to better cope with black hole attacks. In their proposed protocol, a path would be chosen if it involves a greater number of highest trusted nodes, which are ranked by their packet forwarding capability. By doing this, paths with low trust value will be disregarded and data will flow through more secured routes; hence, the reliability of the whole network improved. A similar approach is taken in Eissa et al. (2013) in which a trust-based routing scheme named FrAODV is presented with abundant evidence showing its effectiveness in improving MANET routing environment. In FrAODV scheme, a route is evaluated based on selected features such as trust and identity of the involved nodes before a message is forwarded.

Liebeherr and Dong (2007) proposed a key management and encryption scheme based on the application-layer overlay protocols to maintain the integrity and confidentiality of application data in MANET. The method allows each node to share secrets only with authenticated neighbors to avoid global rekeying operation and lower the risk of global rekeying. Komninos et al. (2006) discussed the solutions of multiple lines of defense from both data link and network layers to prevent a variety of attacks. Joshi (2011) identified a variety of attacks of MANET in the network layers to understand how attacker absorbs network traffic and controls the network traffic flow. It is important to design new protocol not only to establish an optimal and efficient route between nodes, but also to prevent attackers from injecting themselves into the unreliable link between nodes. However, due to the self-organizing, self-configuring nature of MANET, the roles of nodes in MANET change based on the type of the communication needs that create a dynamic and decentralized structure for MANET infrastructure. Such nature creates many challenges to the security and reliability of network infrastructure in MANET.

3.6 *Challenges and future research*

In MANET, mobile devices (nodes) are connected via wireless links. In such network, nodes change locations very often thus posing special difficulty for routing algorithm to address the changing topology. Given that various protocols are available and every protocol is best for a particular setting and makes it difficult to choose which protocol to use in the application, it is critical to develop a set of framework to evaluate these protocols in terms of security policy enforcement and efficient transmission of data.

In addition, mobile technology allows the business and customers to connect at massive scale without time and space boundaries. This capability of mobile commerce will have impacts on the way business happens, for example, mobile technology has already changed the campaign-based consumer marketing strategy to the dialogue-based consumer marketing using social network and encourage consumer to share information. The trusted relationship is not only important for nodes in MANET infrastructure in terms of security and reliability as we discussed earlier, but also critical to the development of mobile commerce applications between business and customers. The connection-oriented nature of mobile communication helps to build a much deep trust channel than other connectionless communication, such as email or blog. However, how to develop and maintain the trust in front of increasing malicious attacks is a challenging task to the application developers and service providers.

References

Abramov, R. and A. Herzberg (2013). TCP Ack storm DoS attacks. *Computers & Security* **33**: 12–27.

Adnane, A. et al. (2013). Trust-based security for the OLSR routing protocol. *Computer Communications* **36**(10–11): 1159–1171.

Agah, A. et al. (2006). Security enforcement in wireless sensor networks: A framework based on non-cooperative games. *Pervasive and Mobile Computing* **2**(2): 137–158.

Aiache, H. et al. (2008). Improving security and performance of an ad hoc network through a multipath routing strategy. *Journal in Computer Virology* **4**(4): 267–278.

Almomani, I. et al. (2013). Logic-based security architecture for systems providing multihop communication. *International Journal of Distributed Sensor Networks* **2013**: 1–17.

Andel, T. R. et al. (2011). Automating the security analysis process of secure ad hoc routing protocols. *Simulation Modelling Practice and Theory* **19**(9): 2032–2049.

Bankovic, Z. et al. (2011). Improving security in WMNs with reputation systems and self-organizing maps. *Journal of Network and Computer Applications* **34**(2): 455–463.

Bar, R. K. et al. (2013). QoS of MANet through trust based AODV routing protocol by exclusion of black hole attack. *Procedia Technology* **10**: 530–537.

Ben Othman, J. and L. Mokdad (2010). Enhancing data security in ad hoc networks based on multipath routing. *Journal of Parallel and Distributed Computing* **70**(3): 309–316.

Bhalaji, N. and A. Shanmugam (2012). Dynamic trust based method to mitigate greyhole attack in mobile ad hoc networks. *Procedia Engineering* **30**: 881–888.

Bialas, A. (2010). Common criteria related security design patterns validation on the intelligent sensor example designed for mine environment. *Sensors (Basel, Switzerland)* **10**(5): 4456–4496.

Boukerche, A. and Y. Ren (2008). A trust-based security system for ubiquitous and pervasive computing environments. *Computer Communications* **31**(18): 4343–4351.

Burmester, M. et al. (2007). Adaptive gossip protocols: Managing security and redundancy in dense ad hoc networks. *Ad Hoc Networks* **5**(3): 313–323.

Chai, Z. et al. (2007). Threshold password authentication against guessing attacks in ad hoc networks. *Ad Hoc Networks* **5**(7): 1046–1054.

Chang, B.-J. et al. (2009). Distributed route repair for increasing reliability and reducing control overhead for multicasting in wireless MANET. *Information Sciences* **179**(11): 1705–1723.

Changda, W. and J. Shiguang (2008). Multilevel security model for ad hoc networks. *Journal of Systems Engineering and Electronics* **19**(2): 391–405.

Chen, B. et al. (2012). Two-terminal reliability of a mobile ad hoc network under the asymptotic spatial distribution of the random waypoint model. *Reliability Engineering & System Safety* **106**: 72–79.

Chen, P.-T. and J. Z. Cheng (2010). Unlocking the promise of mobile value-added services by applying new collaborative business models. *Technological Forecasting and Social Change* **77**(4): 678–693.

Chizari, H. et al. (2012). EF-MPR, a new energy eFficient multi-point relay selection algorithm for MANET. *The Journal of Supercomputing* **59**(2): 744–761.

Cionca, V. et al. (2012). Configuration tool for a wireless sensor network integrated security framework. *Journal of Network and Systems Management* **20**(3): 417–452.

Conti, M. et al. (2013). Clone wars: Distributed detection of clone attacks in mobile WSNs. *Journal of Computer and System Sciences* **80**(3): 654–669.

Cook, J. L. and J. E. Ramirez-Marquez (2008). Reliability analysis of cluster-based ad-hoc networks. *Reliability Engineering & System Safety* **93**(10): 1512–1522.

Czarlinska, A. and D. Kundur (2007). Towards characterizing the effectiveness of random mobility against actuation attacks. *Computer Communications* **30**(13): 2546–2559.

Dahlberg, T. et al. (2008). Past, present and future of mobile payments research: A literature review. *Electronic Commerce Research and Applications* **7**(2): 165–181.

Datta, R. and N. Marchang (2012). Security for mobile ad hoc networks (Chapter 7). *Handbook on Securing Cyber-Physical Critical Infrastructure*, Das, S. K., Kant K., and Zhang, N. (Eds.), Morgan Kaufmann, Waltham, MA, pp. 147–190.

eDigitalResearch (2013). Survey finds 4G will accelerate mobile shopping. From: http://www.edigitalresearch.com/news/item/month/april/year/2013/nid/892198128. Accessed November 10, 2013.

Eissa, T. et al. (2013). Trust-based routing mechanism in MANET: Design and implementation. *Mobile Networks and Applications* **18**(5): 666–677.

Elhdhili, M. E. et al. (2008). CASAN: Clustering algorithm for security in ad hoc networks. *Computer Communications* **31**(13): 2972–2980.

Esparza, O. et al. (2006). Secure brokerage mechanisms for mobile electronic commerce. *Computer Communications* **29**(12): 2308–2321.

Feng, L. et al. (2009). Security bootstrap model of key pre-sharing by polynomial group in mobile ad hoc network. *Journal of Network and Computer Applications* **32**(4): 781–787.

Figge, S. (2004). Situation-dependent services a challenge for mobile network operators. *Journal of Business Research* **57**(12): 1416–1422.

Garcia-Morchon, O. et al. (2013). Cooperative security in distributed networks. *Computer Communications* **36**(12): 1284–1297.

Gopalakrishnan, K. and V. Rhymend Uthariaraj (2012). Collaborative polling based routing security scheme to mitigate the colluding misbehaving nodes in mobile ad hoc networks. *Wireless Personal Communications* **67**(4): 829–857.

Hadjichristofi, G. C. et al. (2011). Routing, security, resource management, and monitoring in ad hoc networks: Implementation and integration. *Computer Networks* **55**(1): 282–299.

Hayat, Z. et al. (2007). Ubiquitous security for ubiquitous computing. *Information Security Technical Report* **12**(3): 172–178.

Hu, F. et al. (2007). Scalable security in wireless sensor and actuator networks (WSANs): Integration re-keying with routing. *Computer Networks* **51**(1): 285–308.

Huang, K.-H. et al. (2009). Efficient migration for mobile computing in distributed networks. *Computer Standards & Interfaces* **31**(1): 40–47.

Hwang, J.-S. et al. (2007). 4G mobile networks—Technology beyond 2.5 G and 3G. *PTC (Pacific Telecommunications Council) Proceedings*, Honolulu, HI.

Itani, W. and A. Kayssi (2004). J2ME application-layer end-to-end security for m-commerce. *Journal of Network and Computer Applications* **27**(1): 13–32.

Jia, Q. et al. (2013). Capability-based defenses against DoS attacks in multi-path MANET communications. *Wireless Personal Communications* **73**(1): 127–148.

Jiang, G. et al. (2010). Agent-based simulation of competitive and collaborative mechanisms for mobile service chains. *Information Sciences* **180**(2): 225–240.

Jin-Hee, C. et al. (2010). Effect of intrusion detection on reliability of mission-oriented mobile group systems in mobile ad hoc networks. *Reliability, IEEE Transactions on* **59**(1): 231–241.

Joshi, P. (2011). Security issues in routing protocols in MANETs at network layer. *Procedia Computer Science* **3**(0): 954–960.

Karyotis, V. and S. Papavassiliou (2007). Risk-based attack strategies for mobile ad hoc networks under probabilistic attack modeling framework. *Computer Networks* **51**(9): 2397–2410.

Komninos, N. et al. (2006). Layered security design for mobile ad hoc networks. *Computers & Security* **25**(2): 121–130.

Komninos, N. et al. (2007). Authentication in a layered security approach for mobile ad hoc networks. *Computers & Security* **26**(5): 373–380.

Kuo, Y.-F. and P.-C. Chen (2006). Selection of mobile value-added services for system operators using fuzzy synthetic evaluation. *Expert Systems with Applications* **30**(4): 612–620.

Kuo, Y.-F. and C.-W. Yu (2006). 3G telecommunication operators challenges and roles: A perspective of mobile commerce value chain. *Technovation* **26**(12): 1347–1356.

Lacey, T. H. et al. (2012). RIPsec using reputation-based multilayer security to protect MANETs. *Computers & Security* **31**(1): 122–136.

Liang, Y.-H. et al. (2012). Cross-layer-based local repair for maximizing goodput and minimizing control messages in multicasting MANET. *Wireless Personal Communications* **66**(1): 1–23.

Liebeherr, J. and G. Dong (2007). An overlay approach to data security in ad-hoc networks. *Ad Hoc Networks* **5**(7): 1055–1072.

Malavenda, C. S. et al. (2012). Delay-tolerant, low-power protocols for large security-critical wireless sensor networks. *Journal of Computer Networks and Communications* **2012**: 1–10.

Mohanapriya, M. and I. Krishnamurthi (2013). Modified DSR protocol for detection and removal of selective black hole attack in MANET. *Computers & Electrical Engineering* **40**(2): 530–538.

Mukherjee, T. et al. (2009). Self-managing energy-efficient multicast support in MANETs under end-to-end reliability constraints. *Computer Networks* **53**(10): 1603–1627.

Padmavathy, N. and S. K. Chaturvedi (2013). Evaluation of mobile ad hoc network reliability using propagation-based link reliability model. *Reliability Engineering & System Safety* **115**: 1–9.

Pai, H.-T. and F. Wu (2011). Prevention of wormhole attacks in mobile commerce based on non-infrastructure wireless networks. *Electronic Commerce Research and Applications* **10**(4): 384–397.

Park, J. H. et al. (2011). Special issue of computer communications on information and future communication security. *Computer Communications* **34**(3): 223–225.

Qureshi, B. et al. (2010). Opportunistic trust based P2P services framework for disconnected MANETs. *Autonomic and Trusted Computing*, Xie, B., Branke, J., Sadjadi, S. M., Zhang, D., Zhou, X., (Eds.), Springer, Berlin, Heidelberg, pp. 154–167.

Qureshi, B. et al. (2012). Trusted information exchange in peer-to-peer mobile social networks. *Concurrency and Computation: Practice and Experience* **24**(17): 2055–2068.

Shim, K. and Y.-R. Lee (2005). Security flaws in authentication and key establishment protocols for mobile communications. *Applied Mathematics and Computation* **169**(1): 62–74.

Singh, A. B. (2012). Mobile banking based money order for India Post: Feasible model and assessing demand potential. *Procedia—Social and Behavioral Sciences* **37**: 466–481.

Stafrace, S. K. and N. Antonopoulos (2010). Military tactics in agent-based sinkhole attack detection for wireless ad hoc networks. *Computer Communications* **33**(5): 619–638.

Sundararajan, T. V. P. et al. (2014). Biologically inspired artificial intrusion detection system for detecting wormhole attack in MANET. *Wireless Networks* **20**(4): 563–578.

Takeuchi, M. et al. (2013). Improving assurance of a sustainable route-split MANET routing by adapting node battery exhaustion. *Telecommunication Systems* **54**(1): 35–45.

Tsalgatidou, A. and E. Pitoura (2001). Business models and transactions in mobile electronic commerce: Requirements and properties. *Computer Networks* **37**(2): 221–236.

Tsuchiya, T. et al. (2006). Counter-based reliability optimization for gossip-based broadcasting. *Computer Communications* **29**(9): 1516–1521.

Veijalainen, J. et al. (2006). Transaction management for m-commerce at a mobile terminal. *Electronic Commerce Research and Applications* **5**(3): 229–245.

Venkatesh, V. et al. (2012). Designing e-government services: Key service attributes and citizens preference structures. *Journal of Operations Management* **30**(1–2): 116–133.

Venkatraman, L. and D. P. Agrawal (2003). Strategies for enhancing routing security in protocols for mobile ad hoc networks. *Journal of Parallel and Distributed Computing* **63**(2): 214–227.

von Mulert, J. et al. (2012). Security threats and solutions in MANETs: A case study using AODV and SAODV. *Journal of Network and Computer Applications* **35**: 1249–1259.

Wang, J. et al. (2011). Building a trusted route in a mobile ad hoc network considering communication reliability and path length. *Journal of Network and Computer Applications* **34**(4): 1138–1149.

Wang, X. et al. (2014). Constructing a MANET based on clusters. *Wireless Personal Communications* **75**(2): 1489–1510.

Wang, X. and H. Qian (2013). A tree-based address configuration for a MANET. *Pervasive and Mobile Computing* **12**(June): 123–137.

Welch, C. (2013). Google's white spaces trial will beam broadband to ten South African schools. From: http://www.theverge.com/2013/3/25/4144946/google-white-spaces-trial-provides-broadband-ten-south-african-schools. Accessed May 15, 2013.

Whitson, J. C. and J. E. Ramirez-Marquez (2009). Resiliency as a component importance measure in network reliability. *Reliability Engineering & System Safety* **94**(10): 1685–1693.

Workman, M. et al. (2008). A structuration agency approach to security policy enforcement in mobile ad hoc networks. *Information Security Journal: A Global Perspective* **17**(5–6): 267–277.

Xiaonan, W. and Q. Huanyan (2013). Cluster-based and distributed IPv6 address configuration scheme for a MANET. *Wireless Personal Communications* **71**(4): 3131–3156.

Xie, X. et al. (2008). Link reliability based hybrid routing for tactical mobile ad hoc network. *Journal of Systems Engineering and Electronics* **19**(2): 259–267.

Yuste, A. J. et al. (2013). Type-2 fuzzy decision support system to optimise MANET integration into infrastructure-based wireless systems. *Expert Systems with Applications* **40**(7): 2552–2567.

Zhao, X. et al. (2013). A novel two-terminal reliability analysis for MANET. *Journal of Applied Mathematics* **2013**: 1–9.

Zhen, Y. et al. (2010). Toward path reliability by using adaptive multi-path routing mechanism for multimedia service in mobile ad-hoc network. *The Journal of China Universities of Posts and Telecommunications* **17**(1): 93–100.

Zyba, G. et al. (2011). Dissemination in opportunistic mobile ad-hoc networks: The power of the crowd. In *Proceedings of IEEE INFOCOM Mini Conference*, San Diego, CA, 2011.

chapter four

Security of wireless ad hoc network

Manmohan Chaturvedi, Preeti Aggarwal, Shilpa Bahl, and Sapna Malik

Contents

4.1 *Introduction and background*

4.1.1 *Network*

A network is collection of entities connected to each other forming a topological structure. The entities can be a personal computer, mobile device, etc. The aim of providing a topological structure to the network is to maintain a path of communication and give direction peer-to-peer communication. A network can be broadly classified as local area network (LAN), metropolitan area network (MAN), and wide area network (WAN).

LAN: It is defined as a network wherein computers are interconnected within a limited area of about a kilometer. This limited area can be an office, school, home, etc. (see Figure 4.1).

MAN: It is defined as a network in which two or more devices or networks are interconnected. The condition associated with this is that the communicating devices or networks are geographically separated but exist in the same metropolitan city. The metropolitan areas span over a geographical area of about 50 km(s) (see Figure 4.2).

WAN: It interconnects a number of LAN and MAN. The communication over WAN can be over a public or private network (see Figure 4.3).

NEED FOR COMPUTER NETWORK

- File sharing
- Printer sharing
- Communication and collaboration
- Organization
- Remote access
- Data protection

4.1.2 *Network topologies*

Different types of topologies through which a network can be defined are as follows:

Bus: This type of network was widely used in the 1980s. In this configuration, every computer (node) shares the networks total bus capacities. In addition, adding more computers will reduce the access speed on the network. Each computer communicates to other computers on the network independently. This is referred to as *peer-to-peer* networking (see Figure 4.4).

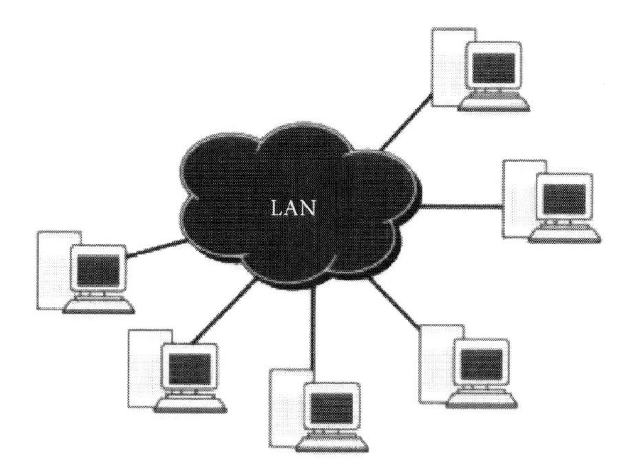

Figure 4.1 Local area network.

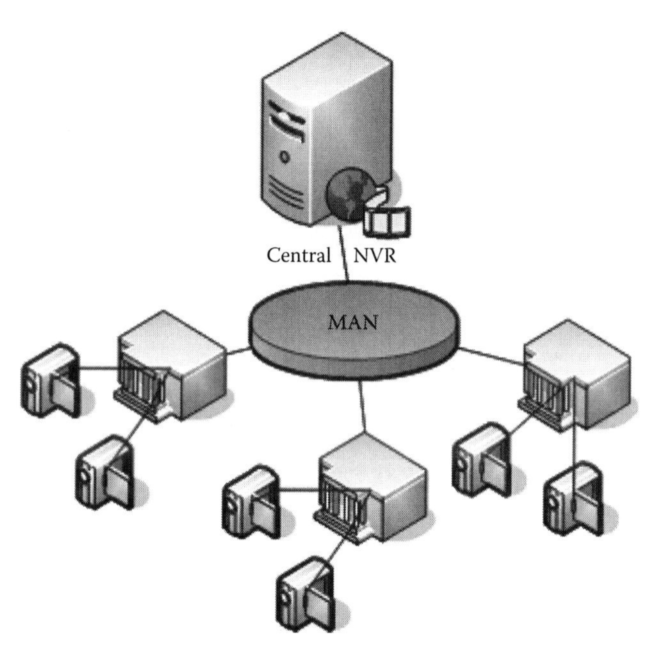

Figure 4.2 Metropolitan area network.

Ring: In this topology, each node is connected to the two nearest nodes so the entire network forms a circle. Data only travel in one direction on this network. There is no central server in ring topology, and it works on the basis of token methodology (see Figure 4.5).

Star: In this topology, every node is connected through a central device such as a hub, switch, or router. Compared to ring and bus topologies, star topology requires that more thought be put into its setup (see Figure 4.6).

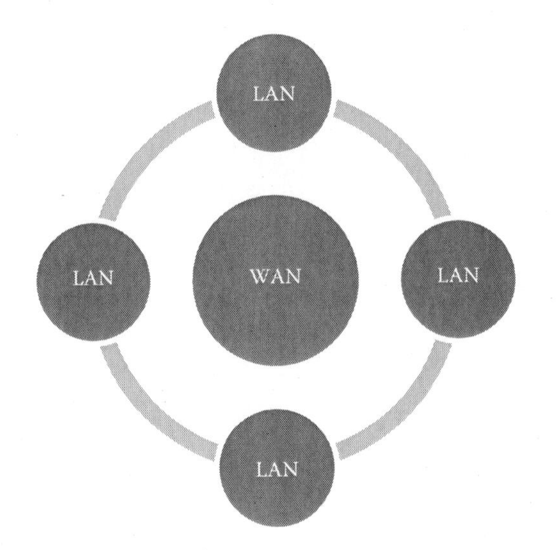

Figure 4.3 Wide area network.

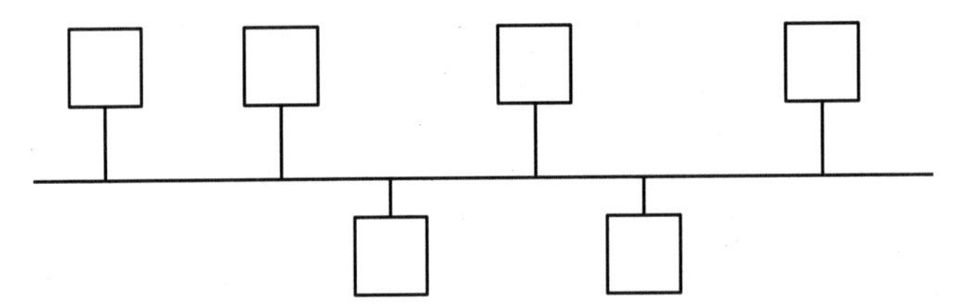

Figure 4.4 Bus topology.

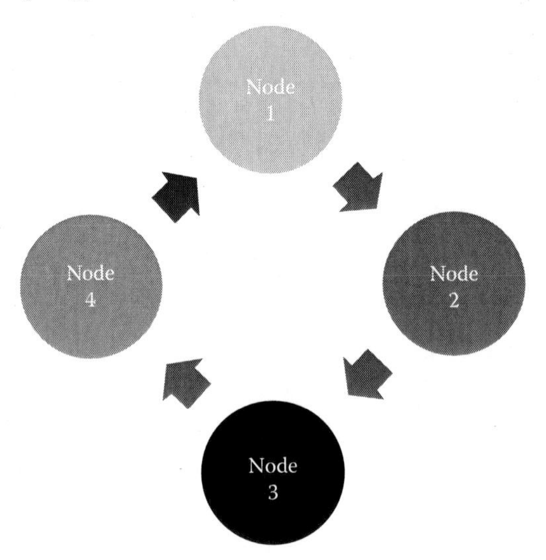

Figure 4.5 Ring topology.

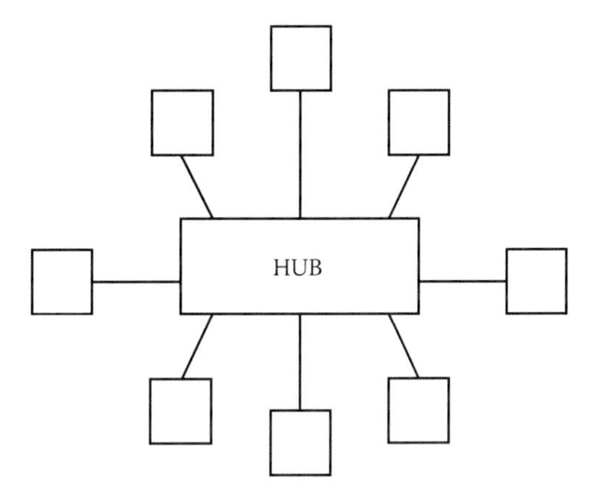

Figure 4.6 Star topology.

Tree: It is a variation of the bus topology. The shape of this type of network is that of a tree with central root branching and subbranching to the extremities of the network. There is no need of removing packets from the medium because when a signal reaches the end of the medium, it is aborted by the terminators (see Figure 4.7).

Mesh: Each node is connected to more than one node to provide an alternative route in case the host is either down or too busy. This topology is excellent for long-distance networking and is used in large internetworking environments with stars, rings, and buses attached to each node (see Figure 4.8).

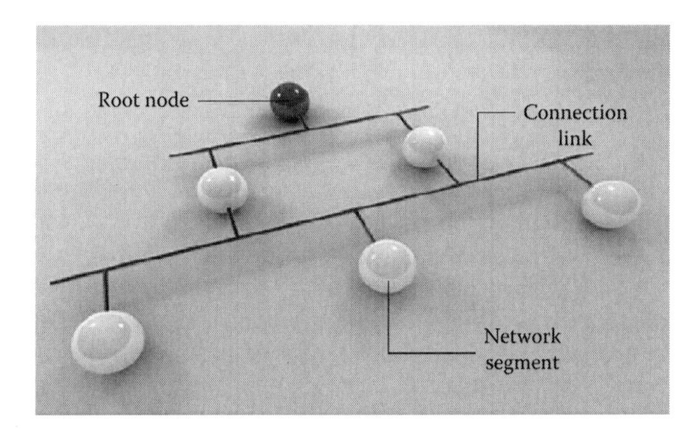

Figure 4.7 Tree topology.

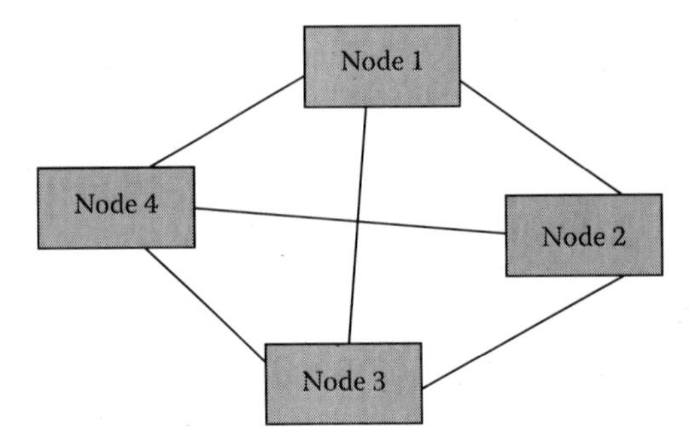

Figure 4.8 Mesh topology.

4.1.3 *Types of network*

4.1.3.1 *Wired network*

Wired network is simply a collection of two or more computers, printers, and other devices linked by Ethernet cables. Wired networks can also be used as part of other wired and wireless networks. Ethernet is the fastest wired network protocol, with connection speeds from 10 megabits per second (Mbps) to 100 Mbps or higher. Typically, the range of a wired network is within a radius of 2000 ft.

4.1.3.2 *Wireless network*

Wireless network, which uses high-frequency radio waves rather than wires to communicate between nodes, is another option for home or business networking. Individuals and organizations can use this option to expand their existing wired network or to go completely wireless. Wireless allows for devices to be shared without networking cable, which increases mobility but decreases range. There are two main types of wireless networking: peer to peer or ad hoc and infrastructure.

4.1.3.3 *Ad hoc wireless network*

Ad hoc or peer-to-peer wireless network consists of a number of computers each equipped with a wireless networking interface card. Each computer can communicate directly with all the other wireless-enabled computers. They can share files and printers this way but may not be able to access wired LAN resources unless one of the computers acts as a bridge to the wired LAN using special software. Mobile ad hoc networks (MANETs) are a kind of ad hoc wireless network that usually has a routable networking environment on top of a Link Layer ad hoc network (Hu et al., 2003).

4.1.4 Comparing wireless network and wired network

The key advantage of wireless networks over wired network emerges from the major difference between the two; that is, one uses network cables and the other uses radio frequencies (see Figure 4.9).

- Though wireless networking is a lot more mobile than wired networking, the range of the network is usually 150–300 indoors and up to 1000 ft outdoors depending on the terrain, but wired network is much more secure than wireless network, and transmission speeds can suffer from outside interference.
- Wired networks are inexpensive compared with wireless network installation, for wireless adapters and access points may cost three or four times as much as Ethernet cable adapters and hubs/switches. Broadband routers cost more, but these are optional components of a wired network, and their higher cost is offset by the benefit of easier installation and built-in security features.
- Wired LANs offer superior performance compared with wireless networks as it degrades because of distance sensitivity; that is, maximum performance will degrade on computers farther away from the access point or other. A traditional wired Ethernet connection offers only 10 Mbps bandwidth, but 100 Mbps Fast Ethernet technology costs a little more and is readily available. Fast Ethernet should be sufficient for file sharing, gaming, and high-speed Internet access. But greater mobility of wireless LANs helps offset the performance disadvantage.
- Wireless LANs are less secure than wired LANs because wireless communication signals travel through the air and can easily be intercepted. But wireless networks can be made as secure as wired networks are for the ease of mobility; that is, wireless networks protect their data through the wired equivalent privacy (WEP) encryption.

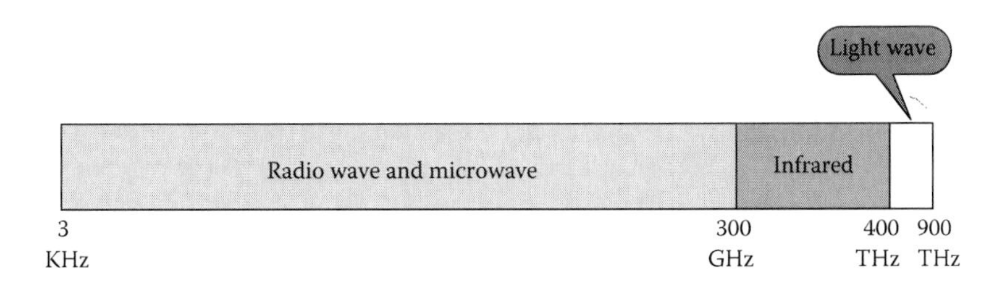

Figure 4.9 Radio frequency distribution.

4.1.5 Wireless network standards

There are four basic types of transmission standards for wireless networking. These types are produced by the Institute of Electrical and Electronic Engineers (IEEE) (Hubaux et al., 2011). These standards define all aspects of radio frequency wireless networking. They have established four transmission standards: 802.11, 802.11a, 802.11b, and 802.11g.

The basic differences between these four types are connection speed and radio frequency. Transmission standards 802.11 and 802.11b are the slowest at 1 or 2 Mbps and 5.5 and 11 Mbps, respectively. They both operate at the 2.4 GHz radio frequency range. Standard 802.11a operates at the 5 GHz frequency range and can transmit up to 54 Mbps, and 802.11g operates at the 2.4 GHz frequency range and can transmit up to 54 Mbps. Actual transmission speeds vary depending on such factors as the number and size of physical barriers within the network and any interference in the radio transmissions.

4.2 Security issues in wireless network

In the arena of technical advancements, it is the need of the hour to ensure safety and security of hardware, software, and data that are either present on the computer entities or on the network.

The three key components of security can be defined as prevention, detection, and reaction (Hu et al., 2003).

Security targets the features of integrity, availability, authenticity, confidentiality, and nonrepudiation (Nogueira et al., 2009), whereas safety ensures physical safety and graceful degradation of the system. These security features are modeled to design the policies for information security (Yang et al., 2004).

Integrity: Integrity of information ensures that the information is accurate and trustworthy.

Availability: Availability ensures that the information is readily accessible to the authorized people.

Authenticity: Authenticity involves two parties: the sender and the receiver. The sender should be genuine, and the receiver can authorized be only if it is the intended receiver. For example, digital signature is one mechanism to authenticate the user.

Confidentiality: Confidentiality targets that the information is not disclosed to unauthorized or unintended users.

Nonrepudiation: Nonrepudiation means that neither the sender nor the receiver can deny sending or receiving the information, respectively, at a later stage.

4.2.1 Security issues

4.2.1.1 Lack of secure boundaries

Mobile networks are generally more prone to physical security threats than are fixed cable networks. They do not have clear defined secure boundaries that can be compared with the clear line of defense in the traditional wired network. Mobile ad hoc networks face these threats due to their nature of freedom to join, leave, and move inside the network. There are increased possibilities of spoofing and denial-of-service (DoS) attacks in these networks.

As in the mobile ad hoc network, there is no need for an adversary to gain physical access to visit the network: once the adversary is in the radio range of any other nodes in the mobile ad hoc network, it can communicate with those nodes in its radio range and thus join the network automatically. As a result, the mobile ad hoc network does not provide the so-called secure boundary to protect the network from some potentially dangerous network access.

4.2.1.2 Threats from compromised nodes inside the network

As there are no secure boundaries defined in a mobile ad hoc network, it causes the occurrences of various link attacks. These link attacks place their emphasis on the links between the nodes and try to perform some malicious behavior to make destruction to the links. Due to some other attacks, they try to gain the control over the nodes themselves by some unrighteous means and then use the compromised nodes to execute further malicious action.

As nodes are free to leave and join, they do not have any strict polices for themselves to prevent possible malicious behaviors from all the nodes it communicate with because of the behavioral diversity of different nodes. The compromised node can frequently change their attack target; thus, it is very difficult to track the malicious behavior of a compromised node, especially in a large-scale ad hoc network. Therefore, threats from compromised nodes inside the network are far more dangerous than the attacks from outside the network, and these attacks are much harder to detect because they come from compromised nodes, which behave well before they are compromised.

4.2.1.3 Lack of centralized management facility

There is no centralized management system in mobile ad hoc networks that can detect the attacks and can control the incoming and outgoing traffic because it is highly dynamic. In mobile ad hoc networks, the anomalous failures like path breakages, transmission impairments, and packet dropping happen frequently. Therefore, malicious failures will be more difficult to detect, especially when adversaries change their attack pattern and their attack target at different periods of time.

Lack of centralized management machinery also affects trust management for the nodes in the ad hoc network as all the nodes have to cooperate for network operations so it becomes difficult to judge trusted and nontrusted nodes.

The other important perspective on the lack of centralized management service is that some algorithms in the mobile ad hoc network rely on the cooperative participation of all nodes and the infrastructure. Because there is no centralized authority, and decision making in mobile ad hoc network is sometimes decentralized, the adversary can make use of this vulnerability and perform some attacks that can break the cooperative algorithm.

4.2.1.4 *Restricted power supply*

In mobile ad hoc networks, due to their mobility, it is common for nodes to rely on batteries as their major source of power.

Due to this limitation, the adversary knows that the target node is battery-restricted; either it can continuously send additional packets to the target and ask it to route those additional packets, or it can induce the target to be trapped in some kind of time-consuming computations.

Sometimes, due to limitation, a node in the mobile ad hoc network may behave in a selfish manner when it finds that there is only limited power supply, and the selfishness can cause some problems when there is a need for this node to cooperate with other nodes to support some functions in the network. Some nodes may genuinely have exhausted their power supply, but there can be some other node that intentionally announces that it runs out of battery power and therefore do not want to cooperate with other nodes in some cooperative operation when actually this node still has enough battery power to support the cooperative operation.

4.2.1.5 *Scalability*

The term *scalability* in the mobile ad hoc network keeps on changing all the time: because of the mobility of nodes in the mobile ad hoc network, you can hardly predict how many nodes there will be in the network in the future, due to which the routing protocols and key management services have to be made compatible with the continuous changing scale of mobile ad hoc network as there are thousands of nodes participating in the network (see Figure 4.10).

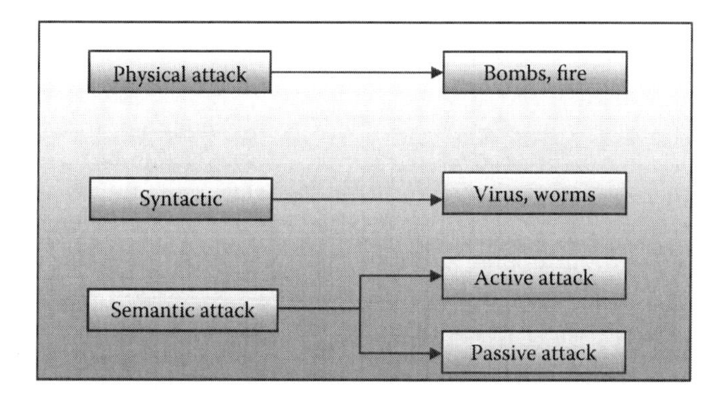

Figure 4.10 Types of attacks.

4.2.2 *Types of attacks*

Attack can be categorized as internal and external. The other category in which attacks can be placed is active and passive attacks.

Viruses: This is a malicious code that requires that you perform some action before it infects your computer like opening an email attachment or going to a particular web page (Nogueira et al., 2009).

Worms: Worms propagate without user intervention and start by exploiting software vulnerability. Similar to viruses, worms can propagate via email, websites, or network-based software. The automated self-propagation of worms distinguishes them from viruses (Nogueira et al., 2009).

Trojan horses: A Trojan horse program is software that does not let the user know its actual consequences. For example, a program that claims it will speed up your computer may actually be sending confidential information to a remote intruder (Nogueira et al., 2009).

Hacker, attacker, intruder: These terms are applied to the people who seek to exploit weaknesses in software and computer systems for their own gain. Although it is difficult to comment on one's intention for doing this because they may or may not cause direct harm to the end user, but DoS definitely deprives the end user to be properly served (Lavania et al., 2011).

Eavesdropping: It is an unethical act of secretly capturing the exchange of information between two ends without the knowledge of the sender and the receiver.

DoS: It is related to attacks against packet forwarding at network layer. Such attacks do not disrupt the routing protocol and poison the routing

Table 4.1 Attacks at Different Network Layers

Layer	Attack	Security measures
Application layer	Virus, worms, malicious codes	Detect and prevent application level abuses
Transport layer	Syn flooding	Authenticate and secure end-to-end communication
Network layer	Routing protocol	Safeguard the routing and forwarding protocols
	Forwarding protocol	
	Wormhole	
	Blackhole	
	Sinkhole	
	Flooding	
	Selective forward	
	Rushing	
	Sybil	
Data link layer	Exhaustion	Protect wireless MAC layer
	Collision	
Physical layer	Signal jamming	Avoid DoS attacks

states at each node. Instead, they cause data packets to be delivered in a way that is intentionally inconsistent with the routing states. For example, the attacker along an established route may drop packets, modify the content of the packets, or duplicate the packets it has already forwarded. Another type of packet forwarding attack is the DoS attack via network-layer packet blasting, in which the attacker injects a large amount of junk packets into the network. These packets waste a significant portion of the network resources and introduce severe wireless channel contention and network congestion (Table 4.1).

4.2.3 *Security issues in protocol stack*

4.2.3.1 *Measures*

There are basically two approaches to protecting ad hoc networks: proactive and reactive. Proactive approach attempts to prevent an attacker from launching attacks in the first place, typically through various cryptographic techniques (Hu et al., 2003).

In contrast, reactive approach seeks to detect security threats a posteriori and react accordingly. Due to the absence of a clear line of defense, a complete security solution for such networks should integrate both approaches and encompass all three components: prevention, detection, and reaction. For example, proactive approach can be used to ensure the correctness of routing states, while reactive approach can be used to protect packet forwarding operations. Through *cryptography,* information is protected by converting it into an unreadable format (cipher text). Only those who possess the secret key can decipher the message (Nogueira et al., 2009). Based on such network-layer vulnerabilities generally fall into one of two categories—routing attacks and packet forwarding attacks—based on the target operation of the attacks (Lavania et al., 2011). Some of the cryptographic algorithms are *RSA, DES,* and *AES,* which are quite popular.

By attacking the routing protocols, attackers can attract traffic toward certain destinations in the nodes under their control and cause the packets to be forwarded along a route that is not optimal or even nonexistent. Attackers can create routing loops in the network and introduce severe network congestion and channel contention in certain areas. Various secure ad hoc routing protocols are *source* routing, *distance vector* routing, and *link state* routing (Li and Joshi, 2008).

4.2.3.2 Defense method against wormhole attacks

In wormhole attacks, an attacker records packets (or bits) at one location in the network, tunnels them (possibly selectively) to another location, and replays them there into the network. The replay of the information will make an illusion, for the nodes that get the replayed packets cannot distinguish them from the genuine routing packets. Due to this, the routing functionality in the mobile ad hoc network will be severely interfered by the wormhole attack (Lavania et al., 2011).

Sunaina and Kumar (2013) have presented the mechanism to detect and to defend the attacks against wormhole. In this mechanism, there are two leashes (an additional information added to a packet designed to restrict the packet's maximum allowed transmission distance): *geographical leashes* and *temporal leashes.* A geographical leash ensures that the recipient of the packet is within a certain distance from the sender. A temporal leash ensures that the packet has an upper bound on its lifetime, which restricts the maximum travel distance, since the packet can travel at most at the speed of light. Either type of leashes can prevent the wormhole attack because it allows the receiver of a packet to detect if the packet travels farther than the leash allows.

4.2.3.3 Watchdog and pathrater

Watchdog and pathrater, as the names themselves suggest, are used to enhance the performance of wireless ad hoc networks in the presence of disruptive nodes. The job of watchdog is to determine the misbehavior of any node participating by copying packets to be forwarded into a buffer and monitoring the behavior of the adjacent node to these packets. Watchdog snoops to decide if the adjacent node forwards the packets without modifications. If the packets that are snooped match with the observing node's buffer, then they are discarded whereas packets that stay in the buffer beyond a time-out period without any successful match are flagged as having been dropped or modified. The node responsible for forwarding the packet is then noted as being suspicious. If the number of violations becomes greater than a certain predetermined threshold, the violating node is marked as being malicious. Information about malicious nodes is passed to the pathrater component for its corresponding job, that is, performing path rating evaluation.

The role of pathrater is to rate the nodes indicated as violated nodes by watchdog according to their reliability. Nodes are initially given a neutral rating, and the node if indicated by the watchdog as misbehaving is given an immediate rating of –100. It is distinguished that misbehavior is detected as packet mishandling/modification, whereas unreliable behavior is detected as link breaks.

SECURITY MEASURES

- Cryptography
- Access control
- Digital signature
- Firewall

References

Hu, Y., Perrig, A., Johnson, D., and Leashes, P. (2003). A defense against wormhole attacks in wireless ad hoc networks. *Proceedings of IEEE INFOCOM'03*, San Francisco, CA.

Hubaux, J.P., Buttyan, L., and Capkun, S. (2011). The quest for security in mobile ad hoc networks. *Proceedings of ACM Symposium on Mobile Ad hoc Networking and Computing*, Paris, France.

Lavania, K.K., Saini, G.L., Kothari Rooshabh, H., and Harshraj Yagnik, A. (2011). Privacy anxiety and challenges in mobile ad hoc wireless networks and its solution. *International Journal of Scientific & Engineering Research*, 2(9), 173–177.

Li, W. and Joshi, A. (2008). Security issues in mobile ad hoc networks—A survey. Department of Computer Science and Electrical Engineering, University of Maryland, Baltimore County, MD.

Nogueira, L.M., dos Santos, A.L., and Pujolle, G.A. (2009). Survey of survivability in mobile ad hoc network, supported by CAPES/Brazil, Grant No. 4253-05-1.

Sunaina, D. and Kumar, S. (2013). Survey on security threats of wireless and mobile ad hoc network. *International Journal of Data and Network Security*, 2(2), 4–8.

Yang, H., Luo, H.Y., Ye, F., Lu, S.W., and Zhang, L. (2004). Security in mobile ad hoc networks: Challenges and solutions. *IEEE Wireless Communications*, 11(1), 38–47.

section three

Mobile electronic commerce systems and the human perspective

chapter five

Mobile social networking service users' trust and loyalty

A structural approach

Norazah Mohd Suki and Norbayah Mohd Suki

Contents

5.1 Introduction: Background and driving forces

Social networking sites allow the individuals to create a profile and connect it to others' profiles, sharing text, images, and photos for the purpose of forming a personal network by applications and groups provided on the Internet (Boyd and Ellison, 2008). Social networking sites can also be accessed via mobile phones as they are portable and integrate the social connections brought about by social networking services (SNSs) and the communication channel. The arrival of social network sites is rapidly changing human interaction (Counts and Fisher, 2010; Zhong et al., 2011) that millions of people worldwide are living much of their lives on social networking sites such as Facebook, MySpace, Twitter, and LinkedIn.

The number of Facebook users continues to increase globally and is quickly becoming one of the most popular tools for social communication and entertainment. It had 167,431,700 active users in the United States and 13,577,760 in Malaysia in 2013 (Socialbakers, 2013). More than 30 billion pieces of content (web links, news stories, blog posts, notes, photo albums, etc.) are shared each month. There are over 900 million objects that people interact with (pages, groups, events, and community pages). Facebook introduced features such as wall, pokes, status, photos, news feed, tag, marketplace, instant messaging, and video. This study is helpful to mobile SNS providers as it discusses the effect of perceived user trust and perceived flow experience on users' loyalty toward their platform. If the mobile SNS provider wishes to retain and increase their customers, special attention will need to be paid toward the information and system quality provided. Apart from that, the management teams of SNS such as Facebook and Twitter will benefit from this study as well because they are capable of altering and providing the perceived enjoyment to their users. Hence, this study aims to examine the structural relationships of (1) perceived information quality and perceived system quality on perceived user trust and perceived flow, (2) perceived user trust on perceived flow, and (3) perceived user trust and perceived flow on mobile SNS users' loyalty simultaneously via structural equation modeling (SEM) approach.

This chapter is organized as follows. The next section review the literature on perceived flow, perceived system quality, perceived information quality, and perceived user trust and users' loyalty with deriving testable hypotheses. Section 5.3 describes the research methodology, while Section 5.4 reports the results of the study. The chapter rounds off with conclusions and direction for future research.

5.2 Factors influencing users' trust and loyalty

This section described the variables that influence mobile SNS users' trust and loyalty, which includes perceived information quality, perceived system quality, and perceived flow.

5.2.1 Perceived information quality and perceived user trust

Continuous exchange of information and access to, and understanding of, information are necessary conditions for establishing a learning process.

Users care about information quality, content richness, and navigation (Ilsever et al., 2007). Information quality is found to be an important trust-building mechanism (Fung and Lee, 1999; Keen et al., 2000). Reliability, relevance, and personalization of information exchange encourage the creation of a trust-based relationship (Yvette and Karine, 2001). Information quality is also the antecedent of overall user satisfaction (Aggelidis and Chatzoglou, 2012; Chang et al., 2012), which influences the perceived value of the e-commerce system (Wang, 2008). Conversely, Zhou (2013) found an insignificant relationship between information quality and user satisfaction. Therefore, it is hypothesized that

H1. Perceived information quality of mobile SNS significantly influences perceived user trust.

5.2.2 Perceived information quality and perceived flow

Information quality presented on the Internet has a significant impact on the user flow experience (Chau et al., 2000) in terms of the pleasure in using mobile SNS when conduct multiple tasks while surfing the Internet to acquire information or entertaining themselves. For instance, users will most likely form negative perceptions about the information quality of mobile SNS platform if the mobile service provider cannot provide accurate, comprehensive, and timely information to its users. Accordingly, it is posited that

H2. Perceived information quality of mobile SNS significantly influences perceived flow.

5.2.3 Perceived system quality and perceived user trust

System quality such as navigational structure and visual appeal influences user trust in mobile commerce technologies (Vance et al., 2008). Website quality tends to be a stronger predictor of trusting beliefs (McKnight et al., 2002). For example, technical aspects of IT artifacts affect users' willingness to trust (Gefen et al., 2006). Prior studies found that system quality influences user satisfaction (Aggelidis and Chatzoglou, 2012; Zhou, 2013) and perceived ease of use (Chen and Hsiao, 2012). Without efficient system quality, provision of quality services is difficult, which in turn diminishes flow experiences (Aladwani and Palvia, 2002). Hence, the following hypothesis is proposed:

H3. Perceived system quality of mobile SNS significantly influences perceived user trust.

5.2.4 *Perceived system quality and perceived flow*

System quality is related to the presence of a fast, reliable connection for navigation (Hsu and Lu, 2004; Nelson and Todd, 2005). The interplay of interactions between the system and the user, such as design factors like user interface, the navigation structure, the time that the system requires to process a request (Garrett, 2003; Chiou, 2005), influences the user's cumulative satisfaction. Deighton and Grayson (1995) affirmed that the flow generated during specific tasks at a website creates memorable experiences that are believed to strengthen the relationship. The optimal experience significantly influences the user's intention to revisit a website. Thus, the following hypothesis is posited:

H4. Perceived system quality of mobile SNS significantly influences perceived flow.

5.2.5 *Perceived user trust and perceived flow*

Trust, that is, belief that using mobile service will be free of security and privacy threats (Doney and Cannon 1997; Wang et al., 2003), will reduce perceived uncertainty and risks, thus reducing the effort spent on monitoring the mobile service provider, subsequently enhancing users' perceived control (Pavlou et al., 2007), and improving their experience. If users trust mobile service providers, they expect positive future experiences (Kim et al., 2009). Trust affects the online travel community users' flow experience (Wu and Chang, 2005). Therefore, it is proposed that

H5. Perceived user trust significantly influences perceived flow.

5.2.6 *Perceived user trust and loyalty*

Trust is one of the key factors when it comes to mobile success in commerce, as when customers register with a mobile SNS, they are giving away private information or personal data. Trust has been shown to be a key determinant of loyalty in offline and online environments (Berry and Parasuraman, 1991). For instance, m-loyalty was found to be influenced by trust (Lin and Wang, 2006). Norazah (2012) noted that customer trust toward the vendor in m-commerce is affected by responsiveness, brand image, and satisfaction toward the vendor in m-commerce. Further, trust and satisfaction positively influence e-loyalty (Cyr et al., 2006). Corbitt et al. (2003) noted a strong positive effect on trust on loyalty to online firms. Accordingly, it is hypothesized that

H6. Perceived user trust significantly influences loyalty.

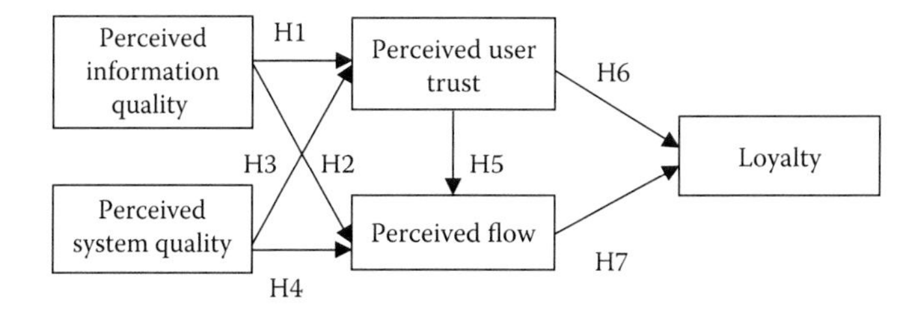

Figure 5.1 Theoretical framework.

5.2.7 *Perceived flow and loyalty*

In a study, it is presented that the flow affects online consumers' purchase and return intention (Hausman and Siekpe, 2009; Zhou and Lu, 2011). Users' satisfaction further determines their continuous use of mobile Internet services (Deng et al., 2010). User finds the flow experience in a certain mobile SNS to be satisfying and will most likely continue using it, which thus promotes loyalty. Understanding the influence of the flow on the trusting belief–loyalty relationship can therefore ensure positive loyalty outcomes (Gupta and Kabadayi, 2010). Thus, it is posited that

H7. Perceived flow significantly influences loyalty.

Figure 5.1 illustrates the proposed theoretical framework based on the aforementioned literature.

5.3 *Methodology*

The completed and usable close-ended questionnaire was circulated to 200 university students studying at the Universiti Malaysia Sabah Labuan International Campus from January 26 to 28, 2011 utilizing the convenience sampling technique as the method does allow the researcher to have any control over the representativeness of the sample. Initially, a total of 250 questionnaires were administered, leading to a response rate of 80%. The students who have at least a social networking site account such as Facebook, MySpace, Twitter, or LinkedIn and have experience in browsing it via mobile phone were randomly chosen in the university campus. The questionnaire comprised three sections. Section A consisted of demographic profile of respondents. Section B consisted of questions related to their mobile SNS experience in terms of the amount of time mobile SNS is used in a week, mobile SNS usage, and the most frequently visited mobile SNS. Section C examined the effect of flow experience on mobile SNS users' loyalty, which aimed to measure their perceived flow, perceived system quality, perceived information quality, and perceived

Table 5.1 Measurement of Variables

Variables	Authors
Perceived information quality	DeLone and McLean (1992)
Perceived system quality	Nelson and Todd (2005); Armstrong and Hagel (1996); Chiou (2005)
Perceived enjoyment	Davis et al. (1992); Koufaris (2002)
Perceived control	Macan et al. (1990)
Attention focus	Intriligator and Cavanagh (2001); Pylyshyn and Storm (1988)
Perceived user trust	Berry and Parasuraman (1991)
Loyalty	Lin and Wang (2006); Cyr et al. (2006)

user trust on mobile SNS users' loyalty. The questionnaire items, as available in Appendix 5.A, were adapted from sources as detailed in Table 5.1 and were measured on a five-point Likert scale ranging from 1 (strongly disagree) to 5 (strongly agree). SEM was utilized to empirically test and estimate the hypothesized relationships by using the AMOS version 20.0 computer software as it has the ability to ensure the consistency of the model with the data and to estimate effects among constructs instantaneously.

5.4 *Data analysis*

Table 5.2 portrays the frequency analysis of the demographic profile of respondents with 49% males and 51% females. More than three-quarter of the respondents aged 21–25 years earn a monthly allowance of lesser than RM300. In terms of mobile SNS experiences, 18% of the sample group use mobile SNS for more than 15 times in a week. More than one-quarter of the respondents have the experience of using mobile SNS for a period of more than 6 months. Facebook is the most frequently visited mobile SNS as compared with Twitter and MySpace.

5.4.1 *Factor analysis*

Perceived flow, which combined three dimensions, perceived enjoyment, perceived control, and attention focus, was developed via factor analysis using varimax rotation. These dimensions are interchangeable, are significantly correlated among each other (Koufaris, 2002; Huang, 2006), are driven by the same underlying construct as it has the same determinants and consequences that reflected perceived flow (Siekpe, 2005; Wang et al., 2007). Table 5.3 exemplifies that item loadings ranging from 0.685

Table 5.2 Socio-Demographic Profile of Respondents

	Frequency	Percentage
Gender		
Male	98	49.0
Female	102	51.0
Age (years)		
<20	26	13.0
21–25	168	84.0
26–30	4	2.0
>31	2	1.0
Monthly allowance		
<RM 300	160	80.0
RM 301–RM 400	13	6.5
RM 401–RM 500	7	3.5
>RM 501	20	10.0
Amount of time mobile SNS is used in a week		
1–5 times	125	62.5
6–10 times	32	16.0
11–15 times	7	3.5
>15 times	36	18.0
Experience of mobile SNS usage		
0–2 months	97	48.5
2–4 months	34	17.0
4–6 months	14	7.0
>6 months	55	27.5
Most frequently visited mobile SNS		
Facebook	147	73.5
MySpace	2	1.0
Twitter	5	2.5
MSN	9	4.5
Yahoo messenger	10	5.0
Others	27	13.5

to 0.848 with the item "I felt that using this mobile SNS is interesting" has a relatively highest loading on perceived flow with Kaiser-Meyer-Olkin Measure of Sampling Adequacy value of 0.896. Results have suppressed small coefficients of absolute value below 0.50 for item "when using this mobile SNS, I felt confused" was deleted as it does not meeting this requirement.

Table 5.3 Factor Analysis

Items	Perceived flow
I felt that using this mobile SNS is interesting	0.848
I felt that using this mobile SNS is enjoyable	0.841
I felt that using this mobile SNS is exciting	0.822
When using this mobile SNS, I was intensely absorbed in the activity	0.815
I felt that using this mobile SNS is fun	0.810
When using this mobile SNS, I concentrated fully on the activity	0.782
When using this mobile SNS, my intention was focused on the activity	0.781
When using this mobile SNS, I felt calm	0.721
When using this mobile SNS, I was deeply engrossed in the activity	0.719
When using this mobile SNS, I felt in control	0.685
Total variance explained	6.275
Percentage of variance explained	57.043
Kaiser-Meyer-Olkin Measure of Sampling Adequacy	0.896

5.4.2 *Structural equation modeling*

SEM is performed via two-step SEM approach: measurement model and structural model.

5.4.2.1 *Measurement model*

The measurement model was checked by inspecting item loading, construct reliability, and average variance extracted (AVE) as summarized in Table 5.4. The Cronbach's alpha values and composite reliability values of all variables exceeded the acceptable level of 0.70, implying all variables are reliable and have high internal consistency. Moreover, each of the standardized loading items is greater than 0.50 on their expected factor, except PC2, PC3, AF4, indicating that the construct validity is acceptable. Convergent validity is also acceptable as all AVE values are greater than the cut-off value 0.50.

Discriminant validity is examined by comparing the shared variances between factors with the squared root of AVE for each construct. The results in Table 5.5 show that the shared variances of the construct with other constructs were lower than the squared root of AVE of the individual factors, confirming discriminant validity. Hence, each construct was statistically different from the others. Correlations between the variables are positively significant at 0.01 level and finds that all of them correlate. Next, the perceived flow had the highest mean of 4.608 whereas the loyalty had the highest standard deviation of 1.459.

Table 5.4 Item Loadings and Validities

	Estimate	Cronbach's alpha	Composite reliability	Average variance extracted
Perceived information		0.962	0.915	0.729
PIQ1	0.804			
PIQ2	0.838			
PIQ3	0.869			
PIQ4	0.902			
Perceived system quality		0.824	0.870	0.628
PSQ1	0.691			
PSQ2	0.737			
PSQ3	0.846			
PSQ4	0.881			
Perceived flow		0.926	0.920	0.601
PE1	0.920			
PE2	0.916			
PE3	0.927			
PE4	0.910			
PC1	0.603			
AF1	0.635			
AF2	0.610			
AF3	0.544			
Perceived user trust		0.865	0.812	0.590
TR1	0.791			
TR2	0.756			
TR3	0.756			
Loyalty		0.853	0.894	0.737
LO1	0.863			
LO2	0.871			
LO3	0.841			

5.4.2.2 *Structural model*

The structural model was evaluated by examining fit indices and variance-explained estimates. The measure of the fit of the tested model was done through examining several goodness-of-fit indices (Table 5.6). The results indicated that the χ^2 of the model was 493.859 with 187 degrees of freedom (χ^2/df = 2.641). The fit indices value for comparative fit index (CFI), goodness of fit index (GFI), and normed fit index (NFI) were above 0.90 and root mean square error of approximation (RMSEA) below 0.08, indicating a satisfactory fit.

Table 5.5 Inter-Construct Correlations

	1	2	3	4	5
1. Perceived information quality	**0.854**				
2. Perceived system quality	0.756(**)	**0.792**			
3. Perceived flow	0.564(**)	0.750(**)	**0.775**		
4. Perceived user trust	0.625(**)	0.608(**)	0.643(**)	**0.768**	
5. Loyalty	0.628(**)	0.705(**)	0.701(**)	0.662(**)	**0.858**
Mean	4.465	4.580	4.608	4.287	4.430
Std. deviation	1.250	1.175	1.382	1.147	1.459
Skewness	−0.483	−0.591	−0.339	−0.219	−0.463
Kurtosis	0.357	0.761	0.023	0.846	0.047

**Correlation is significant at the 0.01 level (2-tailed).
Diagonals represent the square root of the AVE.

Table 5.6 Goodness-of-Fit Indices for Structural Model

	χ^2	df	χ^2/df	CFI	GFI	NFI	RMSEA	PNFI	PCFI
Recommended values	N/A	N/A	<3.0	>0.9	>0.9	>0.9	<0.08	>0.5	>0.5
Hypothesized model values	493.859	187	2.641	0.938	0.962	0.982	0.061	0.746	0.714

The results in Figure 5.2 enumerate that perceived information quality and perceived system quality accounted for 44% of the total variance in perceived user trust ($R^2 = 0.437$) and 75% in perceived flow ($R^2 = 0.751$). There is 76% variance in mobile SNS users' loyalty explained by perceived user trust and perceived flow ($R^2 = 0.759$). Overall, the results signify a sign of adequate model fit between the proposed research model and the empirical data.

The standardized path coefficients for the hypotheses testing are summarized in Table 5.7 and Figure 5.2, implying all hypothesized paths were significant at $p < 0.05$. Specifically, perceived information quality ($\beta_1 = 0.507$, $p < 0.05$) and perceived system quality ($\beta_3 = 0.424$, $p < 0.05$) significantly influence perceived users trust, supporting H1 and H3. Next, perceived flow is also significantly influenced by perceived information quality and perceived system quality of mobile SNS ($\beta_2 = -0.141$; $\beta_4 = 0.685$, $p < 0.05$). Thus, H2 and H4 are sustained. Likewise, significant relationship appears between perceived user trust and perceived flow ($\beta_5 = 0.338$, $p < 0.05$), supporting H5. Moreover, mobile SNS users' loyalty

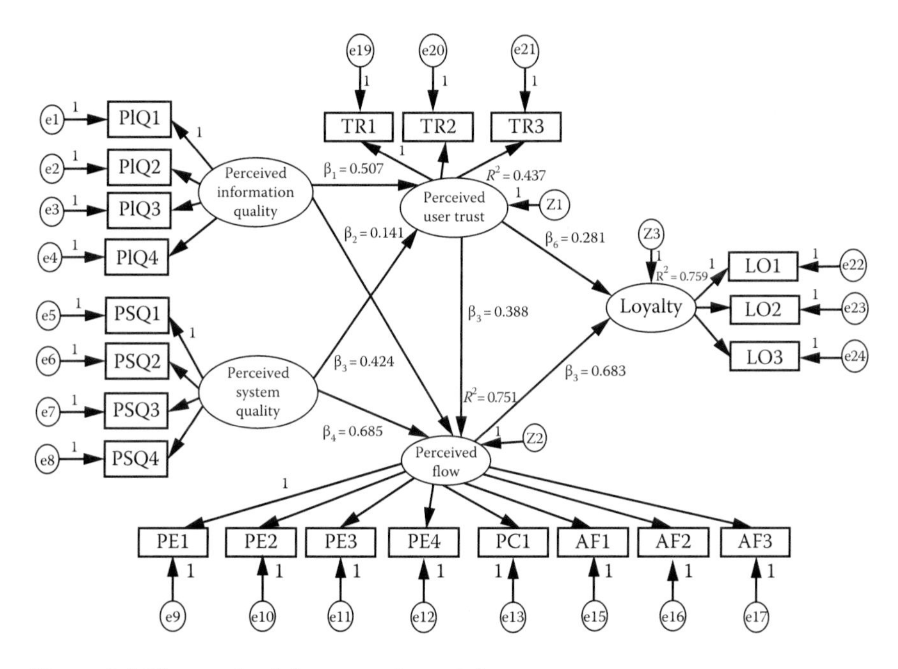

Figure 5.2 The result of the research model.

Table 5.7 Summarized Results of Hypothesis Testing

Causal paths			Estimate	*t*-value	*p*-value	Hypothesis
Perceived information quality	→	Perceived user trust	0.507*	6.623	0.000	H1 supported
Perceived information quality	→	Perceived flow	−0.141*	−2.397	0.017	H2 supported
Perceived system quality	→	Perceived user trust	0.424*	5.389	0.000	H3 supported
Perceived system quality	→	Perceived flow	0.685*	8.720	0.000	H4 supported
Perceived user trust	→	Perceived flow	0.338*	4.620	0.000	H5 supported
Perceived user trust	→	Loyalty	0.281*	4.384	0.000	H6 supported
Perceived flow	→	Loyalty	0.683*	10.533	0.000	H7 supported

*$p < 0.05$.

is significantly affected by both perceived user trust ($\beta_6 = 0.281$, $p < 0.05$) and perceived flow ($\beta_7 = 0.683$, $p < 0.05$), suggesting support for H6 and H7.

The total effects of the variable on the dependent variable of the models refer to the sum of direct and indirect effects. Table 5.8 indicates some strong indirect effects within the relationships between constructs.

Table 5.8 Direct and Indirect Effects

Independent variable	Mediating variable	Dependent variable	Direct effect	Indirect effect	Total effect
Perceived information quality	—	Perceived user trust	0.507	0.000	0.507
Perceived information quality	Perceived user trust	Perceived flow	−0.141	0.171	0.030
Perceived system quality	—	Perceived user trust	0.424	0.000	0.424
Perceived system quality	Perceived user trust	Perceived flow	0.685	0.143	0.828
Perceived user trust	—	Perceived flow	0.338	0.000	0.338
Perceived user trust	Perceived flow	Loyalty	0.281	0.231	0.512
Perceived flow	—	Loyalty	0.683	0.000	0.683

5.5 Discussion

This study examined the structural relationship between perceived information quality, and perceived system quality, perceived user trust, and perceived flow and the effects of (1) perceived information quality and perceived system quality on perceived user trust and perceived flow, (2) perceived user trust on perceived flow, and (3) perceived user trust and perceived flow on mobile SNS users' loyalty simultaneously via SEM approach using maximum likelihood for the estimation method as it provides a consistent approach to parameter estimation problems that can be developed for a large variety of estimation situations. Empirical results provide information necessary to investigate the study hypotheses. Each standardized path coefficients within the model was assessed as significant at the 0.05 level, and the overall model was determined to fit the data well. Perceived information quality highly influences perceived user trust followed by perceived system quality, satisfying H1 and H3. This result is comparable to Wong and Hsu (2008), which notes that the quality of the information posted on the company website or mobile portal has a direct impact on potential customers' perception and trust of its products and services. Sledgianowski and Kulviwat (2009) and Davis (1989) found that the user's thinking as to the usefulness of a system has a great influence and is positively related to the adoption of information technology, thereby developing higher trust in the mobile SNS platform.

Perceived flow is significantly influenced by perceived information quality and perceived system quality of mobile SNS; thus, H2 and H4 are supported. The preceding research (Pilke, 2004) found that informative and pleasant visualization was quite frequently mentioned by participants as an element that facilitates the flow. If the information provided via the mobile SNS is accurate and is of high quality, users will have a better perceived flow. Zhou et al. (2010) noted that mobile SNS platforms that are unreliable and have slow responses or where services are abruptly interrupted will seriously affect users' experience, including enjoyment, attention focus, and control, which are the three reflective dimensions of the flow. Further investigation of the study found that H5 is sustained when users trust the mobile SNS provider and have a better perceived flow compared with those in doubt. Users reported that being confident was important and stressed the value of being able to explore new things in the online environment (Hassanein and Head, 2007).

Corresponding to H6, Gupta and Kabadayi (2010) found that trust beliefs affect users' loyalty toward websites. Thus, it can be said that a high perceived user trust in a certain mobile SNS platform will ensure customer loyalty. Compared with the effect of perceived user trust, perceived flow has a larger effect, indicating perceived flow as the strongest determinant of users' loyalty. Parallel to the significant finding for H7, Palka et al. (2009) stated that providing users with a good experience will promote their continuous use and generate positive word-of-mouth. Kim and Son (2009) revealed the significant effect of net benefits, including perceived usefulness and satisfaction, on user loyalty toward online services. Also, as proposed by Zhou et al. (2010), improving user experience is one of the main ways to enhance the user's loyalty toward mobile SNS. Consequently, if a user finds that the perceived flow from a mobile SNS provider above satisfactory, there are high possibilities that the user will stay loyal.

5.6 Conclusion and perspectives

Empirical results, via SEM, revealed that both perceived information quality and perceived system quality affect perceived flow and perceived user trust, which further influences mobile SNS users' loyalty. Perceived user trust affects perceived flow, and both factors influence user loyalty. Indeed, perceived flow is regarded as the strongest factor that significantly affects users' loyalty on mobile SNS. Mobile SNS providers need to consider user perceived flow to enhance users' loyalty. Thus, mobile SNS providers should implement steps to increase perceived information quality and perceived system quality in order to build and enhance users' trust level and further provide users with a compelling experience. They may need to emphasize loyalty program in an attempt to retain customers in the competitive telecommunication market.

Further research is deemed essential to expand the sample size and investigate at different geographical locations by considering several social groups beyond Malaysia, such as other Asian countries, Europe, and the United States to improve the generalizability of findings as most of the current samples are students from the very same university and the studies lean more toward students and not the general public or adults, as different groups of working professionals would use mobile SNS differently that introduces a big bias. However, this research provides important theoretical contributions as most existing studies only addressed the concern of user acceptance of online social networking sites utilizing the theory of technology acceptance model. This study adopts a comprehensive approach to explain structural relationships of perceived flow, perceived system quality, perceived information quality, and perceived user trust with mobile SNS users' loyalty. In addition, the research model supports the integration of cross-disciplinary studies in virtual community research.

On the practical side, the results provide some tangible recommendations for helping enhance their SNS users' loyalty, thereby bringing continued profitability and business success. It is also important for policy makers to make efforts to outline strategies through its white papers to encourage the use of social networking sites as an avenue to strengthen business activities more competitively by emphasizing the aspects of perceived information quality, perceived system quality, perceived flow, and perceived user trust on mobile SNS users' loyalty.

5.A Appendix

Measurement of Instruments

Perceived Information Quality

The information provided by this mobile SNS is what I need.
The information provided by this mobile SNS is accurate.
The information provided by this mobile SNS is up-to-date.
The information provided by this mobile SNS is comprehensive.

Perceived System Quality

This mobile SNS is reliable.
This mobile SNS provides fast responses to my inquiries.
This mobile SNS is easy to use.
This mobile SNS provides good navigation functions.

Perceived Enjoyment

I felt that using this mobile SNS is fun.
I felt that using this mobile SNS is exciting.
I felt that using this mobile SNS is enjoyable.
I felt that using this mobile SNS is interesting.

Perceived Control

When using this mobile SNS, I felt calm.
When using this mobile SNS, I felt in control.
When using this mobile SNS, I felt confused.

Attention Focus

When using this mobile SNS, I was intensely absorbed in the activity.
When using this mobile SNS, my attention was focused on the activity.
When using this mobile SNS, I concentrated fully on the activity.
When using this mobile SNS, I was deeply engrossed in the activity.

Trust

This mobile SNS has the necessary knowledge and ability to fulfill its tasks.

This mobile SNS will keep its promises.

This mobile SNS is concerned with its users' interests, not just its own benefits.

Loyalty

I will continue using this mobile SNS.
I will recommend this mobile SNS to other users.
When using this mobile SNS, I consider it to be my first choice.

References

Aggelidis, V.P. and P.D. Chatzoglou. 2012. Hospital information systems: Measuring end user computing satisfaction (EUCS). *Journal of Biomedical Informatics* 45 (2012): 566–579.

Aladwani, A.M. and P.C. Palvia. 2002. Developing and validating an instrument for measuring user-perceived web quality. *Information & Management* 39 (6): 467–476.

Armstrong, A.G. and J. Hagel. 1996. The real value of online communities. *Harvard Business Review* 73 (3): 134–141.

Berry, L.L. and A. Parasuraman. 1991. *Marketing Services: Competing through Quality*. New York: The Free Press.

Boyd, D.M. and N.B. Ellison. 2008. Social network sites: Definition, history and scholarship. *Journal of Computer-Mediated Communication* 13 (1): 210–230.

Chang, I.C., Y.C. Li, T.Y. Wu, and D.C. Yen. 2012. Electronic medical record quality and its impact on user satisfaction—Healthcare providers' point of view. *Government Information Quarterly* 29 (2012): 235–242.

Chau, P.K., G. Au, and K.Y. Tam. 2000. Impact of information presentation modes on online shopping: An empirical evaluation of a broadband interactive shopping service. *Journal of Organizational Computing Electronic Commerce* 10 (1): 1–22.

Chen, R.F., and J.L. Hsiao. 2012. An investigation on physicians' acceptance of hospital information systems: A case study. *International Journal of Medical Informatics* 81 (2012): 810–820.

Chiou, J.S. 2005. The antecedents of consumers' loyalty toward Internet service providers. *Information & Management* 41 (6): 685–695.

Corbitt, B.J., T. Thanasankit, and H. Yi. 2003. Trust and e-commerce: A study of consumer perceptions. *Electronic Commerce Research and Applications* 2 (6): 737–758.

Counts, S. and K.E. Fisher. 2010. Mobile social networking as information ground: A case study. *Library & Information Science Research* 32 (2): 98–115.

Cyr, D., M. Head, and A. Ivanoc. 2006. Design aesthetics leading to m-loyalty in mobile commerce. *Information & Management* 43 (8): 950–963.

Davis, F.D. 1989. Perceived usefulness, perceived ease of use, and user acceptance of information technology. *MIS Quarterly* 13 (3): 319–340.

Davis, F.D., R.P. Bagozzi, and P.R. Warshaw. 1992. Extrinsic and intrinsic motivation to use computers in the workplace. *Journal of Applied Social Psychology* 22 (14): 1111–1132.

Deighton, J. and K. Grayson. 1995. Marketing and seduction: Building exchange relationships by managing social consensus. *Journal of Consumer Research* 21 (4): 660–676.

DeLone, W.H. and E.R. McLean. 1992. Information systems success: The quest for the dependent variable. *Information Systems Research* 3 (1): 154–171.

Deng, L., D.E. Turner, R. Gehling, and B. Prince. 2010. User experience, satisfaction, and continual usage intention of IT. *European Journal of Information Systems* 19 (1): 60–75.

Doney, P.M. and J.P. Cannon. 1997. An examination of the nature of trust in buyer-seller relationships. *Journal of Marketing* 61 (2): 35–51.

Fung, R.K.K. and M.K.O. Lee. 1999. EC-Trust (trust in electronic commerce): Exploring the antecedent factors. In Haseman, W.D. and Nazareth, D.L. eds., *Proceedings of the Fifth American Conference on Information Systems*, August 13–15, Milwaukee, WI: Omnipress, pp. 517–519.

Garrett, J.J. 2003. *The Elements of User Experience: User-Centered Design for the Web*. Indianapolis, IN: New Riders.

Gefen, D., P.A. Pavlou, I. Benbasat, D.H. McKnight, K. Stewart, and D.W. Straub. 2006. Should institutional trust matter in information systems research? *Communications of the AIS* 19 (7): 205–222.

Gupta, R. and S. Kabadayi. 2010. The relationship between trusting beliefs and web site loyalty: The moderating role of consumer motives and flow. *Psychology & Marketing* 27 (2): 166–185.

Hassanein, K. and M. Head. 2007. Manipulating social presence through the web interface and its impact on consumer attitude towards online shopping. *International Journal of Human Computer Studies* 65 (8): 689–708.

Hausman, A.V. and J.S. Siekpe. 2009. The effect of web interface features on consumer online purchase intentions. *Journal of Business Research* 62 (1): 5–13.

Hsu, C.L. and H.P. Lu. 2004. Why do people play games? An extended TAM with social influences and flow experience. *Information & Management* 41 (7): 853–868.

Huang, M.H. 2006. Flow, enduring, and situational involvement in the web environment: A tripartite second-order examination. *Psychology & Marketing* 23 (5): 383–411.

Ilsever, J., D. Cyr, and M. Parent. 2007. Extending models of flow and e-loyalty. *Journal of Information Science and Technology* 4 (2): 10–13.

Intriligator, J. and P. Cavanagh. 2001. The spatial resolution of visual attention. *Cognitive Psychology* 43 (3): 171–216.

Keen, P., C. Ballance, S. Chan, and S. Schrump. 2000. *Electronic Commerce Relationships: Trust by Design*. Upper Saddle River, NJ: Prentice Hall.

Kim, G., B. Shin, and H.G. Lee. 2009. Understanding dynamics between initial trust and usage intentions of mobile banking. *Information Systems Journal* 19 (3): 283–311.

Kim, S.S. and J.Y. Son. 2009. Out of dedication or constraint? A dual model of post-adoption phenomena and its empirical test in the context of online services. *MIS Quarterly* 33 (1): 49–70.

Koufaris, M. 2002. Applying the technology acceptance model and flow theory to online consumer behavior. *Information Systems Research* 13 (2): 205–223.

Lin, H. and Y. Wang. 2006. An examination of the determinants of customer loyalty in mobile commerce contexts. *Information & Management* 43 (3): 271–282.

Macan, T.H., C. Shahani, R.L. Dipboye, and A.P. Phillips. 1990. College students' time management: Correlations with academic performance and stress. *Journal of Educational Psychology* 82 (4): 760–768.

McKnight, D.H., V. Choudhury, and C. Kacmar. 2002. The impact of initial consumer trust on intentions to transact with a Web site: A trust building model. *Journal of Strategic Information Systems* 11 (3–4): 297–323.

Nelson, R.R. and P.A. Todd. 2005. Antecedents of information and system quality: An empirical examination within the context of data warehousing. *Journal of Management Information Systems* 21 (4): 199–235.

Norazah, M.S. 2012. Examining factors influencing customer satisfaction and trust towards vendors on the mobile Internet. *Journal of Internet Banking and Commerce* 17 (1): 1–12.

Palka, W., K. Pousttchi, and D.G. Wiedemann. 2009. Mobile word-of-mouth—A grounded theory of mobile viral marketing. *Journal of Information Technology* 24 (2): 172–185.

Pavlou, P.A., H. Liang, and Y. Xue. 2007. Understanding and mitigating uncertainty in online exchange relationships: A principal-agent perspective. *MIS Quarterly* 31 (1): 105–136.

Pilke, E.M. 2004. Flow experience in information technology use. *International Journal of Human-Computer Studies* 61 (3): 347–357.

Pylyshyn, Z.W. and R.W. Storm. 1988. Tracking multiple independent targets: Evidence for a parallel tracking mechanism. *Spatial Vision* 3 (3): 179–197.

Siekpe, J.S. 2005. An examination of the multidimensionality of the flow construct in a computer-mediated environment. *Journal of Electronic Commerce Research* 6 (1): 31–43.

Sledgianowski, D. and S. Kulviwat. 2009. Using social network sites: The effects of playfulness, critical mass and trust in a hedonic context. *Journal of Computer Information Systems* 49 (4): 74–83.

Socialbakers. 2013. Malaysia Facebook statistics. Retrieved from http://www.socialbakers.com/facebook-statistics/malaysia. Accessed January 24, 2013.

Vance, A., E.D.C. Christophe, and D.W. Straub. 2008. Examining trust in information technology artefacts: The effects of system quality and culture. *Journal of Management Information Systems* 24 (4): 73–100.

Wang, L.C., J. Baker, J.A. Wagner, and K. Wakefield. 2007. Can a retail web site be social? *Journal of Marketing* 71 (3): 143–157.

Wang, Y. 2008. Assessing e-commerce systems success: A respecification and validation of the DeLone and McLean model of IS success. *Information System Journal* 18 (1): 529–557.

Wang, Y.S., Y.M. Wang, H.H. Lin, and T.I. Tang. 2003. Determinants of user acceptance of Internet banking: An empirical study. *International Journal of Service Industry Management* 14 (5): 501–519.

Wong, Y.K. and C.J. Hsu. 2008. A confidence-based framework for business to consumer (B2C) mobile commerce adoption. *Personal and Ubiquitous Computing* 12 (1): 77–84.

Wu, J.J. and Y.S. Chang. 2005. Towards understanding members' interactivity, trust, and flow in online travel community. *Industrial Management & Data Systems* 105 (7): 937–954.

Yvette S. and F. Karine. 2001. Information quality: Meeting the needs of the consumer. *International Journal of Information System* 21 (1): 21–37.

Zhong, B., M. Hardin, and T. Sun. 2011. Less effortful thinking leads to more social networking? The associations between the use of social network sites and personality traits. *Computers in Human Behavior* 27 (3): 1265–1271.

Zhou, T. 2013. An empirical examination of continuance intention of mobile payment services. *Decision Support Systems* 54 (2013): 1085–1091.

Zhou, T., H. Li, and Y. Liu. 2010. The effect of flow experience on mobile SNS users' loyalty. *Industrial Management & Data Systems* 110 (6): 930–946.

Zhou, T. and Y. Lu. 2011. Examining mobile instant messaging user loyalty from the perspectives of network externalities and flow experience. *Journal Computers in Human Behavior* 27 (2): 884–888.

chapter six

Comparative study of in-store mobile commerce applications and feature selection, targeted at enhancing the overall shopping experience

Electra Safari and Dimitrios Zissis

Contents

6.1 Introduction

In recent years, we have witnessed the explosive growth of a previously nonexistent market. The widespread use of personal computers and the introduction of high-speed Internet connectivity made the process of sharing business information, maintaining business relationships, and conducting business transactions by means of telecommunications networks a reality. Since the 1990s, a new type of electronic commerce conducted through handheld devices has been rapidly gaining momentum, mostly due to developments in communication technologies and portable computing. This subset of e-commerce, labeled mobile commerce, is attracting significant attention and is shifting the way we conduct business from a wired to a wireless environment, using simply our handheld terminals anytime and anywhere (Ngai and Gunasekaran 2005). Due to its characteristics (among others ubiquity, personalization, and flexibility), mobile commerce promises businesses unprecedented market potential, great productivity, and high profitability (Siau et al. 2003). The number of mobile phone users is rising continuously as the market of mobile technologies is promising speed and convenience. Currently, an overall of 10% of global website views are from mobile devices (smartphones and tablets). On Christmas day 2012, more than 50% of online activity came from mobile devices (Mixpanel 2012). Mobile technology in Japan currently accounts for 35% of all e-commerce transactions. In 5 years' time, more than 50% of all transactions will be through smartphones (IPC 2012).

Various kinds of mobile applications have emerged as a result of these advances and are penetrating our everyday life, changing human behaviors, and generating important social and economic impacts (Xu et al. 2008). Among these efforts, mobile applications for shopping come at the intersection of ubiquitous computing and electronic commerce and are gaining attention from both communities (Xu et al. 2008). Prior work though has mostly focused on the transactional functions of mobile phones and information consumption on-the-go and less on the experiential aspect of shopping and how this can be improved via the use of wireless devices (Xu et al. 2008). Nowadays, there is a fundamental blurring of the boundaries between online and off-line shopping (Gish 2012). Smartphones are fundamentally changing how people shop, browse, use coupons, find locations, and enter local and near field promotions. Shopping behavior is changing and showrooming and snacking (shopping in a spare five minutes) are increasing (Gish 2012). The real and the virtual world are converging into a complete shopping experience. It is becoming typical user behavior for consumers when in stores to go for their handheld devices in search of additional product information, better prices, and location information in an attempt to speed up the overall shopping process and gain a better shopping experience.

Channels are blending and consumers are shopping wherever, whenever, and however they want. According to Forrester, 50% of retail sales are influenced online (IPC 2012). As part of its "Mobile path to Purchase" research, Nielsen and Telmetrics reported that a 46% of mobile users rely exclusively on their mobile device for retail prepurchases in the stores while, interestingly, half of the audience does not need to do a PC research before buying (Sterling 2013).

However, current retailers do not seem to be tuning successfully into this new trend. While in *wired* settings, we have accumulated the necessary knowledge to design and develop commercial systems effectively, the field of m-commerce presents not only unique opportunities but also challenges. It faces a number of businesses, technical, and legal challenges that differ from traditional e-commerce, specifically in relation to location-based services (Tsalgatidou et al. 2000). A thorough analysis of the design space of in-store shopping is a necessity, in addition to the way that mobile device interaction can get embedded in the shopping experience in benefit of the customer (Xu et al. 2008). Moreover, factors that influence actual adoption of mobile shopping applications, such as value creation, user acceptance, and entertainment, must be examined in detail, so that we gain an understanding of the mobile customer's requirements.

In this chapter, we focus on the type of mobile commercial applications that consumers use when shopping in large retail stores or shopping malls, so as to speed up the shopping process and gain an ideal shopping experience. More specifically, we are concerned with the nontransactional functions of commercial applications that can offer convenience and fun during the shopping process. The chapter is structured as follows: we first attempt to define the boundaries of our research by accurately defining the problem space, evaluating not only the benefits and detriments but also the inherent complexities of designing and developing such applications. Within this scope, we perform a literature review with the purpose of eliciting requirements for m-commerce applications. Requirements elicitation is recognized as one of the most critical, knowledge-intensive activities of software development; poor execution of elicitation will almost guarantee that the final project is a complete failure (Gottesdeiner 2002, Hickey and Davis 2003). The process of requirements capturing can be broken down into discovery (elicitation), analysis, modeling and documentation, communication, and validation (Schedlbauer 2011). The volatile nature of m-commerce requires we perform a validation of collected requirements; we employ methods such as interviews, surveys, and on-site interaction to effectively validate the requirement set. Additionally, we perform a comparative study of available in-store commercial applications in a contextual setting, with classification based on user acceptance, in an attempt to identify the characteristics and design features of such an application. From these, a design framework stems, with guidelines and recommendations of considerations that can assist designers and developers of mobile on-site shopping applications with the ultimate goal of designing a better overall shopping experience.

6.2 *Mobile commerce in the shopping field*

Mobile electronic commerce (Tsalgatidou et al. 2000) has recently emerged as a subset of e-commerce and refers to those activities that rely solely or partially on mobile e-commerce transactions. It operates in a different environment than e-commerce, which can be viewed as conducted in the fixed desktop Internet metaphor, mostly due to the different characteristics and constraints of mobile devices, constraints of wireless networks, and ultimately the context, situations, and circumstances in which people use their handheld terminals.

Advances in telecommunication of the last decade have made mobile commerce one of the main channels for purchasing products. Its features, such as ubiquity, reachability, localization, and dissemination (Siau et al. 2003), combined with the recent advanced capabilities of mobile devices have provided the consumer with a great device for decision making. Due to this, we are witnessing a shift from electronic shopping and in-store shopping to a crossbred shopping experience that includes in-store shopping using a mobile device for every step of the shopping process.

Shopping behavior is changing. Channels are blending and consumers are shopping wherever, whenever, and however they want (Gish 2012). According to Forrester, 50% of retail sales are influenced online (IPC 2012). The consumer has found in m-commerce a better way for conducting shopping *on the go* in a more convenient, time saving, and pleasurable way, while simultaneously benefiting from a more personalized service. But mobile commerce benefits are not strictly consumer driven. Businesses can increase their market potential, productivity, and profits, through managing smartly this upward trend (Siau et al. 2003). Retailers now have the chance to influence consumers in a more direct way, by identifying the mobile shopping space and adjusting their services to it. By exploiting mobile commerce capabilities, they are able to track their customer's location and promote special discounts, e-coupons, and messages, identify their special needs and inform about desired products, and direct them inside the store till the point of purchase.

Despite the benefits of mobile commerce for consumers and retailers, there are many constraints induced by the characteristics of wireless communications, the devices, and the context of mobility itself (Tsalgatidou et al. 2000). Wireless communication constraints are attributed to the quality of service of the networking infrastructure consisting of various wireless networks (LAN, WAN, satellite services, etc.) and other protocols, standards, and technologies. This complex networking system is responsible for the efficiency of mobile commerce activities and must be taken into account when designing an m-commerce system so that problems such as frequent disconnection, loss of data, slow download rate, and others are eliminated.

The design and development of mobile applications is highly complex and challenging (Charland and Leroux 2011). It concerns developing for a number of different operating systems and devices with various characteristics in a diverse set of programming languages. In such a fragmented field, designers and developers have to comprehend the restrictions of the technology (e.g., small screen, memory, and CPU) and the particular technical characteristics of mobile operating systems (e.g., iOS, Android, and Windows phone) (Forman 1998). In addition, mobile devices are characterized by a unique synthesis of interaction affordances that can actually transform the user experience when compared to desktop platforms including gesture-based, multitouch interaction with digital content; location awareness and subsequent service and content adaptation; advanced sensing capabilities (with embedded devices such as accelerometer, gyroscope, GPS, and camera); multimedia (photos, sound, and video) capturing and sharing (Corey 2010, Reddy et al. 2010, Ferreira et al. 2012). Therefore, mobile application development requires a whole new way of thinking in respect to interaction design and HCI as well as to software development. Device constraints are enhanced mostly due to the nature of the devices themselves, especially their portability. To be easily carried around, mobile devices are light and small. As a result, they have smaller screens, keypads, memory, disk capacity, and processing power, combined with high power consumption. Mobility constraints are related to the continuous switch of locations of a mobile user and the various scales of bandwidth networks he/she may connect from. User device location tracking can often be interrupted because of often movement and unavailability of networks. Mobility also raises complex security and authentication issues (Tsalgatidou et al. 2000).

In mobile commerce, we also identify several requirements that are not available to traditional electronic commerce and that should be considered in the development of m-commerce systems and applications. Influenced by Varshney and Vetter's (2002) proposed framework for mobile commerce application implementation, we identified three important areas of interest for m-commerce applications from which several requirements derive. Those are as follows:

1. Software (mobile interface and mobile middleware)
2. Hardware (mobile handheld devices such as PDAs and mobile/smartphones)
3. Wireless network (wireless mobile network infrastructure)

We developed a diagram that summarizes the most significant requirements and constraints resulting from each one of the aforementioned areas (Figure 6.1). Software requirements can be classified (ISO/IEC JTC1/SC7 N4098) according to the assigned and inherent (existing or permanent characteristics or features) properties of a software product. Moreover, inherent properties can be classified as either functional properties or quality properties. Functional properties determine what the software is capable of. In our case, these domain-specific functional properties include basically the functions of a mobile commerce system that we want to examine (e.g., locating an item and searching for prices). Quality properties determine how well the software performs. In other words, quality properties indicate the degree to which the software is capable of providing and maintaining specified services. Hardware requirements derive from the nature of mobile devices and

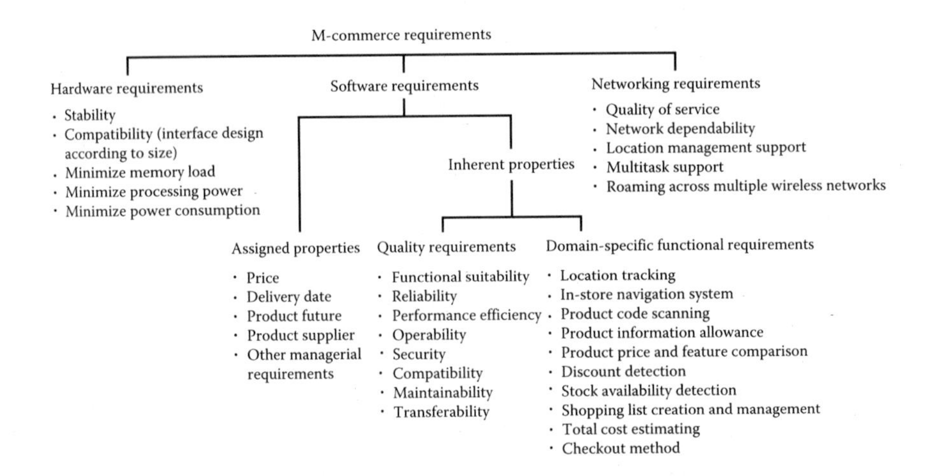

Figure 6.1 Mobile commerce requirements.

their capabilities. Networking properties refer basically to constraints and challenges related to connectivity and bandwidth issues (Varshney and Vetter 2002). Varshney and Vetter (2002) categorized networking requirements for m-commerce applications into five groups, such as location management, multitasking support, network dependability, quality of service, and roaming across multiple networks.

A number of classes of m-commerce applications can be identified, including mobile financial applications, mobile advertising, mobile inventory management, product locating and shopping, proactive service management, wireless reengineering, mobile auctions or reverse auctions, mobile entertainment services and games, mobile offices, mobile distance education, and wireless data centers.

In our research, we focus on "product locating and shopping" class, which includes applications that locate an item in a particular area or location. This also concerns finding an item with certain specifications and its availability across stores (Varshney and Vetter 2002). Although m-commerce activities include trading of goods and services of economic value, this is not strictly limiting; thus, our research does not deal with the transactional aspect of mobile shopping. We are more concerned with the experiential aspect of shopping via handheld mobile devices aiming at simplifying several consumers' tasks. Within this context, we attempt to identify the set of requirements, which should be met when designing an m-commerce application; this set will lead to the specific (design) principles of an m-commerce application targeting at improving the shopping experience. In the related literature review step, we attempt to specific design requirements related to our class of applications.

6.3 Consumer behavior

Today's shoppers generally find the shopping experience at existing retail stores satisfying on the dimensions of convenience, product quality, value provided, and product selection. However, they express their dissatisfaction with the speed of shopping, level of service, available product information, fun of shopping, and security and privacy (Burke 2002). In the last decade, technology has appeared as a viable solution to many of the previous requirements. Mobile phones tend to become a growing shopping channel, filling in gaps in the shopping experience, while providing a better, more convenient, and efficient overall way to shop.

Currently, consumers search several stores to find an item of a certain brand, size, and characteristics to fit their preferences, required features, and price. Using a handheld device and an online product information service, a user is capable of navigating to the exact location of a store or a certain product (Varshney and Vetter 2002). In the case of large shopping stores and malls, consumers attempt to use their mobile devices in store, to speed up the shopping process (using indoor maps for tracking and navigating) so as to gain a pleasant, convenient, and efficient visit in the shop and even avoid staff's *sales pressure*.

Chuck Martin, the CEO of Mobile Future Institute, stated recently that mobile is turning "path to purchase on its head" (2013). Shopping is becoming an iterative rather than a serial process. Consumers no longer go shopping, they always are shopping, he continues. Hackbert (2012) indicates that the phenomenon is growing especially among smartphone users, which he calls app-happy. In his research, he found that 56% of smartphone owners have at least one shopping-related app installed on their phones and 15% have more than six shopping apps currently on their phones.

6.4 Choice of comparative study and environment

In order to create a useful design framework for mobile in-store shopping applications we conducted a three-level study, consisting of a literature review, an online survey, and an on-site comparative study. The research is organized as follows:

First, we perform an extended literature review with the purpose of eliciting requirements for m-commerce applications. Requirements elicitation is recognized as one of the most critical, knowledge-intensive activities of software development; poor execution of elicitation will almost guarantee that the final project is a complete failure (Gottesdeiner 2002, Hickey and Davis 2003). The process of requirements can be broken down into discovery (elicitation), analysis, modeling and documentation, communication, and validation (Schedlbauer 2011). The volatile nature of m-commerce requires we perform a validation of collected requirements.

Second, we employ methods such as interviews, surveys, and on-site interactions with customers to effectively validate the requirement set. We conducted an online survey audience so as to confirm which of the requirements identified earlier were actual consumer needs or could be considered as constraints and identify new ones.

The third part of our work included the comparison of several features in existing mobile applications as well as the evaluation of the requirements gathered in the second step in real-time conditions. To do this, we chose an on-site field study so as to identify additional usability and interaction issues as well as measure user satisfaction throughout the whole shopping process. We conducted our study in the *public* store located in the center of Athens, Greece. The store is structured as a large showroom with multiple floors organized by product category. Its purpose is to encourage people to walk in, experience the various products, and ask questions. We have chosen several existing mobile shopping applications and an appropriate number of users that cover a range of ages and both genders. These applications support in-store shopping activities such as:

- Creating and managing shopping list
- Checking product prices and discounts
- Locating a product in nearby stores
- Finding product information and customer reviews through scanning (barcode, QR code, and image recognition)
- Comparing product features
- Sharing a product with friends

Some of the applications contained transactional functions such as purchasing a product online or mobile payment via a credit card. However, we excluded these features from our research as security and privacy issues could be raised.

6.4.1 *Requirements elicitation*

When designing mobile applications that contribute to the convenience of the shopping process for consumers, there are a number of specific requirements that should be considered, as the shopping experience is related to meeting specific consumer's needs and pleasuring the customer. System and application design alike should be approached from the point of entertainment service provision (Roussos et al. 2003). Roussos et al. (2003) examined this retailing entertainment or *retailtainment* aspect of m-commerce by developing a grocery shopping system. In the first step of their research, they collected several user requirements conducting a field study of a target audience. The study revealed a number of user required features, including features supporting a constant awareness of total shopping costs, access to complete and accurate product descriptions, comparison of value of similar products, personalized and targeted promotions, in-store navigation, and smart checkout. In the second stage of their research, real-time user scenarios were conducted so as to evaluate the system. Participants claimed that using the system transformed the shopping experience into a less stressing experience. However, several privacy constraints were identified. Participants were especially concerned about collection and storage of personal data, even though data protection policies were confirmed.

When shopping in a large store, a shopping mall, or supermarket, user requirements are amplified, as customers' needs grow. Corresponding applications are expected to provide additional functions that can transform the uncomfortable and often disorienting mall experience into a pleasant, convenient, and friendly visit. Some of these additional requirements relate to features such as customer in store guidance and *to-product* guidance. In this context, Asthana et al. (1994) developed a wireless personal shopping assistant based on the capabilities of customers' wireless communication devices. In their work, they support that such a personalized nature of service can add not only to customer's convenience, but could also be a source of direct and indirect revenue for the shopping center. The system must be simple and help the customer navigate within the store and inquire about items. In some cases, it may even track the customers shopping history so as to provide relevant information to them. Asthana et al., as did Roussos et al., identified several constraints that are following the implementation of this system and its functions such as privacy issues, or the size, weight, power consumption, and frequency bandwidth for the handheld device.

Xu et al. (2008) also tried to explore the experiential aspect of mobile shopping in their design approach of a vision-based mobile interface for in-store shopping. After conducting a 1-month long diary study, they formulated a list of design principles for mobile shopping applications. First, they support that these applications should not demand a high and continuous level of attention, so as not to interfere with the shopping experience itself. Second, the portability of the mobile phone platform and its connection to other resources need to support the shopping activity over a longer term and across different locations. Moreover, mobile shopping applications need to simplify access to product information (in comparison to desktop usage) and provide highly relevant content in relation to location, time, and personalization opportunities. Mobile shopping applications also need to focus on supporting organizational and communicational functions, as these are considered important by consumers. Lastly, they mention that online and in-store shopping experiences need to be bridged as the user's cognitive load of shopping tasks, when switching from one context to the other, should be minimized. Interestingly, after examining their application via a number of user real-time scenarios, they also found that participants were more than interested in sharing and saving functions such as "adding a product to a wish list" and "send to a friend."

Adelmann (2007), in his study on mobile phone interactions with everyday products, claims that the existence of a good middleware is the solution for a fast and convenient interaction between the handheld devices and the products. He supports that "when being on the go, a simple and fast user interaction is essential." For succeeding in this, he states that a good recognition rate must be provided by the application. Most of the times, recognition conditions are bad because of limited lighting, low camera resolution, possible hand movement when scanning, or even the quality of the image. When in the shopping field, users have neither the convenience to scan products appropriately so that the system will return them the related information nor the time to manually enter data into their phone. As a result, a mobile shopping application should ease and accelerate the interaction process, relying mostly on tag-based interaction whenever possible and using good recognition algorithms that can cope with complex lighting conditions (Adelmann 2007).

In the related literature, less attention is paid to the role of aesthetics as being an influential factor in the use of mobile commerce applications. Cyr et al. (2006) tried to explore the possibility of visual design aesthetics leading to *mobile loyalty* of users. They discovered that an aesthetic design could significantly impact perceived usefulness, ease of use, and enjoyment all of which ultimately influence user's loyalty toward a mobile service. User acceptance of new and well-designed technologies and interface is crucial to the adoption of mobile commerce systems.

Stemming from our review of relative articles, we made a list with the specific design requirements for mobile shopping applications in the following:

- Should not demand a high, continuous level of attention
- Should minimize cognitive load of shopping tasks
- Have simple and fast interaction, hands-free operation, minimize data entry whenever possible
- Have an aesthetically pleasing design
- Support organizational and communicational functions (create lists, sharing functions)
- Provide an accurate item recognition and interaction function
- Calculate the total shopping cost constantly
- Provide access to product information and descriptions
- Locate a product's specific position
- Navigate the customer inside the store

- Compare the value of similar products
- Provide personalized and targeted product promotions
- Provide smart checkout method

Many constraints and restrictions identified in the relative research were related not only to privacy and security issues, pointing to a lack, but also to the necessity of trust between the customers and commercial institutions. Additionally, issues that derive from the nature of handheld devices and wireless communications were mentioned, such as the size, weight, power consumption, frequency bandwidth, coverage area, and connectivity.

6.4.2 Communication and requirements validation

We identified m-commerce requirements for shopping applications and within this context we performed an investigation, using online questionnaires and on-site interviews, to validate the requirements set and identify any additional unique requirements. The methods selected for this survey included user feature testing, questionnaires, and wrap-up interviews with customers at a shopping mall in Athens, Greece.

The questionnaire comprising a total of 13 questions: the first 5 were demographic, and the rest focused on requirement identification and validation. We conducted our questionnaire online directed to 158 people of ages between 10 and 70 years enquiring about their experience with wireless devices, their usage of electronic or mobile commerce, if they had ever tried an m-commerce application while shopping in a store and what exact functions they had used during their shopping trip. We also asked participants to propose any features they believed would improve their shopping experience and also to identify elements that could deter it.

A total number of 158 participants took part in this survey; they were between 10 and 70 years of age: 66.46% of user between 16 and 24, 29.75% of users were between 25 and 45, and 3.80% between 46 and 70. As expected, a huge number of the audience (75.95%) were smartphone users and a 15.19% also used tablet devices (Figure 6.2). We enquired about device ownership and attempted to identify market fragmentation (Figures 6.3 and 6.4). It was interesting to see that even in a small fragment of the overall market, device fragmentation percentages seem to be very close to other published reports (Android as a market leader, followed by Apple and Windows).

To examine their history with electronic commerce we asked them if they ever purchased an item from the Internet (Figure 6.5). A large percent responded positively as having shopped occasionally at popular commercial websites such as eBay.com (52.90%), Skroutz.gr (31.16%), Amazon.com and E-shop.gr (39.13%), and other. Interestingly, only 40 users out of 158 have ever used their mobile devices to perform electronic shopping,

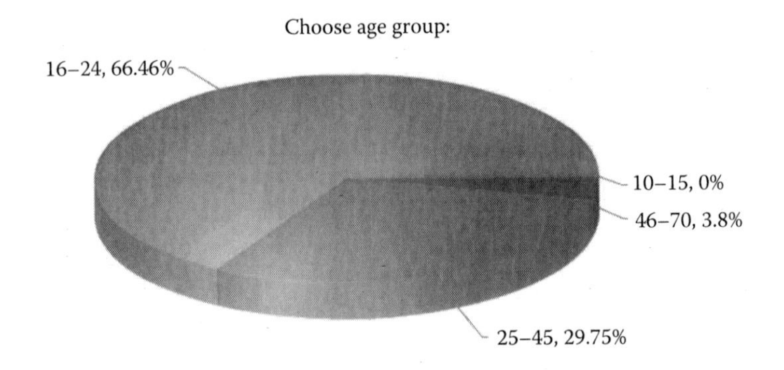

Figure 6.2 User age group.

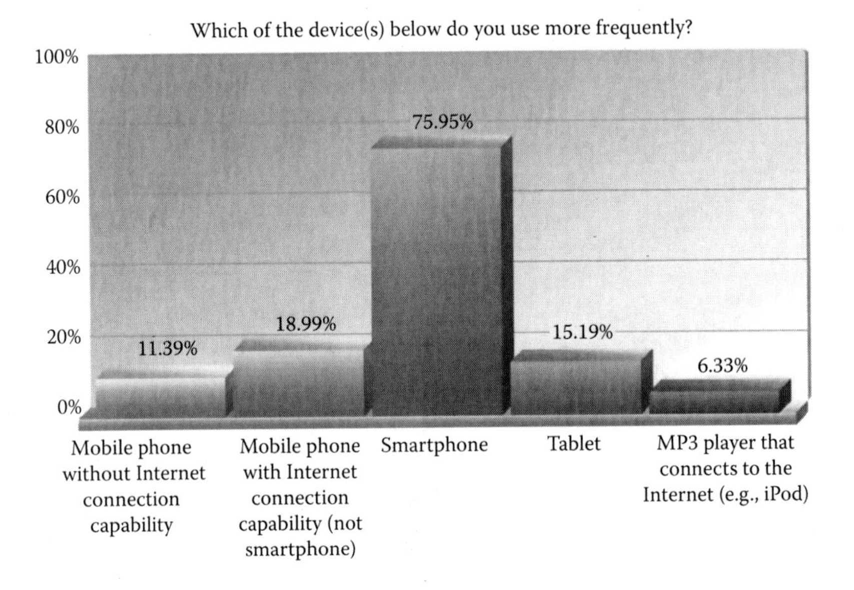

Figure 6.3 User device of preference.

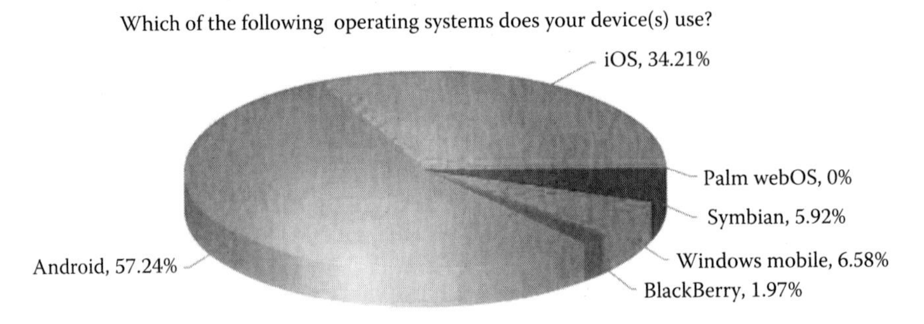

Figure 6.4 Mobile device operating system fragmentation.

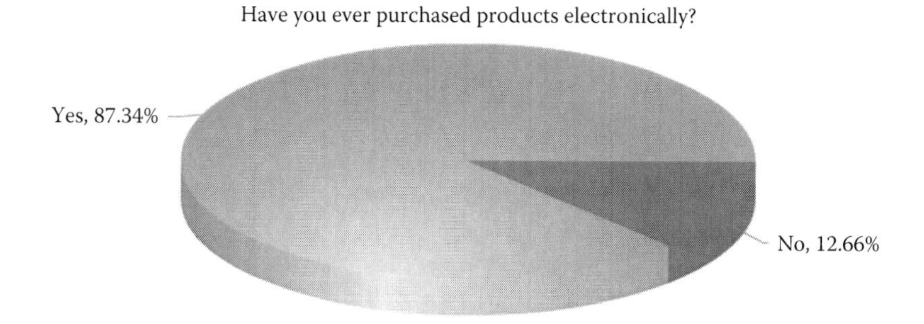

Figure 6.5 Electronic purchasing.

either by visiting a related website, by downloading a mobile shopping application or both (Figure 6.6).

We then asked them if they had ever used a mobile shopping application to simplify their shopping experience and if not what were the reasons (Figure 6.7). A 25.95% claimed to have used a shopping application while inside a store so as to compare prices or products with similar features (65.96%), create a list with products (42.55%), calculate total shopping cost (27.66%), check product availability in the store's stock (25.53%), find discounts (25.53%), and search for the same product in nearby stores (23.40%) (Figure 6.8). However, a large part of the audience that responded negatively in the previous question claimed to often use the Internet to research certain products before visiting a traditional store or did not know that such was possible. Some of the respondents comments were related to the fact that most of these applications are time consuming and can get frustrating due to continuous network disconnections; thus prefer using a desktop or laptop before visiting a shop. Others of course

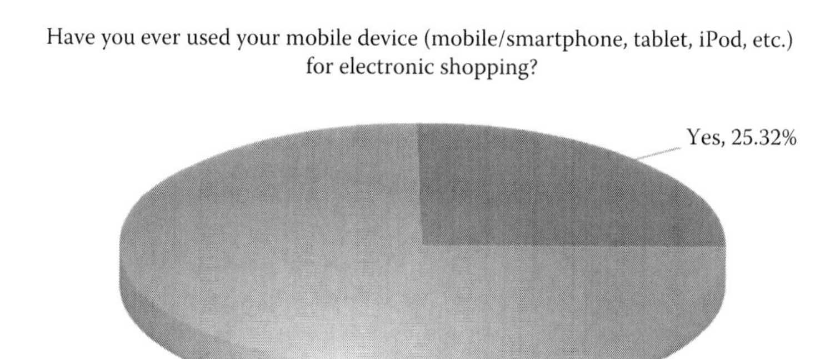

Figure 6.6 Percent of customers that have ever used their mobile phone/tablet to perform e-commerce-related interactions.

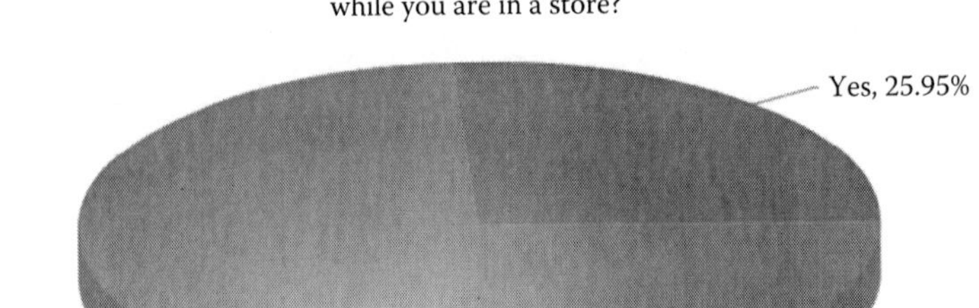

Figure 6.7 Percentage of users ever used an m-commerce application while in store.

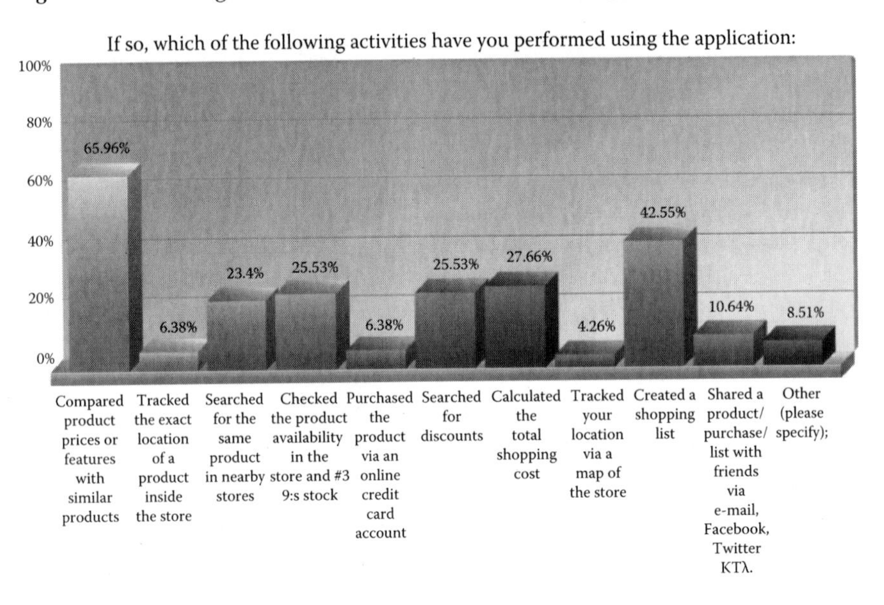

Figure 6.8 Most used/important features in m-commerce applications.

supported that mobile applications are a more fun and entertaining way of shopping. A big drawback identified by a large number of users was that many stores do not always provide free Internet access for customers.

We also asked the audience to identify which feature was more important for a mobile shopping application in their opinion (Figure 6.9). Functions such as comparing prices and product features, checking stock availability, searching product in nearby stores, finding discounts, and calculating total cost were most highly rated. Instead, interestingly, activities such as locating a product and navigating in the store, purchasing a product and sharing product/list/purchase with friends, or creating a

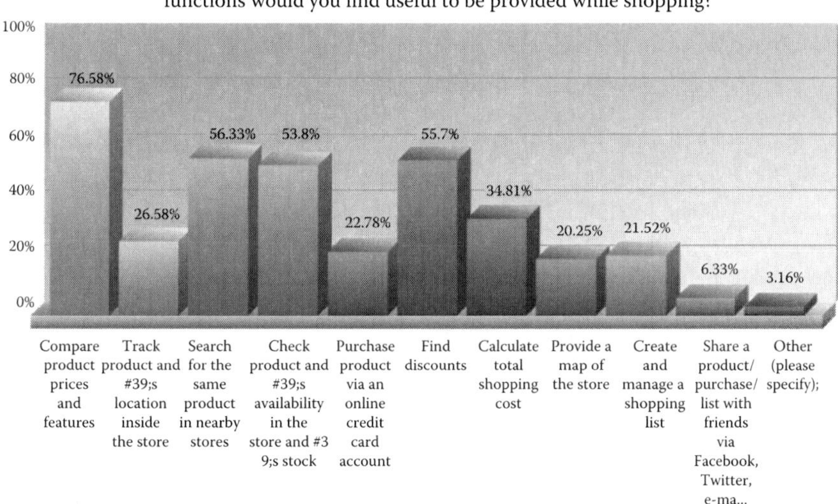

Figure 6.9 Desired user features for in-store m-commerce applications.

shopping list did not attract so much interest. Some of the respondents also suggested that the system should provide customer reviews and information about the country of origin, product guarantees, history of purchases in that store, or even recommend to them related products that can be used in the same way with the one they are searching for. It is interesting to view the differences between most used and required features for m-commerce in shop applications. Although comparing product features remains the top choice in both occasions, finding discounts and products in other stores are the second top required features.

Enquiring about what features customers find more important in relation to mobile shopping, respondents preferred simple and easy interactions, accuracy and speed in results provided, good visual design, and the elimination or automation of steps to result (provide a wide range of search features, accept speech input instead of manual data entry, not to surpass four steps otherwise it gets tiring). They also found important not having to download another application so that the first will be in the position to operate (this is often required by application using external libraries for QR code or other image recognition capabilities). Some respondents commented that the application should be able to save their preferences so that they do not need to repeat them next time and to use their 3G connections wisely in case the store does not provide free WiFi.

We finally asked the audience what usually prevents them from using a mobile application. Most of the respondents agreed that the time they have to wait and the number of steps (interactions) required before getting to the necessary result is an important barrier when *on the go*, while the price of an application (in case it is not available for free), the amount of power consumed, screen size, keyboard size, and security and privacy issues remain important. A smaller percentage of responses referred to issues such as limited network capabilities and frequent disconnections or the frustration caused by having to change networks when moving between locations. Some respondents added that the time spent, the complexity of such systems, and the memory consumption are not worth using such applications most of the times.

Comparing the survey results with the requirements and constraints gathered in the first stage of our study, we formulated a new list with the most important functional, quality, and networking objectives of a mobile shopping application system:

Functional objectives:

- Check product prices/discounts
- Find additional information about a product (characteristics, country of origin, and guarantees)
- Compare the value of similar products (provide product reviews and suggest similar products)
- Calculate the total shopping cost
- Check the availability of a product in the store's stock
- Search for the same product in nearby stores
- Locate a product in the store
- Purchase a product (provide checkout method)
- Create and manage shopping list (create a list, save item to wish list)

Quality objectives:

- Simple and fast interaction, ease of use and search, automate or eliminate tasks
- Interface design compatible with the size and capabilities of wireless device
- Information accuracy (high recognition rate, broader database search)
- Speed in providing results (high recognition algorithm, quick database search)
- Should not demand the installation of another application so as to operate
- Aesthetic interface design
- Security and privacy of personal data
- Minimize power consumption
- Minimize memory consumption
- Support organizational and communicational functions (create list, save products and past findings, share information)

Table 6.1 Mapping Functional Requirements to Design Features

Functional requirements	Features
Check product prices/ discounts	Scan the barcode or the whole product/say or type the name/code of the product
Find additional product information	Scan, say, or type product to find information about its characteristics, guarantee, country of origin, etc., via an automated database search
Compare value of similar products	Compare a product with a list of similar ones/ provide consumer reviews/suggest similar products
Calculate total shopping cost	Automate calculation of total cost of selected products, for example, scan product to add prices/calculate existing price discounts
Search product in nearby stores	On map location tracking of nearby stores that sell the same product in different prices
Check product availability in the store's stock	Scan, say, or type product to check via the store's database if it is available in the store's stock
Locate product in the store	Indoor searchable store map
Purchase product	Scan and store loyalty cards, create an account that collects data of past purchases
Create and manage shopping list	Provide features such as "create new list," "edit current list," "add to wish list"

Networking objectives:

- Maximum connectivity
- Maximum frequency bandwidth
- Maximum coverage area
- Efficient location-based services
- Availability of resources provided in mobile devices

Stemming from the aforementioned objectives, a design framework can be formulated (Table 6.1) presenting features that attempt to address predefined user requirements. This will be evaluated during the on-site evaluation.

6.4.3 On-site evaluation and comparative study

In an attempt to evaluate if the objectives gathered earlier were met in existing high-ranking mobile shopping applications and to identify additional characteristics and design features, we performed an on-site evaluation of such application features in a contextual setting. For this purpose, we chose to conduct an on-site study in the *public* store, located in the *The Mall Athens* shopping center (a large shopping mall in Athens, Greece).

We selected a number of applications based on user ratings, availability, and features so as to cover as many as previously identified (Table 6.2). We interviewed a number of customers (20 customers) and provided a device for them to interact with features of selected applications. Free Internet access was available by the store.

At first, we required that the customers download the specific applications to their own smartphones and spend some time using each feature except from functions such as "add to cart" or "purchase product." During the overall process, we interacted with the users, asking various questions and observing user patterns. Answers collected approximately matched the questionnaires results. Participants were mostly interested in features such as checking online product prices, finding discounts and reviews or local stores that were selling the same product inside the mall, as they supported that it is the only thing that the staff cannot actually answer. They were also equally satisfied by either having to scan the barcode of a product or the product itself as the recognition was most of the times accurate and fast. However, they negatively commented on the fact that some applications still required a manual entry of the product's name, for example, the title of a book. Finding details for a product was one of the features that did not gain much attention, as participants claimed that the store had a very good description of every product in the showroom.

When questioned about the "create a list" feature, most of the participants replied that it could be really useful as they tend to forget their shopping notes at home or lose them. They also found it useful to be able to save some products in a new list which they could organize later. A participant commented that "I would definitely use it when window shopping for Christmas presents, so as to make a remember-to-buy list for later." Possibly, this feature could be used at later time to review product information on a personal computer and find bargains.

Most of the participants though found the "share to friends" feature not that interesting, supporting that they would not want to share information regarding a personal purchase on Facebook or other social sites, but they would rather like to share a product with a specific friend so as to get his opinion (possibly by email). They did not seem to make use of the *total calculation cost* as they believed it would require too many steps to be completed.

None of the selected applications gave us the chance to test features such as "check product availability" and "product location tracking," so we asked participants if they would possibly use one of these features. Interestingly, one of the participants answered that "I would be amazed to have the chance to find if a product is available in stock, as sometimes staff lies if it is on bargain or offer." Most of the participants found location tracking an interesting feature, as long as it is used for finding a product available in the shopping center.

Table 6.2 Feature Testing in On-Site Study

Features	APP 1	APP 2	APP 3	APP 4	APP 5	APP 6
Check product prices/ discounts	Barcode scanning/ manual data entry for search	Barcode scanning/ manual data entry for search	Image recognition		Manual data entry for search	
Give additional product information (size, weight, color, etc.)	Database search	Database search	Google search		Internet search	
Share a purchase with friends	Share item or list via Facebook	Share item or list via email, Facebook, Dropbox, etc.			Share item via Facebook, twitter or email	
Compare product features (get customer reviews, suggest similar products)					Item comparison through database search of similar items	
Calculate total shopping cost						Calculator feature (add items' prices)
Purchase product		Scan and store loyalty card			Online credit card account or gift card	
Locate product in nearby stores	On map location of local stores	On map location of local stores				
Create and manage shopping list	Shopping list feature (save to wish list, create new list)	Shopping list feature (save to wish list, create new list)		Shopping list feature (name a new list/ add items)		

As we did not ask them to perform any transactional functions, we only requested their opinion about m-commerce transactions. We enquired if they would trust an application with their bank account or credit card details so as to electronically purchase an item. Responses were equally balanced. Some participants replied that they would have no problem as long as the application was certified by the store, otherwise they would like to able to use a voucher or token bought from the store just for this purpose. Others, although skeptical, supported that paying without waiting in the queue would be an advantage.

During the interviews, we noticed that most of the participants preferred those applications that offered the fewest functions or even one specific feature, as they were simpler to use. We enquired about the information architecture of each application, which one guided them better and if they were able to understand what they had to press. Most of them were disappointed by the general interface design of the applications at hand. They thought that some applications were better at providing information but had the *look and feel* of traditional websites as they had to do a lot of steps to get a result. Application interface design and aesthetics were closely linked to store and brand quality. Consistency in functionality, information, and visualization across channels is important to customers (Song et al. 2013), but careful information design is necessary for mobile information outlets. One of the participants replied that "for almost all shopping related tasks three clicks (interactions) should be more than enough to get to the required result, as after that I am bored." They instead liked those applications that had only one clear function and as a result not so much information screen cluttering. They commented on the fact that most of the applications required some experience before use, as their basic functions could not be easily detected when operating the application for the first time. A participant reported that "an application must ease your life rather than make it more complicated" after being asked about one of the applications. We finally asked them if the visual design played a role in spending more time using one application rather than the rest. Most of the participants commented that an aesthetic design would definitely prompt them to try the application (a person supported that the use of colors in an application was really helpful), but again the functionality and accuracy were the most important factors.

6.5 Conclusion

In this chapter, we presented our study of in-store mobile commerce applications and feature selection with the ultimate goal of designing a better shopping experience. We attempted to identify a complete set of requirements that should be met when designing an m-commerce application;

this set will lead to the specific (design) principles of an m-commerce application targeting at improving the overall shopping experience.

Considering the premature stage and the challenges faced in the field of m-shopping, we performed a three-level research with the purpose of evaluating and discovering unique requirements. From the literature review, we managed to collect a number of user requirements and several constraints deriving from related studies in the mobile shopping field. We then conducted an online survey to validate our findings according to the audiences' preferences and needs, which resulted in a list of functional, quality, and networking requirements. In an attempt to ensure if the objectives gathered are met in existing mobile shopping applications and to identify design features addressing these requirements, we performed an on-site evaluation of such application features in a contextual setting.

In addition to the functional properties, our study explored the aspect of middleware quality. The importance of good visual interface and efficient information architecture was indicated in the on-site comparative study, where consumers had to interact with the several application features. In the field study, the importance of simplicity and responsiveness was emphasized. Users tend to be satisfied from an application that returns accurate results fast and without complexity. They also prefer applications with clearly defined functions and an interface that provides straightforward navigation. As stated in Alqahtani and Goodwin (2012), the most important thing when designing mobile applications is to design the application in such a way that it does not distract the user from the main purpose of the application. Users also agreed that an aesthetic visual design, referring to colors, shapes, font types, etc., actually influences them in choosing one application instead of the other and it could also be used for navigation purposes. However, in the end they based the value of mobile shopping applications in their capacity to simply offer required results on request.

References

Adelmann, R. 2007. *Mobile Phone Based Interaction with Everyday Products—On the Go*. Zurich, Switzerland: ETH Zurich, Institute Pervasive Computing.

Alqahtani, A. S. and Goodwin, R. 2012. E-commerce Smartphone Application. *International Journal of Advanced Computer Science and Applications* (*IJACSA*), 3(8), 1–6.

Asthana, A., Cravatts, M., and Kryzanowski, P. 1994. *An Indoor Wireless System for Personalized Shopping Assistance*. Murray Hill, NJ: AT&T Bell Laboratories.

Burke, R. R. 2002. Technology and the customer interface: What consumer want in the physical and virtual store. *Journal of the Academy of Marketing Science*, 30(4), 411–432.

Charland, A. and Leroux, B. 2011. Mobile application development: Web vs. native. *Communications of the ACM*, 5(54), 49–53.

Corey, G. 2010. Nine ways to murder your battery (these are only some of the ways). In *Battcon'10*, Hollywood, FL.
Cyr, D., Head, M. M., and Ivanov, A. 2006. Design aesthetics leading to m-loyalty in mobile commerce.
Ferreira, D., Kostakos, V., and Dey, A. K. 2012. Lessons learned from large-scale user studies: Using android market as a source of data. *IJMHCI*, 4(3), 16.
Forman, G. H. 1998. Wanted: Programming support for ensuring responsiveness despite resource variability and volatility. HP labs technical reports. Available at http://www.hpl.hp.com/techreports/98/HPL-98-15.html. Accessed February 1, 2014.
Gish, C. (eBay MarketPlace). 2012. Enabling electronic commerce: Social, local, mobile. In *IPC, Stanford Graduate School of Business Symposium*, Stanford, CA.
Gottesdeiner, E. 2002. *Requirements by Collaboration*. Addison-Wesley.
Hackbert, P. H. 2012. *The Smartphone as an Economic Development Tool: A Preliminary Investigation*. Phoenix, AZ: Berea College, Society of Business Research.
Hickey, A. M. and Davis, A. M. 2003. Requirements elicitation and elicitation technique selection: Model for two knowledge-intensive software development processes. In *Proceedings of the 36th Annual Hawaii International Conference on System Sciences*, Honolulu, Hawaii, January 6–9 2003, 10pp. doi: 10.1109/HICSS.2003.1174229.
IPC. 2012. Summary—Enabling electronic commerce: Social, local, mobile. *IPC, Stanford Graduate School of Business Symposium*, Stanford, CA.
Martin, C. June 11, 2013. The mobile shopping life cycle. Retrieved from: Harvard Business Review Blog Network: http://blogs.hbr.org/cs/2013/06/the_mobile_shopping_life_cycle.html. Accessed February 2, 2014.
MixPanel. 2012. Celebrating Christmas with Data, https://mixpanel.com/blog/2012/12/27/celebrating-christmas-with-data/. Accessed February 2, 2014.
Ngai, E. W. T. and Gunasekaran, A. 2007. A review for mobile commerce research and applications. *Decision Support Systems*, 43(1), 3–15.
Reddy, M., Mun, S., Burke, M., Estrin, J., Hansen, D., and Srivastava, M., 2010. Using mobile phones to determine transportation modes. *ACM Transactions on Sensor Networks*, 6(2).
Roussos, G., Kourouthanasis, P., and Moussouri, T. 2003. Designing appliances for mobile commerce and retailtainment. Birkbeck College, Eltrun-Athens University of Economics and Business, RCMG University of Leicester, Leicester, U.K.
Schedlbauer, 2011. The quest for good requirements, Corporate Education Group, http://www.corpedgroup.com/resources/ba/TheQuestforGoodReqs.asp. Accessed February 2, 2014.
Siau, K., Lim, E.-P., and Shen, Z. 2003. Mobile commerce: Current states and future trends. In *Mobile E-Commerce on Mobile Phones* Do van Thanh (Ed.), pp. 1–17, IGI Global, Hershey, PA.
Song, J., Kim, M., Baker, J., and Kim, J. 2013. Fostering a seamless customer experience in cross-channel electronic commerce. In *2013 SIGBPS Workshop on Business Processes and Service*, Milan, Italy.
Sterling, G. PC Free: 46 percent used only mobile devices in purchase process according to new study, marketing land. http://marketingland.com/pc-free-46-used-only-mobile-devices-in-purchase-process-according-to-new-study-41846. Accessed on February 2, 2014.

Tsalgatidou, A., Veijalainen, J., and Pitoura, E. 2000. Challenges in mobile electronic commerce. In *Third International Conference on Innovation through E-Commerce*, Manchester, U.K., pp. 1–12.

Varshney, U. and Vetter, R. 2002. *Mobile Commerce: Framework, Applications and Networking Support*. Dordrecht, the Netherlands: Kluwer Academic Publishers.

Xu, Y., Spasojevic, M., Gao, J., and Jacob, M. 2008. Designing a vision-based mobile interface for in-store shopping. In *Fifth Nordic Conference of the Human–Computer Interaction: Building Bridges*, Lund, Sweden, pp. 393–402. New York: ACM.

chapter seven

Moving toward a mobile website for www.india.gov.in

Rakhi Tripathi

Contents

7.1 Introduction

India is deploying e-governance solution to provide online public services and record keeping, including online bill payment, taxes, land records, income certificates, loans, driving licenses, birth and death certificates, and various government entitlement programs.

India.gov.in is the Indian government's web portal for citizens. It presents information resources and online services from government sources, accessible from a single point. It is also known as the National Portal of India. The objective is to provide a single-window access to the information and services such as passport, driving licenses, company registration, and so on, being provided by the Indian government for the citizens and other stakeholders.

Many urban centers have set up web portals to facilitate the citizen-to-government interface necessary to support the delivery of services. An electronic interface as a one-stop destination for public access to information on various aspects of government functions well as a single window for the delivery of government services. However, certain centers such as rural areas and cities that lack the infrastructure necessary for every citizen to have web access are disadvantaged because in the absence of web connectivity, provisioning of government services is done manually. In regions that do not have Internet access, citizens are forced to wait in long lines for each service or information request.

On the contrary, India is witnessing an exponential growth of Internet users on mobile devices that access government information using a cell phone over the cell carrier network. People in rural India are spending more than those in urban areas, according to a study by Accenture.

The aforementioned problem and recent developments conclude that India is moving toward m-government. M-government involves a number of functions, for example, short message service (SMS), applications, and mobile website. This study focuses on the mobile version of National Portal of India. At present, National Portal of India does not have a mobile version; hence, the access to the National Portal of India and its services is limited on a mobile device. This chapter is an attempt to analyze what are the factors involved in the mobile site of India.gov.in and what are the levels of mobile websites.

7.2 Literature review

Since the time mobile computing came into the picture, a few developed government organizations have started moving toward m-government. Various authors have defined m-government in their own way. Table 7.1 presents definitions of m-government from different researchers over one decade. It can be noted in Table 7.1 that over the years the definitions have also evolved. Earlier m-government was more of a subset of e-government, and now it is an independent identity. According to Kailasam (2012), m-governance is not a replacement for e-governance, rather it complements e-governance.

Extensive research on m-government has been done over the last decade. Appolis et al. (2012) investigated m-government services and the barriers to adoption. Surveys were done to ascertain how citizens interact and access government services and information via their mobile phones. Initially, there were studies on issues and challenges related to m-government. Rossel et al. (2006) explored mobile e-government issues by analyzing their historical evolution. Carroll (2005) focused on issues that would enhance or delay the widespread acceptance of m-government services among the users of public sector organizations. Around the same time, Kumar and Prasad (2007) discussed some technical and policy consideration of mobile technology in the context of e-government, which can be better known as m-government. Antovski and Gusev (2005) have discussed not only the issues related to m-government but

Table 7.1 Definitions of M-Government

Author	Year	Definition
Kushchu	2003	The strategy and its implementation involving the utilization of all kinds of wireless and mobile technology, services, applications and devices for improving benefits to the parties involved in e-government including citizens, businesses and all government units.
Arazyan	2002	A functional subset of all inclusive e-government.
Carroll	2005	M-Government is largely a matter of getting public sector IT systems geared to interoperability with citizen's mobile services.
El-Kiki et al. and Ntaliani et al.	2005	M-government is considered as a subset of e-government comprising another channel to provide governmental information and services.
Sheng and Trimi	2008	M-government is defined as the use of mobile and wireless communication technology in providing government service.
Ntaliani et al.	2007	Mobile government (m-government) refers to the use of ICTs by government institutions with the help of mobile technologies to deliver electronic services to the public. This definition is derived from the definition of electronic government because m-government is its subset of electronic government.
Wang	2010	M-government focuses on a complete government system supported by a perfect m-government project. It does not simply encompass the use of m-government tool for the public services, but also a complex instrument for the development and implementation of strategies.
Signo	2012	M-government provides multidirectional advantage with the effective use of wireless devices.

also interoperability, security, openness, flexibility, and scalability, which according to the researchers were the principles of m-government.

After the issues and challenges, it can be noted in the literature that the research on m-government shifted to framework and implementation. Sheng and Trimi (2008) proposed a framework based on the theory of task–technology fit to understand mobile technology (MT) and its implications for m-government applications. Amailef and Lu (2008) presented a framework of mobile-based emergency response system that has five main components (register, monitoring, analysis, decision support, and warning) aiming to provide a new function and service to m-government. The legislative and regulatory considerations during design and implementation of m-government initiatives and during any postimplementation were studied by McMillan (2010). Rannu et al. (2010) tried to understand the mobile phone as a value-adding device in government services. In 2009, Fasanghari and Samimi developed six stages (five phases) m-government model. In this model, all of the required phases for implying mobile technologies in government service delivery to citizens are considered. El-Kiki and Lawrence (2007) highlighted and defined the barriers to the success of m-government service projects.

In the past few years, the research question shifted from issues and framework to adoption of mobile government. Appolis et al. (2012) studied the availability of m-government services, and the barriers to adoption of m-government were investigated. Abdelghaffar and Magdy (2012) introduced a conceptual model for youth adoption of m-government services in developing countries. Zamzami and Mahmud (2012) examined m-government services in three main areas: mobile interface design from different government sectors, the information quality of the applications, and whether this information has met user satisfaction.

7.3 *M-government and India*

India is aggressively moving toward m-government though it is at nascent stage. SMS is in the lead among other m-government facilities, the reason being that people in India, especially in rural India, use a normal cell phone handset with basic features of SMS, call, watch, music, etc., and not a smartphone where they can use a website or a mobile application. Following are some successful cases where m-government has become an integral part of citizens of India:

- Farmers of Haryana can use SMS for agriculture problems. SMS for immediate redressal of agriculture-related problems of the farmers in the state, introduced by the Haryana Agriculture Department in 2007, has been a success with the farmers asking a wide range of questions from experts. This free of cost service was launched on February 02, 2007, and the farmers could send an SMS for immediate redressal of their queries. The concerned specialists and officials answer questions sent to the department through this service within 24–48 h telephonically.
- Mobile phones have been used by fishermen in Kerala for a decade. Fishing is a vital industry in this part of south-west India, employing over one million people. The Harvard economist Robert Jensen has studied the effect that the spread of mobile communications has had on the fishermen's livelihoods. The arrival of mobile phones in the region meant that fishermen could compare prices and demand for fish across the area. This led to

the setting of a single rate for sardines along the Kerala coast, eliminated any wasted catches, and saw a fall in consumer prices. This more efficient market benefited everyone. Consumer prices fell by an average of 4%, and there was an 8% rise in the average fisherman's profits, meaning that mobile phones usually paid for themselves within the first 2 months.
- Dr. SMS is a novel project of the Kerala State IT Mission—the technology implementation wing of the Government of Kerala in India, launched on May 29, 2008, with the aim to increase the access of health care services by common people through simple and innovative use of mobile telephony. The project aims at improving the health of the citizens of Kerala by
  - Improving access to health care resources by making available authentic information on the same
  - Providing timely information on medical and diagnostic facilities
  - Providing informational alerts about emerging diseases

The main idea behind this project has been to enable people to use their mobile phones to receive information on health resources. The project aims at providing users with a comprehensive list of medical facilities available in a particular locality including hospitals with infrastructure and expertise in various medical specialties (cardiology, gynecology, etc.), specialized doctors in the locality, all through the simple facility of the SMS. Since an SMS is used to gain information on medical services, the project has been named Dr. SMS.

- Education sector in India: the most popular use of mobile technology in the education space through SMS. This has been used to impart English language training, conducting tests, sending alerts for examination dates, results, etc. Multimedia messaging service (MMS) are also used but are restricted to conducting training and sending educational messages.
- Over the years, banking has transcended from the traditional brick and mortar model to one where banking services are available anytime anywhere; the portability of mobile phones thus makes mobile banking a much desired phenomenon by banks and customers. Checking account balances is the most popular banking service used by urban Indians with almost 40 million users, followed by checking last three transactions with 28 million users, and status of cheques with 21 million users.

Bhatnagar (2009) explores conditions that can promote the use of mobile phones for successful delivery of government services to those at the Bottom of the Pyramid in India. Dr. D.C. Misra, I.A.S. (Retd.) in 2010 conveyed that efforts in the direction of m-government so far in India have been sporadic and piecemeal. No holistic view of m-government has so far been taken as a result of which the vast potential of m-government continues to be unrealized.

7.4 Research methodology

After looking into the literature and investigating the growth of m-government in India, it can be noted that m-government is limited to SMS mostly. According to Telecom Regulatory Authority of India (TRAI), November 5, 2013 (http://www.trai.gov.in/):

- Total cell phone subscribers in India are 870.58 million where the monthly average addition is around 20 million.
- Urban subscribers of cell phone in India are 520.21 million.
- Rural subscribers of cell phone in India are 350.37 million.
- There are 51 million smartphone users in urban India today, an 89% increase from 2012, when there were just 27 million users. The study also reveals that the biggest spike is in the youngest age group between 16 and 18 years, where numbers have gone from 5% in 2012 to 22% this year, a fourfold increase.
- Subscribers who access the Internet through cell phones are 143.2 millions.

This shows that India has a decent base for m-government services. With 143 million mobile Internet users (which are increasing rapidly), an initial mobile website of the government can do wonders for the citizens such as providing a unified view of all government services, providing relevant information, etc. The existing portal of India does not have a mobile version.

7.5 Portal stage model

A portal before becoming a one-stop undergoes different growth stages. There have been over 15 e-government stage models that have been developed in last 10 years. Lee (2010) has done a comprehensive study on this, in which he has compared and analyzed the e-government growth models and their respective stages.

Before planning for a mobile website, it is important to analyze the portal india.gov.in. The portal of Government of India is currently at the transaction level where in certain cases transactions can be made. Also at the back-office, there is integration though not fully vertical or horizontal.

Now, it becomes essential to identify the stages of a mobile website of india.gov.in. A mobile website needs to achieve all these stages in order to give a one-stop satisfaction to the citizen where all the government services are provided to the citizen at a single stop. Maranny (2011) proposed and validated a stage maturity model for m-government implementation process. The stage maturity model of m-government is intended to measure the performance of m-government implementation and to provide the roadmaps and recommendations for the future directions.

Tripathi et al. (2011) came up with five stages of growth. These stages are applicable for mobile website as well:

- Information: Stage governments are focused on establishing an online presence for the government and is mostly limited to online presentations of government information.
- Interaction: A more sophisticated level of formal interactions between citizens and service providers is facilitated through e-mail and "post comments" area.
- Transaction: empowers citizens to deal with their governments online anytime.
- Integration: *Vertical integration* refers to local, state, and federal governments connected for different functions or services of government. At stage three, federal, state, and local counterpart systems are expected to connect or, at least, communicate to each other. At *horizontal integration* stage, databases across different functional areas communicate with each other and ideally share information, so that information obtained by one department will propagate throughout all government functions.
- Transformation: One-stop mobile portal for India.gov.in.

As of now, it would be better to start with the first stage, that is, information stage, the reason being that citizens can easily acquire the knowledge about government services. For example, a citizen can get the information about passport issue, tenders, latest events, etc.

7.6 Mobile websites: Features, different from desktop websites, India's mobile site?

It is determined that the mobile website providing information will be appropriate for the time being. There are certain features that a mobile website must have when compared with the desktop website. Table 7.2 highlights major differences between mobile websites and normal websites.

Table 7.2 Features of Mobile Website versus Desktop Website

Basis	Mobile websites versus websites
Content prioritization	In comparison to the design of websites for desktop computers—typically, for a 1024 × 768 screen resolution—the biggest challenge in designing a website for a smartphone with a 320 × 480 screen resolution is how to cope with this dramatic difference in screen size without sacrificing the user experience. While desktop websites often contain a wide range of content, mobile sites usually include only the most crucial functions and features—particularly those that leverage time and location. Mobile site designs should give priority to the features, and content users are most likely to need when viewing a site using a mobile device.
Vertical instead of horizontal navigation	Horizontal navigation is a widely accepted means of structuring and presenting content on desktop websites. Users scan a navigation bar from left to right, and then click a link to go to a different section of a site. When a navigation bar is at the top of a page, users can typically more easily focus on page content rather than being visually attracted to the navigation bar on the side. However, vertical navigation has replaced horizontal navigation on more than 90% of the mobile sites we analyzed.
Bars, tabs and hypertext	Hypertext is the signature component of the Internet and the web. However, we see much less hypertext on mobile pages. It is not that pages are no longer linked, but that links instead appear in the form of bars, tabs, and buttons. The reason for this is the optimization of mobile design for users' operation of mobile devices with their fingers. Hypertext is ideal when users click links using a mouse on a computer, but tapping links using your fingers on a touchscreen mobile device is not easy. Users can too easily activate a link they did not intend to tap and accidentally land on an undesired page. This can lead to a bad user experience. Fitts's law tells us that the time required to acquire a target area is a function of the distance to and the size of the target. Bigger objects such as bars, tabs, or buttons allow users to tap with more precision. It is essential to make the actionable objects on mobile sites big and easily noticeable.

(*Continued*)

Table 7.2 (Continued) Features of Mobile Website versus Desktop Website

Basis	Mobile websites versus websites
Text and graphics	On web pages, we can use graphics for many different purposes such as promoting, marketing, or navigating. However, designers often remove promotional or marketing graphics from the designs of mobile sites. The company logo remains for navigational purposes. Users can tap it to go to the home page. There are several reasons for this transition from many to few graphics. One reason is that some mobile devices do not support the software we traditionally use for desktop website design. Other reasons include the small screen sizes of mobile devices and the limited available screen real estate in which to display content as well as the slow download speeds on mobile devices.
Contextual and global navigation	Desktop websites typically use various forms of navigation. Some of them are global and remain consistent across a site, while others are contextual and change depending on where users are on a site. However, while global navigation is common on mobile sites, contextual navigation is not. The main reason for the reduction of global and contextual navigation on mobile sites is the limited screen real estate on mobile devices. However, a lack of global and contextual navigation may cause users to find themselves in the middle of nowhere, not knowing where they are. Therefore, it is essential to reduce hierarchy when organizing the content on mobile sites, so users do not have to dig too deeply to get things done. They should be able to achieve what they want to accomplish before becoming lost.
Breadcrumbs	On desktop websites, breadcrumbs are an effective way of reassuring users they are on the right page and allow them to backtrack on their navigational path. They make sense for large, hierarchical websites with lots of different contents at multiple levels in a hierarchy. However, breadcrumbs rarely appear on mobile sites, and there is usually no necessity for them. Limited space is one reason breadcrumbs are uncommon on mobile sites. But the main factor is that the design of mobile sites prevents users from having to go too deep into a hierarchy to find what they are looking for. Again, users should be able to achieve what they want to accomplish on a site before they start feeling lost.

(*Continued*)

Table 7.2 (Continued) Features of Mobile Website versus Desktop Website

Basis	Mobile websites versus websites
Progress indicator	On desktop websites, when users must progress through multiple steps to complete a process—whether they are making a purchase or filling out a long registration form—there is often a progress indicator at the top of the page to guide users through the process. Such progress indicators do not appear on mobile sites. Again, limited space is the main reason. Use alternative approaches to make users aware of their progress without a progress indicator. For example, instead of using buttons with implicit actions such as Next or Continue, use buttons with explicit labels that inform users exactly what the next step is—for example, Proceed to Checkout or Specify Shipping & Payment. Users still receive information about where they are in a process and what to expect at the next step.
Integration with phone functions	Smartphones or mobiles with Internet are communications devices, so making phone calls is their most basic function. While mobile platforms place many limitations on design and content, they also open up new opportunities that traditional websites cannot provide. For example, there is better integration with phone functions such as direct calling and text messaging, which lets mobile sites facilitate ordering products by phone or sending promotional text messages. Usually, mobile sites let users select a phone number, then call or text that number—without having to type the number.
Footers	There are two types of footers that are in common use on desktop websites. One type of footer provides links to content that users might expect to see on a site's home page, but has a lower priority than the primary content on the home page—for example, Careers or Sitemap. Another type of footer provides quick links to content users typically need to view most often. These quick links are often grouped in lists in a footer, so that users have access to them across a site. Mobile sites employ footers that provide access to content users often looking for on a home page, keeping its links to a minimum, but they do not use footers containing quick links.

(*Continued*)

Table 7.2 (Continued) Features of Mobile Website versus Desktop Website

Basis	Mobile websites versus websites
Localized and personalized search	Another area of opportunity that is unique to mobile sites is the use of geo-location services or support. While this technology has been available for some time, only in the last 5 years has it gained traction in the consumer marketplace. Now, it is commonplace for mobile applications and websites to take advantage of this functionality by integrating it into value-adding services such as mobile search. Many mobile devices can automatically detect where users are and give them local search results. This capability offers powerful opportunities for businesses to promote their products or services based on a person's proximity to their place of business and their immediate intent.
Screen resolution	Perhaps the most obvious, screen resolution is of course hugely different for a mobile browser than for a regular PC/Mac browser. The typical screen resolution is anywhere from around 120 pixels wide for a mobile device, whereas the most popular resolution is a minimum of 1024 pixels wide for regular monitors. Content designed for a regular browser simply will not display properly in a mobile browser *out of the box*. Changes have to be made if the browsing experience is to be any good at all.
Navigation	Navigation links are more often than not consigned to the left vertical portion of the screen when viewing a website on a PC/Mac-based browser. In mobile web browsing this is not the case—the *norm* is to have page navigation near to, or right at the very top of each page, for ease of navigation using a mobile browser. Given the limited page width, the key to viewing page content on a mobile browser lies in the vertical axis—everything has to flow vertically, rather than horizontally.
Bandwidth	Images and other page elements can contribute an unnecessarily high amount to the overall page size. This might not seem much of an issue in today's high-bandwidth world, but for mobile web browsers this can be a real problem, particularly given the fact that many mobile users pay a premium fee for data transfer to their mobile. Every byte counts, so ensuring a mobile site treats elements such as images appropriately can make a huge difference.

(*Continued*)

Table 7.2 (Continued) Features of Mobile Website versus Desktop Website

Basis	Mobile websites versus websites
Image sizes	Images that are designed for viewing on large monitors can of course look garishly large on a mobile device screen. Although image sizes can of course be resized on-the-fly by some modern mobile web browsers (in dimensional terms), this often leads to distorted images and, again, a poor quality browsing experience.
Contrast and color schemes	When looking at a regular site using a large, well lit, quality LCD screen all of the subtleties and color shades used in a website can be appreciated. Conversely, when bustling along a busy street, one hand holding a mobile phone, eyes squinting at a small screen, a whole new set of issues come into play. Suddenly, design elements become less important—the clarity of the information becomes paramount (simply to be able to read it!).
Touch vs. click	Finger tapping is less precise than mouse clicking. Unlike a precise mouse pointer on your desktop, the human finger is a blob, and finger tapping requires large targets on the screen for hyperlinks. Expect to see larger rectangular tap targets (tiles) on mobile web pages, and fewer text-based hyperlinks. Additionally, menus will often be replaced with large buttons and large tabs to accommodate the imprecision of finger taps.

7.7 *Making government websites mobile compliant*

The Government of India will support a *one web* approach to government websites. "One web means making, as far as is reasonable, the same information and services available to users, irrespective of the device or the browser they are using." This also confirms to the guidelines laid down in the Right to Information Act (RTI) of Government of India. This means that the mobile device compatibility of all the government websites should be ensured. It has been estimated that mobile users in India will reach one billion subscribers by 2015. Although not all of the devices used by the citizens may be enabled currently for full web access, it does not underscore the importance of web content on mobile devices. It is proposed through this recommendation that standards for mobile accessibility of central and state government websites be established and that best practice guidelines for enabling mobile access to government sites be followed. Provided in the following are some of the guidelines that may be adopted:

1. Central and state government websites should consistently enable mobile site access as new sites are designed and implemented. Mobile Web Best Practices 1.0 from the World Wide Web Consortium (W3C) at http://www.w3.org/TR/2008/REC-mobile-bp-20080729 may be referred to while formulating standards for enabling mobile web access.
2. All government agencies should use mobile optimized content as a primary method for device support, with device specific style sheets as appropriate. It may be noted that in some smartphone instances (e.g., iPhone, Windows Mobile Phones, PalmPre, Android, and so on) not much effort may be required to render a usable website on the device.
3. Mobile website access should be supported by CSS specific files for major smartphone devices using any of several device detection methods such as the Wireless Universal Resource File (WURFL) at http://wurfl.sourceforge.net to redirect to specific mobile implementations. Other types of web-enabled mobile devices, when detected, should degrade gracefully to text-based CSS implementations for less capable devices.
4. Specialized website addresses for mobile sites should be avoided in preference to using standard web addresses. As an example, a single URL address, www.india.gov.in, should render the website on a computer browser or a mobile device through a dedicated script that is capable of detecting the device used and redirect to the corresponding content structure.
5. The government websites should avoid the use of large graphics, image buttons, or graphic rich form elements, and assume constrained resources for bandwidth, screen size, color, and resolution peculiar to mobile devices.

   The government websites, before being made live for citizen use, should be tested for mobile website implementations in as many targeted devices and simulators as is practicable.

7.8 Analysis

Considering the facts such as adoption of large number of cell phones in India; mobile website's different stages; different features of mobile and desktop websites; and one web, some useful insights can be highlighted:

- India is at the nascent stage of m-government. However, there is an exponential growth in cell phone subscribers in India. Not only urban but rural India is moving toward mobile phones, hence making the base for m-government.

- There is a positive increase in the number of Internet subscribers on wireless devices in India. This shows that apart from SMS and calls, a government website can be very useful for citizens.
- Mobile website of the government has to undergo different stages, from providing basic information to the citizens to a one-stop destination for its citizens. As it can be noted that India is at early stages of m-government and is moving actively toward m-government, the mobile website of government must provide information to the citizens.
- The moment information website for mobile is planned, content of the website becomes the outmost priority. With a limited screen size, the content of the mobile website must be clear so that it is easily understood by the citizens.
- The bandwidth of India as of now is comparatively low with developed countries. Hence, the size of the mobile website should be small. Usage of pictures and videos (one in india.gov.in) should be minimal.

7.9 Conclusion and future work

India has shown a remarkable development in e-government over the years. www.india.gov.in has shifted from mere having a web presence to an integration level. Concurrently, India has witnessed significant growth in not only Internet users and mobile users but also those users who access the Internet from their mobiles. This grouping has raised an unquenchable thirst for a mobile version of government website in India. The mobile website will not only be a lighter version but also provide significant information to the citizens at one place.

This chapter analyzes the current scenario that forms the basis of a mobile version of www.india.gov.in. Before coming up with a final design of mobile version, it is important to check the current status and predict the growth of mobile subscribers in urban as well as in rural India. Then it becomes vital to plan the level of the mobile website. In this study, it has been found that the information level is the most appropriate one to start with.

This study can be applied to different government organizations as well. Moreover, a similar study can be done for commercial websites. As a future work, a complete mobile website design of www.india.gov.in can be developed that will include both technical and managerial issues.

References

Abdelghaffar, H. and Magdy, Y. (2012). The adoption of mobile government services in developing countries: The case of Egypt. *International Journal of Information and Communication Technology Research*, 2(4), 333–341.

Amailef, K. and Lu, J. (2011). A mobile-based emergency response system for intelligent m-government services. *Journal of Enterprise Information Management*, 24(4), 338–359.

Antovski, L. and Gusev, M. (2005). *M-Government Framework, EURO mGOV 2005*, Brighton, U.K., pp. 36–44.

Appolis, K., Alexander, B., Parker, M., and Wills, G. (2012). Availability and adoption of m-Government services in South Africa. *Proceedings of the 14th Annual Conference on World Wide Web Applications*, Durban, South Africa.

Arazyan, H. (2002). M-government: Definitions and perspectives. Retrieved from: September 2013, www.developmentgateway.org

Bhatnagar, S. (2009). Exploring conditions for delivery of successful M-government services to the Bottom of the Pyramid (BOP) in India. International Development Research Centre, Ottawa, Ontario, Canada and the Department for International Development, London, U.K.

Carroll, J. (2005). Risky business: Will citizens accept m-government in the long term. Paper presented at the *From E-Government to M-Government*, University of Sussex, Brighton, U.K.

El-Kiki, T. and Lawrence, E. (2007). Emerging mobile government services: Strategies for success. *20th Bled eConference*, Bled, Slovenia.

El-Kiki, T., Lawrence, E., and Steele, R. (2005). A management framework for mobile government services. Paper presented at the *COLLECTOR*, Sydney, New South Wales, Australia.

Fasanghari, M. and Samimi, H. (2009). A novel framework for m-government implementation. *International Conference on Future Computer and Communication*, Kuala Lumpur, Malaysia, January 2009, DOI:10.1109/ICFCC.2009.146 ISBN: 978-0-7695-3591-3.

Kailasam, R. (2012). m-Governance: Leveraging mobile technology to extend the reach of e-Governance. Accessed September 2013, Available at http://papers.eletsonline.com/2012/01/18/m-governance-leveraging-mobile-technology-to-extend-the-reach-of-e-governance.

Kushchu, I. and Kuscu, H. (2003). From E-government to M-government: Facing the inevitable? *Proceedings of European Conference on E-Government (ECEG 2003)*, Trinity College, Dublin, Ireland.

Kumar, M. and Prasad Sinha, O. (2007). M-Government—Mobile technology for e-government. *International Conference on E-government*, Hyderabad, India.

Lee, J. (2010). 10 Year retrospect on stage models of e-Government: A qualitative meta-synthesis. *Government Information Quarterly*, 27(3), 220–230.

Maranny, E.A. (2011). Stage maturity model of m-Government (SMM m-Gov): Improving e-Government performance by utilizing m-Government features. University of Twenty, Enschede, The Netherlands, Accessed September 2013, Available at http://essay.utwente.nl/62691/.

McMillan, S. (2010). Legal and regulatory frameworks for mobile government. *Proceedings of mLife 2010 Conferences*, October 27–29, Brighton, U.K.

Misra, D.C. (2010). Make M-government an integral part of E-government: An agenda for action, compendium, *National Forum on Mobile Applications for Inclusive Growth and Sustainable Development*, New Delhi, India, April 7–8, 2010, pp. 78–86.

Ntaliani, M., Costopoulou, C., and Karetsos, S. (2007). Mobile government: A challenge for agriculture. *Government Information Quarterly*, 25(4), 699–716.

Rannu, R., Saksing, S., and Mahlakõiv, M. (2010). Mobile government: 2010 and beyond. Mobil Solutions Ltd., Electronic references retrieved from: August 2013, http://www.mobisolutions.com.

Rossel, P., Finger, M., and Misuraca, G. (2006). Mobile e-government options: Between technology-driven and user-centric. *The Electronic Journal of e-Government*, 4(2), 79–86.

Sheng, H. and Trimi, S. (2008). Emerging trends in m-government. *Communications of the ACM*, 51(5), 53–58.

Signo, C.M. (2012). PNP text 2920 service: A communication perspective on m-government: Philippine context. Retrieved from: September 2013, http://www. cprsouth.org/wp-content/uploads/drupal/Cristina_Signo. pdf.

Tripathi, R., Gupta, M.P., and Bhattacharya, J. (2011). Identifying factors of integration for an interoperable government portal: A study in Indian context. *International Journal of Electronic Government Research*, 7(1), 64–88.

Wang, C. (2010). Annual report on China's e-government development. Social Sciences Academic Press, Beijing, China, pp. 231–233.

Zamzami, I. and Mahmud, M. (2012). Mobile interface for m-government services: A framework for information quality evaluation. *International Journal of Scientific & Engineering Research*, 3(8), 39–43.

chapter eight

Security in mobile electronic commerce

James Scott Magruder

Contents

8.1 M-commerce

The buying of products or services through a computer network (including the Internet) is called e-commerce (Wikipedia, the free encyclopedia 2013, 1). Initially, this activity was performed on a stand-alone, hardwired PC. Then wireless systems were developed. This allowed movement of the device used in e-commerce. This gave way to m-commerce. M-commerce ensures the ability to conduct electronic commerce using a mobile device no matter where the customer is geographically (assuming the mobile device can connect to an appropriate network). The term "mobile commerce" originally meant "'the delivery of electronic commerce capabilities directly into the consumer's hand, anywhere, via wireless technology' (Morris 1997)" (Wikipedia, the free encyclopedia 2013, 2).

"With the proliferation of smartphones over the past couple of years, it is easy to forget that mobile commerce (m-commerce) began back in 1997 when the first two mobile phone-enabled Coca-Cola vending machines were installed in Helsinki, Finland. Those machines accepted payment via SMS text messages" (Sahota 2011).

"Mobile commerce, as a term, has taken on a dual meaning, describing both making online payments or purchases from a mobile device (mobile e-commerce) and using a mobile device to make payments at a physical store's point of sale" (Roggio 2013). "Over a billion people use the PC platform to interact with the Internet and yet security breaches are still an issue for the apparently 'mature' technology" (Staff Writer 2011). mobiThinking gave the figure of 5.3 billion subscribers in the mobile market (the web page is dated 2011) (Staff Writer 2011). The PC platform numbers may be going down (replaced perhaps by the mobile market), but the mobile market will continue to grow in numbers. As the mobile market continues to grow, those who would try to hack into mobile devices will also grow in numbers (Staff Writer 2011).

This means that the security for mobile devices used for m-commerce must improve substantially. Authentication has been a major problem for some time in networking. Both sides of the connection must be authenticated as to who they say they are. The website must indicate their true validity (usually via a digital certificate (Symantec)). The customer must validate their credentials as well. This is often done by the customer logging in to their account on the website. After an account has been set up, the log-in procedure can be automated via software (app) so the username and password combination is automatically sent to the website when the customer uses a (mobile) device to access the website. Obviously, this *convenience* may be a problem if the mobile device is lost.

"Mobile devices have become computers in their own right, with a huge array of applications, significant processing capacity, and the ability to handle high bandwidth connections. They are the primary communications device for many, for both personal and business purposes" (Silva 2012). So the advantages/disadvantages of computers now apply to mobile devices.

Typically, in e-commerce/m-commerce, a (mobile) device will communicate with a website on a network (Internet) and search the website for a desired product. The catalog is *served up* via a database of the organization's products. Once a product(s) is chosen, the customer logs in to the website (if not previously done) and fills out a form, and the product is sent to the customer. The product could be some sort of service, software, or a physical product. The service or software may be electronically sent to the customer. The physical product would be shipped later. The products that the customer looks at in the catalog can be saved as part of the customer's account on that website. This information could be used by the organization as part of a marketing plan toward the customer, or it could be sold to another organization. One or more databases contain the website's catalog and the customer's information.

Cookies may be used to pass information between the web server and the browser and also used to send the user a customized page or to identify the user (Webopedia). The cookie law used in the EU "covers the use by business of information stored on users' 'terminal equipment' and this covers mobile sites and apps as well as desktop sites" (Charlton 2012). Cookies may be associated with browser use and tracking e-mails. They may also have to be considered in mobile apps and *other tech channels*

(Charlton 2012). Getting the user to consent to use a cookie when they are online looking at products and/or making a purchase may be more problematic due to the fact that the screens are smaller. This may also cause the user to stop their current process and go to another site (Charlton 2012).

8.2 Changing business models

Do business models change for mobile e-commerce? Yes and no. It is possible to take a business model used in the *regular business world* and implement it online (as a website) where nonmobile and mobile devices can access the website. Accessing the website via a mobile device is much more convenient for the customer. It may, however, not be better for the customer or the organization. Security issues come into play when the customer accesses a website with a mobile device.

Some basic business model categories are given in a web page by Rappa (2010):

- Brokerage
- Advertising
- Infomediary
- Merchant
- Manufacturer (direct)
- Affiliate
- Community
- Subscription
- Utility (Rappa 2010)

It is interesting to note that most of these business models have been implemented in the *regular* business world. Most of these business models put information into a database and allow the customer to access that database remotely.

Joseph lists other types of business models for e-commerce (Joseph 2014):

- Vanity
- Storefront
- Subscription
- Business-to-business
- Affiliate marketing (Joseph 2014)

The vanity business model allows the developer to *blog* their opinions and earn some income through advertisements placed on the website (Joseph 2014). Affiliate marketing allows an organization to sell some products of another organization and get commission for their efforts. Again, most of these business models have been implemented in the *regular* business world. The main difference between these business models and their *regular business* counterparts is that the former are accessed with a mobile device.

Many *regular* business models are extended to an online environment. This may make for a more convenient and (with a correct website design) satisfying customer experience. Happy customers are usually the ones who return to the website.

There is at least one area where business models have changed (or may be changing): payment in the mobile environment (Tode 2013). "Several trends promise to disrupt the payments space this year, with mobile playing an important role as key players look to deliver greater value for merchants and customers, according to a new report from Forrester" (Tode 2013). Competition in the mobile payment environment may well produce new business models for this industry. Some are being tested now (Tode 2013).

For example, LevelUP is linking a quick response (QR) code to a credit/debit card. This is to make the transaction free, fast, and simple (Tode 2013). LevelUp "uses marketing fees to replace interchange fees, with merchants paying LevelUp for delivering new and returning customers to their business" (Tode 2013).

"*QR code* (abbreviated from *Quick Response Code*) is the trademark for a type of matrix barcode (or two-dimensional barcode) first designed for the automotive industry in Japan. A barcode is an optically machine-readable label that is attached to an item and that records information related to that item" (Wikipedia 2013, 3). QR codes cannot be read by humans and must be read by a QR code reader. There are smartphone apps to read QR codes. However, the QR code reader app should always read the contents of the QR code in a sandbox as there may be malware concealed within the QR code information. The QR code information can contain URL information and (under certain circumstances) automatically take the smartphone to a malicious web page (Mazhar 2011). There are QR code generators available on the Internet (just search on "QR code generators"). So it may be possible for an individual to develop a malicious QR code and print it out via a printer. If an unsuspecting individual scans this malicious QR code, their mobile device may become infected with malware.

"*Next generation smartphones* or tablets equipped with fingerprint scanners may provide shoppers with greater security and boost both online and in-store mobile commerce" (Roggio 2013). "The argument that biometric security—i.e., using human traits for identification—will affect mobile commerce includes the idea that biometrics, like fingerprint scanners, will be both more secure and easier to use than existing mobile payment security methods" (Roggio 2013). This fingerprint scanner will make logging in to the mobile device easier and mobile commerce shopping easier (remotely or in a physical store) and more secure (Roggio 2013). However, a report already indicates this security technology may have been hacked. A simple process to defeat this security apparatus was used (Finkle 2013).

Mobile devices tend to be smaller and therefore more mobile. The screen size may present a problem with regular websites. Due to this, m-commerce websites may have to be adapted for the mobile device's smaller screen. This may mean less graphics and better use of fonts (Janssen 2010–2014). Some organizations may have a website for regular viewing and also one for mobile device viewing (Butkovich 2013). When a customer visits the website, the server could sense if the user is using a mobile device and redirect the request to the appropriate site (both of which could be on the same server).

Tracking customers inside physical stores is an experiment that some retailers are pursuing. If a customer chooses a product, then the customer may receive a message describing an associated product (Brandon 2013). "Location-based tracking is getting much more precise. New technologies like magnetic field detection, Bluetooth Low Energy, sonic pulses, and even transmissions from the in-store lights can tell when you enter a store, where you go, and how you shop" (Brandon 2013). This tracking may be done without allowing a customer to *opt out*. A good deal may be costing us our privacy (Brandon 2013).

8.3 Security

This may be the most important aspect of mobile electronic commerce. Without appropriate security, the organization may not remain in business for a long period of time. Security hacks to the organization's network and database may cause the organization's name to be disgraced and lawsuits may be filed against the organization. Security is intertwined throughout the development and usage of mobile electronic commerce. Some of the security considerations are mentioned earlier indicating how much security is such a big part of this process.

It should be remembered that even mobile (wireless) devices must send their messages through networks. If portions of these networks (routers, servers, etc.) are compromised, then the information sent may also become compromised. Encryption should take care of this problem, that is, even if the message is captured by a hacker (via a sniffer), it could not be read. This is not as true as it has been in the past. It should also be remembered that mobile devices transmit their messages via radio waves. Anyone with the correct equipment can intercept the signal. Whether it can be decrypted is another issue.

Bring your own device (BYOD) presents additional security problems. There are potential problems when personal data and the corporation's data are saved on a mobile device. These two sets of data should be separated on the mobile device. Corporate network endpoints must be protected with the appropriate security. Access from outside of the corporate network that is not authorized should be stopped by a firewall. Allowing mobile devices an access to the corporate network may be circumventing the network's security. If the mobile devices have malware on them, the corporate network may become infected with this malware (Kaspersky 1997–2013).

The connection between a mobile device and an e-commerce website should always be secure. This is especially the case when the customer is filling out forms, giving credit card numbers, etc. Usually, there is an icon in the mobile devices browser that indicates whether the connection is secure or unsecure. It is up to the customer to determine if the connection is a secure one before putting sensitive information into the connection. If the connection is not a secure one, then the information should not be sent via the mobile device (Butkovich 2013). A secure connection over a network requires that both sides of the conversation use secure software. A browser could use secure software, but if the e-commerce website does not use secure software, then the connection is not secure.

"Gaining the trust of online customers is vital for the success of any company that transfers sensitive data over the Web" (VeriSign 2010). Without this trust, customers may be unwilling to use the company's website for mobile commerce. Malware could be on the organization's website and may be downloaded to the customer's mobile device.

Of course, trust goes both ways. The organization must be able to trust the customer that has just logged in to their m-commerce website that the customer is who they say they are and can legitimately pay for purchased products/services.

Phishing (FBI 2009) is an attack that has been around a long time and can be implemented in mobile and nonmobile environments. This is an attack where a user is asked for sensitive information (log-in information, social security number, etc.). This information should not be given out by the user. The request is often made via the phone or e-mail. Rashid (2011) gives some examples and indicates how they can be recognized. Phishing attacks can be made in different operating systems and different hardware. This is an insidious aspect of phishing attacks.

Spear phishing (FBI 2009) may be thought of as a targeted phishing attack. Instead of general information in the attack, a spear phishing attack contains more information the targeted user may recognize and therefore conclude the request is legitimate (FBI 2009). Users should be aware that this type of sensitive information should never be requested electronically. The IT department does not need a user to give them a user's log-in name and password. A bank does not request a customer's account number via an e-mail.

A new targeted attack is called water holing (WEBROOT 2013). The attackers determine the website(s) that a targeted group of people often go to. Then they attempt to compromise the website(s) with malware and create a *drive-by download* (WEBROOT 2013) from the website. When the targeted group of people visit the website(s), they may inadvertently download the malware (WEBROOT 2013).

Some malware attacks require that the user perform some action. The user must click on a URL, open an e-mail attachment, fill out a form, etc. Other malware attacks are *automatic* and do not require an action on the users' part. Usually, these types of malware are more specific to operating systems and hardware. They also must be downloaded to the target device and then executed separately. The user must usually be tricked into executing the malware.

The network hardware used in the mobile environment runs in a similar fashion to the nonmobile network environment. This hardware has operating systems or other software needed to run correctly. Operating systems and applications (apps) can be attacked. As an example of how the mobile environment continues to become more dangerous for the customer/user, there is a report that the first-ever malware for the Firefox operating system will be demonstrated at a conference in New Delhi this year (Kumar 2013, 1). "FirefoxOS is a mobile operating system based on Linux and Mozilla's Gecko technology, whose environment is dedicated to apps created with just HTML, CSS, and JavaScript" (Kumar 2013, 1). All of the apps in the operating system are web apps. There are two types of apps for this operating system. One type is downloaded to the mobile device, and the other is on a website (Kumar 2013, 1). Which type is better? Robles (2011) attempts to answer this question. Basically, the answer is "It depends." It depends on the organization, the app, and the customers the organization wants to serve. For the same app, both may be appropriate. For different apps, one or the other may be more appropriate (Robles 2011). Which one is more secure? This question has not been answered. This answer may also *depend* on the given situation.

On the server side of the network, a recent example is about Oracle. "Oracle on Tuesday fixed 127 security issues in Java, its database, and other products, patching some flaws that could let attackers take over systems" (Constantin 2013). Systems have become so complex that the aforementioned number of security issues is not necessarily unusual. These are just two recent examples of the dangerous networked environment. There are many others, and nothing should be concluded from the use of these companies as examples. The point is that that there are individuals (and groups) that will try and take advantage of flaws in complex systems for their own purposes. This makes the Internet a dangerous place. The risk of using it must be managed appropriately by the user/customer when they use the Internet to access a resource.

In addition, the user/customer expects the organization to make their website as secure as *reasonably* possible.

Symantec indicates that digital certificates are the answers to users and devices authentication. These certificates are highly secure, stable, and scalable. They can be used to move data securely over a network (including the Internet and mobile network). They can also be used to verify that devices and users are, in fact, who they say they are (Symantec).

8.4 Surveillance

In addition to security issues in the mobile environment, mobile devices may be under surveillance. Spiegel has seen documents, which are classified, that indicate the National Security Agency (NSA) has set up working groups to determine procedures to access three mobile platforms that are very popular. The NSA wants to be able to access the user's location, their contact lists, and SMS (short messaging service) (Musil 2013). One of the devices, the BlackBerry, is considered to be very secure. The other two popular devices are Android and iPhone (Rosenbach et al. 2013).

Between 2010 and 2011, the NSA tracked US mobile phones in a pilot project that at the time was secret. Mobile phone location data were collected. Basically, it was a test project to see if this information could be captured and stored in the system (Gallagher 2013).

General Keith Alexander, who is the leader of the NSA, stated that there was no current reason to capture mobile device location data at this time. However, he also stated that there may be reasons for such data collection in the future (Serwer 2013).

Instant messaging and e-mail contact lists from hundreds of millions of accounts are reportedly being collected by the NSA. This may include Americans as well. It is currently illegal for the NSA to acquire contact lists in bulk from devices in the United States. However, they can collect this information from other worldwide locations (Homeland Security FoxNews.com).

One way to try to remain anonymous on the Internet is to use Tor. Tor uses relays in a distributed network that is *within* the Internet. Volunteers run the relays. These relays are distributed throughout the world. The website that the user visits will not be able to determine the location of the user. In addition, no one can determine the websites that the user visits (Tor Project: Anonymity Online). Android can access the Tor network using a package called Orbot (Tor Project: Android Instructions). It will probably be available on other mobile devices soon (if it is not available already).

Since Tor is basically a (distributed) network within the Internet, it has entry and exit points in the Internet. So users may not be able to be tracked within the Tor network, but they may be able to be tracked at these entry and exit points. The NSA has used Google Ads (Rosenblatt 2013) and browser cookies (Kumar 2013, 2) to try and keep up with where users enter and exit the Tor network. The NSA has also used a vulnerability in the Firefox browser to try and track these users. They also run their own relays (nodes) (Kumar 2013, 2). Using browser cookies to track Tor users may not be as easy as it may seem. Goodin discusses the process and states that these types of attacks may be possible to use, but only under certain circumstances (Goodin 2013).

It is also reported that the NSA is monitoring millions of Americans' Internet habits (histories on the Internet) even though these Americans are not suspected to be involved in an activity that is deemed criminal. Marina is the codename for the program used by the NSA. The data may be maintained for a year (RT.com 2013). They may also be monitoring contact lists maintained in mobile devices (and nonmobile devices.) (FoxNews.com, Politics 2013) This is another aspect of the federal government's surveillance that may trouble users in mobile and nonmobile environments. Part of these habits/histories may include making payments for international transactions using credit cards. The NSA may have a separate portion of the organization to track these transactions (Butcher 2013). Of course, it may very well be important for the NSA to track some of these transactions as they may be involved with international terror plots. The problem is which of these transactions may be involved in these types of plots and which ones are not.

Imagine that you are the head of a technology company in America that makes mobile devices and/or software for mobile devices. Your company wants to sell these products in America and other countries as well. Would recent NSA revelations cause an international customer to pause before they buy your products? Would this affect your companies' public relations with the international community? Potentially, yes (Ingersoll 2013).

8.5 Conclusion

It should be noted that all of this technology is a *double-edged sword*. It can be used for a good purpose or for a nefarious purpose. Today, the technology is perhaps used for a good purpose. Tomorrow, the same technology may be used for a nefarious purpose. It is this second situation that should give us pause. Will the benefits of the mobile technology outweigh the nefarious costs of the technology? We will see.

Between the security aspects of mobile electronic commerce and the potential surveillance that may occur on mobile devices, it is very clear that consumers (users of mobile devices and the Internet) must be very careful in using this technology. Be safe out there.

References

Brandon, J. 2013. Retail stores plan elaborate ways to track you, Smarter America, FoxNews.com. Accessed July 26, 2013.

Butcher, M. 2013. NSA allegedly spies on international credit card transactions, TechCrunch, http://techcrunch.com/2013/09/15/nsa-allegedly-spies-on-international- credit-card-transactions/. Accessed September 16, 2013.

Butkovich, S. 2013. 5 Big mistakes to avoid in mobile e-commerce, SLI Systems Search, Learn & Improve, http://www.sli-systems.com/blog/2013/06/5-mistakes-to-avoid-in-mobile-ecommerce.html. Accessed October 2, 2013.

Charlton, G. 2012. How will the EU cookie law affect mobile marketing? http://econsultancy.com/us/blog/9773-how-will-the-eu-cookie-law-affect-mobile- marketing. Accessed October 18, 2013.

Constantin, L. 2013. IDG news service, Oracle plugs severe security holes that put systems at hijack risk, InfoWorld, http://www.infoworld.com/d/security/oracle-plugs-severe-security-holes-put-systems- hijack-risk-228874. Accessed October 17, 2013.

FBI. 2009. Spear phishers angling to steal your financial info, http://www.fbi.gov/news/stories/2009/april/spearphishing_040109. Accessed October 20, 2013.

Finkle, J. 2013. German group claims to have hacked Apple iPhone fingerprint scanner, http://www.reuters.com/article/2013/09/23/us-iphone-hackers-idUSBRE98M01X20130923. Accessed October 19, 2013.

FoxNews.com, Politics. 2013. NSA reportedly collecting millions of personal online contact lists worldwide, the Associated Press contributed to this report, http://www.foxnews.com/politics/2013/10/14/nsa-reportedly-collecting-millions- personal-online-contact-lists-worldwide/. Accessed October 15, 2013.

Gallagher, R. 2013. NSA admits tracking US mobile phones, Theage.com.au. http://www.theage.com.au/it-pro/security-it/nsa-admits-tracking-us-mobile-phones- 20131003-hv1vz.html. Accessed October 3, 2013.

Goodin, D. 2013. How the NSA might use Hotmail, Yahoo or other cookies to identify Tor users, http://arstechnica.com/security/2013/10/how-the-nsa-might-use-hotmail-or- yahoo-cookies-to-identify-tor-users/. Accessed October 24, 2013.

Homeland Security FoxNews.com. 2013. NSA reportedly collecting millions of personal online contact lists worldwide, http://www.foxnews.com/politics/2013/10/14/nsa-reportedly-collecting-millions-personal-online-contact-lists-worldwide/. Accessed October 15, 2013.

Ingersoll, G. 2013. NSA's mobile hack is another disaster for American tech, Business Insider Military & Defense, http://www.businessinsider.com/the-nsa-is-a-huge- disaster-for-tech-2013–9. Accessed October 19, 2013.

Janssen, C. 2010–2014. Mobile e-commerce (M-Commerce), techopedia, http://www.techopedia.com/definition/1540/mobile-e-commerce-m-commerce. Accessed October 2, 2013.

Joseph, C. 2014. Demand media, Types of eCommerce Business Models, http://smallbusiness.chron.com/types-ecommerce-business-models-2447.html. Accessed October 18, 2013.

Kaspersky. 1997–2013. Security technologies for mobile and BYOD, www.kaspersky.com/business-security/securing-mobile-endpoints 2.1 and 2.2 on page 3. Accessed October 21, 2013.

Kumar, M. 2013, 1. First ever Malware for Firefox Mobile OS developed by Reasher, http://thehackernews.com/2013/10/exclusive-security-researcher-developed.html. Accessed October 19, 2013.
Kumar, M. 2013, 2. NSA using browser cookies to track Tor users, The Hacker News Security in a serious way, http://thehackernews.com/2013/10/nsa-using-browser- cookies-to-track-tor.html, Accessed October 7, 2013.
Mazhar, A. 2011. WorstTech!, infected QR codes brings security threats, http://www.worsttech.com/security/infected-qr-codes-brings-security-threats-1109123.html. Accessed October 19, 2013.
Morris, C. 1997. Quoted in Wikipedia, the free encyclopedia Mobile_commerce, http://cryptome.org/jya/glomob.htm.
Musil, S. 2013. NSA can reportedly tap smartphone users' data, Politics and Law—CNET News, http://news.cnet.com/8301-13578_3-57601883-38/nsa-can-reportedly-tap-smartphone-users-data/. Accessed September 10, 2013.
Rappa, M. 2010. Managing the digital enterprise. 5. Business models on the web, http://digitalenterprise.org/models/models_text.html. Accessed October 18, 2013.
Rashid, F.Y. 2011. IT security & network security news & reviews: Phishing attacks keep proliferating: How to recognize them. eWeek, http://www.eweek.com/c/a/Security/Phishing-Attacks-Keep-Proliferating-How-to-Recognize-Them-747404/. Accessed September 10, 2013.
Robles, P. 2011. Native apps versus mobile websites: Three simple rules. Tweet, http://econsultancy.com/us/blog/8178-native-apps-versus-mobile-websites-three-simple-rules. Accessed October 17, 2013.
Roggio, A. 2013. Apple fingerprint scanner may boost mobile commerce, *Practical Ecommerce Insights for Online Merchants*, http://www.practicalecommerce.com/articles/58192-Apple-Fingerprint-Scanner-May-Boost-Mobile-Commerce. Accessed October 2, 2013.
Rosenbach, M., Laura, P., and Holger, S. 2013. iSpy: How the NSA accesses smartphone data, http://www.spiegel.de/international/world/how-the-nsa-spies-on-smartphones-including-the-blackberry-a-921161.html. Accessed September 10, 2013.
Rosenblatt, S. 2013. NSA tracks Google ads to find Tor users, CNET>News>Security & Privacy, http://news.cnet.com/8301-1009_3-57606178-83/nsa-tracks-google-ads-to-find-tor-users/. Accessed October 7, 2013.
RT.com. 2013. New Snowden leak: NSA is monitoring the internet histories of millions of Americans, Edited: October 01, 2013, http://rt.com/usa/nsa-leak-internet-history-549/. Accessed October 3, 2013.
Sahota, D. 2011. Mobile commerce: Why IT chiefs must be in the driving seat, http://www.computing.co.uk/ctg/feature/2027819/mobile-commerce-business-drivers. Accessed October 11, 2013.
Serwer, A. 2013. NSA doesn't want your mobile location data–yet, http://www.msnbc.com/msnbc/nsa-doesnt-want-your-mobile-location-data. Accessed October 19, 2013.
Silva, P. 2012. Tech brief secure mobile access to corporate applications, Technical marketing manager, security, f5, http://resources.idgenterprise.com/original/AST-0094437_mobile-access-to-applications-tb.pdf. Accessed October 17, 2013.

Staff Writer. 2011. Security developments for mobile e-commerce, transactionage, http://www.transactionage.com/2011/04/13/security-developments-for-mobile-e-commerce/. Accessed October 2, 2013.

Symantec. Meeting mobile and BYOD security challenges with digital certificates, http://www.cio.com/white-paper/741092/Meeting_Mobile_and_BYOD_Security_Challenges_with_Digital_Certificates, page 1. Accessed October 22, 2013.

Tode, C. 2013. Business model for mobile payments still to be determined: Forrester, http://www.mobilecommercedaily.com/the-business-model-for-mobile-payments-still-to-be-determined-forrester. Accessed October 18, 2013.

Tor Project: Android Instructions. Tor on Android, https://www.torproject.org/docs/android.html.en. Accessed October 8, 2013.

Tor Project: Anonymity Online. Anonymity Online, https://www.torproject.org/. Accessed December 23, 2013.

VeriSign. 2010. Security and trust: The backbone of doing business over the internet, https://www.mercurymagazines.com/pdf/VERISIGN3.pdf. Accessed October 21, 2013.

Webopedia. cookie-Web cookies, "What are cookies (cookie)?—A word definition from the Webopedia, http://www.webopedia.com/TERM/C/cookie.html. Accessed October 18, 2013.

WEBROOT. 2013. Phishing 2.0 Why phishing is back as the No. 1 web threat, and how web security can protect your company, http://www.webroot.com/shared/pdf/Phishing_and_Web_Security_WP_Mar13.pdf. Accessed October 15, 2013.

Wikipedia, the free encyclopedia, 2013, 1. E-commerce, http://en.wikipedia.org/eiki/E-commerce.

Wikipedia, the free encyclopedia, 2013, 2. Mobile commerce, http://en.wikipedia.org/eiki/Moile_commerce.

Wikipedia, the free encyclopedia. 2013, 3. QR code, http://en.wikipedia.org/wiki/QR_code.

section four

Mobile electronic commerce development

chapter nine

Enhancing electronic commerce with hybrid mobile application development architecture

Edward T. Chen

Contents

9.1 Introduction

Mobile device technology is evolving at an unbelievable rate. The use of cell phones to only make phone calls has become a thing of the past. Cell phones have been replaced with mobile devices, and they have been packaged with enough hardware to run applications that were previously confined to desktop computers. Everywhere you turn there are new devices, applications, and ideas flooding the mobile market. It did not take long for corporate America to take notice of the shift in consumer interest and take advantage of the mobile market (Smith and Chaffey, 2005; Chang and Chen, 2009; Serrano et al., 2013).

Mobile applications have been poised to be the singular most influential new software technology in the last two decades. Next to Microsoft Windows, no other software innovation has garnered the attention, success, and financial backing that mobile applications have recently experienced. Navigating the mobile application landscape is challenging for organizations (Gavalas et al., 2011; Ko, 2013; Nicolaou, 2013). While organizations may still be struggling to recognize the full potential of mobile applications, many have come to the realization that mobile applications are a necessary evil and one that customers have come to expect (Deng et al., 2010; Li and Yeh, 2010).

As with any new technology, there have been growing pains, and organizations have had to adapt quickly to the rapidly changing mobile landscape. What works today is not guaranteed to work tomorrow. This mantra is apparent in the mobile application field. Mobile application development as an industry is still in its infancy. As new device operating systems and software packages hit the market, mobile applications as well as organizations are struggling to keep up with the proliferation of devices, backwards-compatible applications, and consumer demand (Mahatanankoon et al., 2005). Even with the wide range of challenges and growing pains, the mobile application space is currently the most innovative and exciting area of technology today (Nicolaou, 2013; Serrano et al., 2013).

With the current batch of mobile applications, we are witnessing the end of the first generation of mobile applications. This first generation of applications due to hardware, software, and vendor constraints were limited to run only on a single operating system. For example, Instagram, one of the most successful mobile applications today, can only be installed on devices that operate Apple's iOS. This group of mobile applications is classified as native applications. They are built to run on code that is native to the device (Charland and Leroux, 2011; Serrano et al., 2013).

The first generation of mobile application was an important step for the electronic commerce and many lessons were learned from it. However, organizations in general do not want to develop different instances of the same application for each mobile platform that exists in today's market (Hasswa et al., 2007). Each mobile platform has its own development language, nuances, personnel commitments, and hardware requirements (Zheng and Ni, 2005). Whether the application is developed in-house or outsourced, it is generally an expensive endeavor. When combined with maintenance costs, short application life cycle, and other fringe expenses, some organizations are questioning whether native mobile applications are the right choice for them (Charland and Leroux, 2011; Serrano et al., 2013).

Hybrid applications combine the development speed and flexibility of web-based applications with the feature-rich environment and performance that only a native application can provide. This second generation of application architecture is positioned to become the next big innovation in mobile application development (Fu et al., 2013). Hybrid mobile applications are written using standard web technologies such as HTML5, CSS, and JavaScript (Nixon, 2011; Ko, 2013; Nenava and Choudhary, 2013). This code is then bundled with third-party mobile development frameworks, and then packaged to run on a variety of mobile device operating systems. The core application code is written in web technologies. Thus, it can be modularized and embalmed to run on any mobile device such as those devices running Apple iOS, Google Android, or Microsoft Windows Phone. The mobile development framework is what makes this all possible. It comprises a thin layer of code that enables the web-based code to access native device functionality such as geolocation services, camera functions, contact lists, and notification features on the mobile device (Haklay et al., 2008; Nguyen-Vuong et al., 2008; Soutschek, 2010).

As with any new technology there are challenges that must be overcome. Hybrid mobile applications are no different. This chapter will explore the strengths and weaknesses of hybrid applications and how their unique set of characteristics affect organizational decisions. This chapter will also cover some of the more popular development frameworks, environments, and how they are making hybrid mobile development more streamlined. Finally, the chapter will explore the future of hybrid mobile applications and how advances in HTML5 adoption and economic realities affect hybrid development (Unhelkar and Murugesan, 2010; Nixon, 2011; Nenava and Choudhary, 2013).

9.2 What is hybrid mobile application development architecture?

The current mobile application landscape comprises three types of application architectures: native, hybrid, and mobile web. To fully understand what a hybrid application is, it is helpful to have a general knowledge of the other two application architectures. Every mobile application architecture has its own set of advantages and disadvantages. It is important for organizations and IT professionals to understand the instance where each architecture is best suited for (Fu et al., 2013; Ko, 2013).

Native applications are currently the most widely utilized application architecture. The name "native" comes from the fact that each application is written using the native application code for the specific operating system that the application will run on. For example, applications written to operate on an Apple device will be coded in the Objective C language. Applications coded to run on Google Android will be written in Java, while Microsoft Windows Phone applications are coded in Microsoft. Net. Native code is generally compiled into an optimized code set which allows it to run faster than an interpreted language such as JavaScript (Charland and Leroux, 2011; Newton and Dell, 2011). The requirement that each native application be uniquely coded for its runtime environment puts both application developers and business planners in a difficult position. Each instance of a mobile application generally requires separate resources, development software, planning, and testing. Many organizations have come to realization that the overall cost associated with the development and maintenance of native applications exceeds any positive organizational impact the app might provide (Nicolaou, 2013).

Performance is where native applications really shine as well as in the collection of features they have access to. Being written in the same coding language as the operating system provides native applications with access to more application programming interfaces (APIs). Most mobile operating systems prohibit access to certain APIs from nonnative applications. This gives native app developers and designers a larger feature pool to choose from when designing and developing their applications. The performance of the applications is also a distinct advantage that native applications have when compared to other application architectures. Nonnative application code requires an additional step during the runtime to translate the nonnative code to native code so that it can be run on the specific device. This code translation in most circumstances leads to a degradation in performance (Google Developer, 2012; Fu et al., 2013).

The next mobile application architecture is mobile web apps. Mobile web apps are applications that operate strictly in a browser and do not require any native code to function. These types of applications are coded using the same languages that many websites utilize, such as HTML, CSS, and JavaScript. Mobile web applications differ from traditional websites in that they are optimized to run on devices with smaller screens, and their code is usually paired down to accommodate wireless data transfer limitations and speeds (Charland and Leroux, 2011; Serrano et al., 2013).

Mobile web applications in most cases are not installed on the user's device. Instead, the application is run in the client's mobile web browser. Popular mobile web browsers include Safari on Apple devices, Firefox for Android, and Opera Mobile. Excluding cookies, browser security parameters and settings prevent mobile web apps from storing code and information on user's devices (Nixon, 2011; Nicolaou, 2013).

Mobile web applications are coded so that they are not dependent on any particular operating system or runtime environment. They provide the same or very similar application code to each device regardless of manufacturer, device, or operating system. The write once, deploy anywhere concept behind mobile web apps makes them a very affordable and efficient option when compared to the highly focused and uniquely complex native application development. Mobile web applications are unspecific by design. They are rarely able to take advantage of native device APIs.

Therefore, they cannot provide the same level of usability that a full featured native application might provide. The characteristic that mobile web apps cannot utilize native functionality forces mobile web apps to largely comprise simple and rudimentary functionality. The native functionality limitation paired with the fact that mobile web apps must run in a browser significantly limits the use cases for mobile web apps. This limitation has excluded them as a viable application development option from many organizations' mobile application architecture discussions (Fu et al., 2013).

Organizations and developers struggle between choosing the high cost and discontinuity of native application development. They are sometimes puzzled with the finding of the more cost effective but less capable mobile web application development. Organizations and developers have agreed on one thing in common. They know that they both want a good and robust application. An application that performs well is stable and reliable and has limited usability issues. Until the hybrid mobile application was born, those requirements have always led organizations to choose native application development (Newton and Dell, 2011). Hybrid mobile applications provide developers and organizations with a best of both worlds solution. Hybrid mobile applications comprise three essential parts: the web-based application code, a hybrid application framework, and a native application wrapper.

Hybrid application code is written using the same, traditional, web-based languages that a mobile web application would use, such as HTML, CSS, and JavaScript. This allows the application to be written once and then deployed on different devices and operating systems with very little, if any, code changes. To take advantage of native device features and functionality, the web-based application code is then coupled with a hybrid application framework. The framework is a thin layer of native code that maps JavaScript functions to native APIs, and allows the web-based code to call native functions on the device. For example, the framework can expose camera and geolocation APIs, allowing developers to incorporate native functionality into their apps. Without the hybrid framework, a hybrid application would simply behave like a mobile web app.

The final step for developing a hybrid mobile application involves encapsulating the applications code in a native wrapper. The wrapper is simply a basic native application that contains the hybrid application framework and the application's web-based code. The native application wrapper allows developers and organizations to offer their applications on various application stores and marketplaces. For example, an Apple iPhone hybrid mobile application would have to be wrapped in a native iOS application so that it could be offered for download in Apple's AppStore. If an organization and a developer wanted to offer the same iPhone application on an Android device, they would simply take the web-based application code and the hybrid framework and copy them into a native Android application. Once packaged and deployed onto a device, the Android app and the iPhone app would both be utilizing the same web-based application code. The hybrid application framework would handle the native functionality integration on various devices (Unhelkar and Murugesan, 2010; West and Mace, 2010). The app would function as designed on both devices with very little customization. That is the extra value hybrid mobile applications provide. Many of the hybrid application frameworks support the wrapping of applications for iOS, Android, Windows Phone, Blackberry, and WebOS. The hybrid frameworks will be covered in more detail later in the chapter.

9.3 *Advantages and disadvantages of hybrid mobile application*

Hybrid mobile applications may appear to be the easy choice for organizations looking to develop a new mobile application, but they are not the right choice for every situation. The task of choosing a mobile application architecture is no simple task. There is no widely accepted method of selecting the correct mobile application architecture. This section will outline a few considerations that organizations should take into account when determining if the hybrid architecture is right for them.

Organizations, in large part, work very hard to keep pace with what their competition is doing. As one organization tries to differentiate itself from its competition, the competition will recognize the organization is trying to do something innovative and different. They will likely try to copy the innovative approaches. This "follow the leader" mentality is especially evident in the mobile application space, where numerous organizations have developed a mobile application simply because their competition has one. In the past, this methodology might have worked, but mobile application development is different nowadays. Despite the potential and promise of mobile applications, there is a need to justify their viability, usefulness, and values to the stakeholders (Keeney, 1999; Nah et al., 2005; Kuo et al., 2009). Some organizations have found that the total cost of ownership (TCO) for mobile applications exceeds any advantages that might be gained by having a mobile application. Instead of abandoning mobile applications all together, organizations should look at hybrid mobile applications and investigate how they could possibly be leveraged to decrease their mobile application's TCO.

One of hybrid mobile application's main advantages is the idea of code sharing and reuse. Hybrid applications are written using portable web technologies. They are easily transferable between operating systems and devices. This characteristic of hybrid application development allows organizations to have a smaller mobile development team. This special task team can focus on a standard platform of web coding languages and processes.

It also allows developers to focus their efforts on other challenges including operating system fragmentation, data management, and network connectivity (Gavalas et al., 2011). Developers with web development skills are both easier to find and more cost effective in today's market. In contrast, to complete a native application, companies would have to either staff more developers with niche job skills, hire contractors, or be forced to outsource to a third-party vendor to complete the work. All of these options are not ideal from an organizational perspective and can lead to over staffing, loss of project control, loss of core competencies, high maintenance costs, and fragmented applications (Ko, 2013).

Another advantage of hybrid mobile applications is that while they are developed using standard web-based technologies they do not require that the device be connected to an external data source for them to be functional. Similar to a website, mobile web apps require an Internet connection so that they can download the required application code. Hybrid mobile applications are packaged into a native application wrapper and installed on a user's device. This gives the application access to all of its code regardless of external connection or device bandwidth speed.

As stated earlier, hybrid mobile applications are not the perfect choice for every application. For instances where performance is vital to the success of the application, careful consideration and performance testing must be completed before an organization selects hybrid as its mobile development architecture. There is an extra level of code interpolation that hybrid applications have to complete that native applications do not. This extra step causes a slight decrease in application performance. For many organizations, developers and product owners are sensitive to the loss in application performance. The loss is trivial when contrasted with the benefits of hybrid mobile applications. Hence, the performance loss is deemed acceptable (Williams et al., 2011).

Another area of concern that application developers must take into account when developing hybrid mobile applications is application security. Contrary to public perception, mobile devices are very secure, and there are a number of things that software companies can do to enhance the basic security functions built into the devices themselves (Dwivedi et al., 2010). Programmers and developers need to be aware of the built-in device security functions as well as the particular security vulnerabilities that are unique to hybrid applications. Some of those vulnerabilities surround local storage concerns, cross-domain policy changes, and erroneous i-frame implementations. These security vulnerabilities are not limited to hybrid mobile applications, but instead are general HTML5 security vulnerabilities. As HTML5 adoption increases, so the attention will be surrounding these security vulnerabilities (Nixon, 2011).

Application security as simple as password protection and authentication is incorporated into almost all new mobile devices and is easy to implement. The layering and encouragement of this kind of security mechanism, while not linked to mobile applications, can significantly increase the overall security of not only an organization's application but also the user's mobile device as a whole (Dwivedi et al., 2010).

9.4 Practices of hybrid mobile application development

9.4.1 Cambridge iReport

The city of Cambridge Massachusetts wanted to develop a mobile application that allows its citizens to report on conditions around the city that they felt needed attention by the cities services. Engaged Cambridge citizens now have an application they can download to their iOS or Android device that allows them to conveniently and easily report an unplowed side street, an overflowing trashcan, unsightly graffiti, or any other problems found within the city limits of Cambridge (Angel, 2011).

The iReport application is a great example of how a hybrid mobile application can be functional, useful, and successful while keeping a low TCO. Isite, the company that developed the application for the city of Cambridge, utilized PhoneGap, a popular hybrid mobile application framework (Wargo, 2012). By utilizing PhoneGap, Isite was able to control costs through code reuse while simultaneously providing an application that would run on both the iOS and Android platforms. The application integrates with the cities online resolution system and allows the application users to seamlessly upload photos and descriptions of specific issues, all while integrating with the native geolocation feature of the user's device. Overall, the application has been well received, getting a 4.5 out of 5 star user rating on Apple iTunes.

9.4.2 Dolphin Tale

As Warner Bros Pictures prepared for the debut of their newest full-length feature film *Dolphin Tale*, they were looking for an innovative way to introduce the movie to potential moviegoers. The organization decided that a nifty and viral way to promote the movie would be to develop a mobile game. Warner Bros Pictures selected Trailer Park, a Hollywood entertainment-marketing firm, with the development. Operating on a tight schedule and budget, Trailer Park knew that they would have to utilize a hybrid mobile framework to deliver an application that could be deployed on both the iOS and Android platforms.

Trailer Park selected Corona software development kit (SDK) as their hybrid framework for the *Dolphin Tale* project. Trailer Park stated that was the only way they were able to build the app and launch it on time and budget. Outside of the advantages hybrid applications offer, such as speed to market and standardized technologies, organizations also utilize hybrid frameworks like Corona SDK. Corona SDK allows them to build hybrid mobile applications in a controlled and supported environment. Corona SDK provides organizations that use their software with support options. Developers have access to Corona's support network during development for trouble shooting and solution. This level of service affects the organization's decision making in selecting which framework or architecture to develop mobile applications (Fu et al., 2013).

9.4.3 Lotte Awards

The hybrid mobile application development and implementation by Lotte Card highlights some of the main reasons why organizations are choosing hybrid development over native development. The Lotte Awards application is a complex and diverse credit card monitoring application that contains over 100 screens, providing users with essential cardholder information, and has unique features such as augmented reality and mobile coupons that can be scanned.

Lotte Card partnered with FAS, an IT solutions provider located in Korea and Japan, to develop the hybrid application. The entire development of the project took only 2 months. The group's accelerated project delivery schedule was made possible by the utilization of the IBM WorkLight Mobile Platform (WorkLight, 2011). Similar to other hybrid application frameworks and platforms, WorkLight provides developers with a rich set of tools and features that aid mobile application development as well as in the process of combining native and web code. The speed-to-market gained through the use of the WorkLight platform as well as the web-based technology that surrounds hybrid mobile applications spotlights the efficiencies that hybrid mobile development provides. The success that the Lotte Awards application has garnered in the iPhone and Android marketplace has Lotte Card considering a new venture into a tablet-specific application (IBM, 2012).

9.5 Frameworks and tools of mobile application development

The world of hybrid mobile applications is full of new and emerging groups and organizations that are trying to put their stamp on the future of hybrid development. This section of the chapter will outline some of the significant players in hybrid mobile applications and what their tools or frameworks are.

9.5.1 PhoneGap

Originally a product of Nitobi, PhoneGap was started in 2009 and was one of the first hybrid mobile development frameworks. Its innovative features, open source availability, online community, and high early adoption by mobile developers gave PhoneGap much success and fanfare. PhoneGap was believed to have so much potential for mobile application development. Adobe Inc. purchased PhoneGap's parent company Nitobi in late 2011.

Adobe's plans for PhoneGap were recently announced in a blog post on the PhoneGap website. There were some concerns from the PhoneGap community that Adobe would start charging for PhoneGap licenses or potentially change the general purpose of the software. Adobe has communicated that they intend on keeping PhoneGap free for the public in an open source version. Apache Cordova will be the new name for the open source and free version of PhoneGap. PhoneGap itself will remain the name of Adobe's version of the software. With the PhoneGap acquisition, Adobe will likely incorporate it into some of their other software offerings as an enhancement to their already successful software packages.

PhoneGap currently supports the Apple iOS, Google Android, HP WebOS, Microsoft Windows Phone, Nokia Symbian OS, and RIM BlackBerry operating systems. Some of the device features that PhoneGap supports include camera operations, notifications, device local storage, network status, geolocation, and contacts. PhoneGap keeps an accelerated release schedule with releases being distributed on an almost monthly basis. This allows them to quickly provide customers with both bug fixes and feature enhancements (Wargo, 2012).

PhoneGap has also recently introduced a build service which allows developers the ability to upload web-based code and in return receive packages that are ready to be loaded on specific devices. PhoneGap Build could be a significant time saver for developers who would otherwise have to maintain development environments for each specific device operating system (Wargo, 2012; Piao and Kim, 2013).

9.5.2 Corona SDK

With the tagline, "Write once, build to both iOS and Android," it is easy to understand why Corona has experienced considerable success in the mobile application development domain. While Corona seems to be gaining in popularity, it still remains to be seen if it can differentiate itself enough from the competition. One way that Corona is differentiating itself is by having its own scripting language, Lua. Lua is layered over the native C++/Objective C code and acts as quick translation scripting language that claims to allow Corona developers the ability to code faster and debug easier. The drawback to having a proprietary scripting language is that in some instances the language limits the customized behavior that developers can introduce into their applications. For this reason, some agencies and developers have decided to choose other standardized hybrid frameworks instead of Corona SDK.

Corona SDK is a for-profit project and therefore users of the framework are charged a yearly subscription fee. The fee covers users to develop and deploy mobile applications to both iOS and Android. A fee to use any software is not always a deterrent because the licensing fees usually include some kind of product support mechanism. Large organizations and new developers favor the support aspects in the SDK. It supplies them with a point of contact when and if there are issues during the development life cycle. Corona SDK offers a rich set of device and software APIs that provide developers with almost endless functionality options. Some of the APIs include native device functionality support, Facebook integration, ad network integration, map features, in-App purchases, and many others.

9.5.3 WorkLight

Organizations are looking for a total package solution to hybrid mobile application development with direct update capabilities and push notifications. IBM WorkLight falls in this category. WorkLight's client list spans a wide range of industries including travel, retail, financial services, technology, and hospitality. Through the use of industry-accepted technologies and tools, WorkLight allows developers and organizations to author hybrid, mobile web, and native mobile applications.

The technology behind WorkLight is not just a framework but also includes an integrated development environment (IDE) and a proprietary middleware consisting of server hardware and software that allows organizations to effectively manage their mobile applications. WorkLight must have been on the right path, so IBM announced that they had reached an agreement to acquire WorkLight (IBM Communications, 2011).

WorkLight has a different approach on hybrid mobile applications and breaks them up into two categories: hybrid web and hybrid mixed. Hybrid web are the kind of applications discussed throughout this chapter. Hybrid mixed applications are simply extensions of hybrid web applications. They also include additional native code. For example, an organization might decide that their hybrid mobile application should allow for near field communication (NFC). When they attempt to incorporate the functionality, they discover that their hybrid framework does not yet support NFC. Instead of relying on the framework, the organization goes ahead and custom codes a native solution that is then integrated into their hybrid application and a hybrid mixed application is created. Since the new functionality was written in native code, the resulting application becomes less portable and requires additional development time if ported to run on additional operating systems.

WorkLight understands that the different mobile application requirements that surround each organization are unique. Their product provides a development environment for each of the mobile application architectures including mobile web and full native development. This development flexibility paired with their IBM support model should make WorkLight a serious development tool consideration for any organization entering the mobile application development landscape.

9.5.4 Appcelerator

Founded in 2006, Appcelerator is a company that focuses on powering their clients with tools to author innovative and intuitive mobile applications that can be deployed to multiple devices from a single code base. Their client list is impressive and includes organizations such as NBC, eBay, CISCO, ING, and Reuters. Operating as a full service mobile development organization, Appcelerator offers a wide array of products such as their flagship product Titanium Mobile, their development environment Titanium Studio, Titanium Analytics, and Titanium Desktop.

Like other hybrid mobile application frameworks, Appcelerator allows developers who use their software tools to write application code in standard web-based languages. When completed, the application is then packaged to be deployed on a variety of mobile device platforms. There have been some concerns raised by potential customers and developers that users of the Appcelerator software will have to learn the Titanium API in addition to the unique set of Appcelerator processes.

9.6 Future of hybrid mobile applications

The future of hybrid mobile applications is bright. Organizations have quickly learned that native development is time consuming, highly fragmented, and costly. It is likely that strict native application development will be left entirely to those applications that require significant performance capability such as immersive mobile gaming applications and applications requiring acute graphical interfaces. At the same time, hybrid mobile applications that rely heavily on standardized web technologies, native integration, and portability are going to garner increased attention from organizational decision makers and developers (Ko, 2013; Nenava and Choudhary, 2013).

The future of hybrid mobile applications is not solely hinged on technology adoption and feature development. Mobile application development as a whole is also largely influenced by economic factors. Paid applications and advertising revenue are a driving force behind mobile applications. Hybrid's emergence onto the application landscape is sure to raise a few eyebrows. Apple was the first on the scene with its revolutionary application development platform and delivery channel (West and Mace, 2010). Google soon followed with their Apps Marketplace. Both Apple and Google have proven that there is significant money to be made in the mobile space and they are not likely to leave the economics to chance. One of the key characteristics of a hybrid mobile application is its use of HTML5 technologies. Therefore, the success of hybrid mobile applications is loosely coupled to the overall adoption and success of HTML5. All indications point to HTML5 becoming an even more significant player in the future of web development (Nixon, 2011; Ko, 2013).

There are some technology experts who believe that eventually HTML5 mobile web apps will replace both native and hybrid applications as the de facto mobile application architecture (Gavalas et al., 2011; Fu et al., 2013; Ko, 2013). Browsers, both mobile and desktop, would have to create APIs for a standardized manner. Thus, web pages could communicate to the devices that they are operating on to get specific information. For example, a web page running on a mobile device might be able to access the devices camera and include a photo in the site. We have already seen some of this technology shift occurring in browsers. Many of them have begun to make available geo-location services and APIs with the caveat that the application user must knowingly allow this information to be shared (Haklay et al., 2008; Soutschek, 2010).

It is likely a matter of time that leading-edge browsers such as Google Chrome will begin to make other device APIs available to users and web developers (Charland and Leroux, 2011). As the list of available APIs grows, it is predicted that at some point developers will become less reliant on native and hybrid dependencies and be able to develop entirely for the browser. In reality, browser technology is notorious for their slow response to emerging technologies. As new advances in device features enter the market, they become available to native and hybrid applications much sooner than the browser developers are able to implement their APIs. For this reason alone, it is expected that hybrid mobile architecture will be relevant to mobile application development in electronic commerce for the foreseeable future.

9.7 Conclusion

After examining the research available on hybrid mobile architecture and the growth of hybrid mobile application development, we categorize two specific findings immediately illustrated in the forefront. The first finding is that the hybrid mobile architecture is probably the right mobile application architecture for most mobile application projects excluding performance centric games (Edoh-Alove et al., 2013). Most mobile application requirements that an organization would likely include can be fulfilled through the use of the hybrid architecture. These hybrid mobile applications are quicker to develop, can be deployed to multiple platforms, have a low TCO, offer more staffing options, and deliver comparable performance. The first finding points out that the hybrid architecture is becoming the intelligent choice for organizations to develop mobile applications.

The second finding is centered on the idea that while the hybrid mobile architecture is currently viewed as an innovative and progressive technology, it is simply a necessary progression of the mobile application evolution. Hybrid mobile applications are, by design, a combination of two established technology platforms. The hybrid architecture is strongly coupled and is dependent on external technologies beyond its control. This makes the hybrid architecture volatile and likely to experience significant changes as the mobile application landscape shifts to determine its technological winners and losers.

When the mobile market, developers, and organizations choose the industry platform that will drive electronic commerce, the hybrid mobile application development architecture apparently provides a platform for the next generation of mobile application innovation. As seen in the fast change in other application development methodologies in various fields, the hybrid mobile architecture is simply a stopgap in the evolution of mobile application architecture. The industry will come up with new architectures to meet the demand of the dynamic electronic commerce market.

There are numerous potential opportunities for future research in the mobile application architecture spectrum. A few interesting research topics might include surveying both users and organizations for their feedback on how a chosen hybrid application performs in comparison to a similar native or mobile web application. It would be also interesting to determine some high-level savings estimates that organizations reap from switching to hybrid mobile development. Results from empirical studies of various mobile application development architectures such as quantitative analyses of savings and TCOs could help organization's decision making in choosing appropriate infrastructure and architecture. This chapter analyzes the advantages and disadvantages of hybrid method with several practical business cases. Researchers and practitioners can use them as a set of guidelines to develop research framework as well as adoption of the hybrid mobile architecture.

References

Angel, K.D. (2011). ISITE design and the city of Cambridge launch a mobile citizen reporting application. Retrieved from: October 1, 2013, http://www.isitedesign.com/news/isite-design-and-city-cambridge-launch-mobile-citizen-reporting-application.

Chang, H.H. and Chen, S.W. (2009). Consumer perception of interface quality, security, and loyalty in electronic commerce. *Information and Management, 46*(7), 411–417.

Charland, A. and Leroux, B. (2011). Mobile application development: Web vs. native. *Communications of the ACM, 54*(5), 49–53.

Deng, Z.H., Lu, Y.B., Wei, K.K., and Zhang, J.L. (2010). Understanding customer satisfaction and loyalty: An empirical study of mobile instant messages in China. *International Journal of Information Management, 30*(4), 289–300.

Dwivedi, H., Clark, C., and Thiel, D. (2010). *Mobile Application Security*. New York: McGraw-Hill.

Edoh-Alove, E., Hubert, F., and Badard, T. (2013). A web service for managing spatial context dedicated to serious games on and for smartphones. *Journal of Geographic Information System, 5*(2), 148–160.

Fu, Z.Q., Wang, H.R., Wang, J.Y., and Li, Y. (2013). A novel wireless network based mobile e-business general architecture. *Journal of Theoretical and Applied Information Technology, 49*(2), 756–763.

Gavalas, D., Bellavista, P., Cao, J., and Issarny, V. (2011). Mobile applications: Status and trends. *The Journal of Systems and Software, 84*(11), 1823–1826.

Google Developers. (2012). Web performance best practices. Retrieved from: September 10, 2013, http://developers.google.com/speed/docs/best-practices/rules_intro.

Haklay, M., Singleton, A., and Parker, C. (2008). Web mapping 2.0: The Neogeography of the GeoWeb. *Geography Compass, 2*(6), 2011–2039.

Hasswa, A., Nasser, N., and Hassanein, H. (2007). A seamless context-aware architecture for fourth generation wireless networks. *Wireless Personal Communications, 43*(3), 1035–1049.

IBM. (2012). Case studies: Lotte Card Co., Ltd. Augmenting offerings with the IBM WorkLight platform. Retrieved from: September 10, 2013, https://www-01.ibm.com/software/success/cssdb.nsf/CS/CPAR-8TNQF4.

IBM Communications. (2011). IBM completes acquisition of WorkLight (Press Release). Retrieved from: September 10, 2013, http://www-03.ibm.com/press/us/en/pressrelease/36919.wss.

Keeney, R.L. (1999). The value of internet commerce to the customer. *Management Science, 45*(4), 533–542.

Ko, C.R. (2013). Research trends and its determinants in mobile commerce research (1999–2012). *Asian Journal of Innovation and Policy, 2*(2), 150–172.

Kuo, Y.F., Wu, C.M., and Deng, W.J. (2009). The relationships among service quality, perceived value, customer satisfaction, and post-purchase intention in mobile value-added services. *Computers in Human Behavior, 25*(4), 887–896.

Li, Y.M. and Yeh, Y.S. (2010). Increasing trust in mobile commerce through design aesthetics. *Computers in Human Behavior, 26*(4), 673–684.

Mahatanankoon, P., Wen, H.J., and Lim, B. (2005). Consumer-based m-commerce: Exploring consumer perception of mobile applications. *Computer Standards & Interfaces, 27*(4), 347–357.

Nah, F., Siau, K., and Sheng, H. (2005). The value of mobile applications: A utility company study. *Communications of the ACM, 48*(2), 85–90.

Nenava, S. and Choudhary, V. (2013). Hybrid personalized recommendation approach for improving mobile e-commerce. *International Journal of Computer Science & Engineering Technology (IJCSET), 4*(5), 546–552.

Newton, D. and Dell, A. (2011). Evaluating the effectiveness of apps for mobile devices. *Journal of Special Education Technology, 26*(4), 59–63.

Nguyen-Vuong, Q.T., Agoulmine, N., and Ghamri-Doudane, Y. (2008). A user-centric and context-aware solution to interface management and access network selection in heterogeneous wireless environments. *Computer Networks, 52*(18), 3358–3372.

Nicolaou, A. (2013). Best practices on the move: Building web apps for mobile devices. *Communications of the ACM, 56*(8), 45–51.

Nixon, R. (2011). *HTML5 for iOS and Android*. San Francisco, CA: McGraw-Hill.

Piao, G. and Kim, W. (2013). Introduction to iPad application development with PhoneGap. *International Journal of Innovation, Management and Technology, 4*(1), 47–51.

Serrano, N., Hernantes, J., and Gallardo, G. (2013). Mobile web apps. *IEEE Software, 30*(5), 22–27.

Smith, P.R. and Chaffey, D. (2005). *E-Marketing Excellence: At the Heart of e-Business*. Oxford, U.K.: Butterworth Heinemann.

Soutschek, M. (2010). Usage of Geoweb technologies in tourism. *HMD: Practice of Economy Informatics, 47*(276), 77–87.

Unhelkar, B. and Murugesan, S. (2010). The enterprise mobile applications development framework. *Mobile Computing, 12*(3), 33–39.

Wargo, J. (2012). *PhoneGap Essentials: Building Cross-Platform Mobile Apps*. Upper Saddle River, NJ: Addison-Wesley Professional.

West, J. and Mace, M. (2010). Browsing as the killer app: Explaining the rapid success of Apple's iPhone. *Telecommunications Policy, 34*(5–6), 270–286.

Williams, A., Ekins, S., Clark, A., Jack, J., and Apodaca, R. (2011). Mobile apps for chemistry in the world of drug discovery. *Drug Discovery Today, 16*(21–22), 928–939.

WorkLight, Inc. (2011). Lotte credit card releases Asia's first financial mobile app with augmented reality for iPhone and android devices (Press Release). Retrieved from: September 10, 2013, http://www.prweb.com/releases/worklight/mobile-platform/prweb5210334.htm.

Zheng, P. and Ni, L. (2005). *Smart Phone & Next Generation Mobile Computing*. (Adams, R., Ed.) San Francisco, CA: Morgan Kaufmann.

chapter ten

Using the Apache Cordova open source platform to develop native mobile applications

Sam S. Gill

Contents

10.1 Introduction: Background and driving forces

In June 2013, I was approached by a friend who asked me to help him develop a mobile application for his business. He specified that he would like his business application to run on multiple platforms: Android* (phones and tablets), iOS† (iPhone, iPad), and perhaps Windows 8‡ as well. At the time, I had been teaching Android and Windows 8 application development courses extensively. Contemplating my friend's request, I realized that pursuing what I had been teaching my students would lead me to have to learn Objective-C§ for development on the iOS platform.

* http://en.wikipedia.org/wiki/Android_(operating_system).
† http://en.wikipedia.org/wiki/IOS.
‡ http://en.wikipedia.org/wiki/Windows_8.
§ http://en.wikipedia.org/wiki/Objective-C.

It would also mean that I would need to use three different platforms for development: Eclipse* for Android, Visual Studio† for Windows, and Cocoa Touch‡ for iOS. Furthermore, I would have to maintain several different code bases for the application, one for each of the platforms that could lead to many issues in developing applications that would have the same appearance, have the same functionality, and can be easily maintained. I needed a better solution. First challenge: choosing the development platform.

10.2 Nitobi

In April 2009, an Internet start-up, Nitobi Software,§ was launched to create rich web applications and developer toolkits in Ajax and JavaScript. The company produced a product, PhoneGap,¶ a developer package that enables developers to create mobile web applications. PhoneGap is an open source platform that allows developers to build cross-platform mobile applications with HTML5 and JavaScript, and distribute these applications to a variety of platforms. By using HTML5 and JavaScript, Nitobi eliminated the need for creating specific applications for different mobile and tablet platforms.

In October 2011, Nitobi Software was acquired by Adobe.** Under Adobe stewardship, PhoneGap was separated from its underlying software which was named Apache Cordova and submitted as an incubation project to the Apache Foundation.†† In October 2012, Apache Cordova graduated to become a top-level project within the Apache Software Foundation (ASF).‡‡ In a March 2012 blog, Brian Leroux explained the difference between PhoneGap and Apache Cordova:

> PhoneGap is a distribution of Apache Cordova. You can think of Apache Cordova as the engine that powers PhoneGap, similar to how WebKit is the engine that powers Chrome or Safari. (Browser geeks, please allow me the affordance of this analogy and I'll buy you a beer later.)

* http://en.wikipedia.org/wiki/Eclipse_(software).
† http://en.wikipedia.org/wiki/Visual_Studio.
‡ http://en.wikipedia.org/wiki/Cocoa_Touch.
§ http://www.crunchbase.com/company/nitobi-software.
¶ http://phonegap.com/.
** http://techcrunch.com/2011/10/03/adobe-acquires-developer-of-html5-mobile-app-framework-phonegap-nitobi/.
†† http://en.wikipedia.org/wiki/Apache_Cordova.
‡‡ http://cordova.apache.org/#about.

> Over time, the PhoneGap distribution may contain additional tools that tie into other Adobe services, which would not be appropriate for an Apache project. For example, PhoneGap Build and Adobe Shadow together make a whole lot of strategic sense. PhoneGap will always remain free, open source software and will always be a free distribution of Apache Cordova.*

10.3 *Apache Cordova*

Apache Cordova is an open source platform for building native cross-platform mobile applications using HTML 5, CSS 3, and JavaScript, thus avoiding each mobile platforms' native development language. Apache Cordova facilitates packaging and deployment of mobile applications as native applications on the target platform. On each target platform, mobile applications execute within wrappers and rely on standards-compliant application programming interface (API) bindings to access a device's sensors, data, and network status. Mobile applications are rendered by the web engine of the target machine: WebKit† for Apple's Safari and Google's Chrome as well as Trident‡ for Internet Explorer.

The Cordova documentation has detailed information about the APIs and how to use them.§ Apache Cordova provides support for most mobile device features including Accelerometer, Camera, Capture, Compass, Connection, Contacts, Device, Events, File, Geolocation, Globalization, InAppBrowser, Media, Notification, Splashscreen, and Storage.

Apache Cordova has different levels of support for the various platforms¶:

- The platforms that are fully supported by the Apache Cordova project are iOS, Android, BlackBerry, Windows Phone, and Windows 8. The support consists of a standard low-level native plug-in bridge API (and related utilities).
- The platforms that are currently partially supported by the Apache Cordova project (but may be upgraded to full support at a later date) are Tizen, Qt, Firefox OS, Ubuntu Mobile (Qt), and Windows (Win32).
- The platforms that are being retired from support by the Apache Cordova project are Symbian, webOS, and Bada.

* http://phonegap.com/2012/03/19/phonegap-cordova-and-what%E2%80%99s-in-a-name/.
† http://en.wikipedia.org/wiki/WebKit.
‡ http://en.wikipedia.org/wiki/Trident_(layout_engine).
§ http://cordova.apache.org/docs/en/3.1.0/guide_overview_index.md.html#Overview.
¶ http://wiki.apache.org/cordova/PlatformSupport.

10.4 PhoneGap

PhoneGap is an open source free distribution of Apache Cordova.* Since PhoneGap is free, developers and companies can use PhoneGap for mobile applications that are free, commercial, open source, or any combination of these. The PhoneGap implementation provides a different set then the Apache Cordova supported API: Accelerometer, Camera, Compass, Contacts, File, Geolocation, Media, Network, Notification (Alert), Notification (Sound), Notification (Vibration), and Storage. In general, PhoneGap extends the Apache Cordova API by having more detailed notification capabilities.

PhoneGap is an HTML 5 application platform that allows developers to author native applications with web technologies and get access to APIs and app stores. Applications are built as normal HTML pages and packaged to run as a native application within a UIWebView or WebView (a chromeless browser, referred to hereafter as a webview).

10.5 Status check

The discovery of Apache Cordova and PhoneGap brought us closer to our goal of creating a platform agnostic native mobile application based on open source standards. Table 10.1 shows that Apache Cordova or its implementation PhoneGap has solved part of the framework for creating native mobile applications that can run on any platform.

Table 10.1 also shows that the selection of Apache Cordova or PhoneGap still leaves two issues unresolved: A unified user interface library and a development environment.

Building a ubiquitous native mobile application can be facilitated if a user interface library can be identified that addresses several requirements:

- The library should be built on top of JavaScript.
- The library should support rendering on all iOS, Android, and Windows 8 platforms.

Table 10.1 Framework for Creating Native Agnostic Mobile Applications

	User interface
Development environment	User interface library
	Apache Cordova or PhoneGap
	HTML 5, CSS 3, JavaScript
	WebKit or Trident
	Mobile Devices

* http://phonegap.com/about/faq/.

Figure 10.1 SPA navigation.

- The library should provide a single-page application (SPA)* framework. SPA is a mobile application that fits on a single view with the goal of providing a fluid user experience. In an SPA, either all necessary code—HTML, JavaScript, and CSS—is retrieved with a single page load, or the appropriate views are dynamically loaded and added to the page as necessary as shown in Figure 10.1.
- Since the basic platform for developing ubiquitous mobile applications is based on HTML 5, CSS 3, and JavaScript, it is important to have a clear separation of the development of the graphical user interface (View) from the development of the business logic or back-end logic (Model) and the components that mediate between the View and the Model (View–Model). The Model–View–ViewModel (MVVM)† is a design pattern that enhances the development experience by creating applications that are maintainable.

The second component that is missing from the mobile development framework is a developer friendly development environment. Both Apache Cordova and PhoneGap utilize a command-line interface (CLI) to create the mobile applications and to deploy them to the various native mobile platforms. The CLI allows developers to create new projects, build them on different platforms, and run them within an emulator. From a developer productivity perspective, the CLI is insufficient. In addition, the development environment must provide emulation capabilities for all the mobile platforms to facilitate testing of ubiquitous native mobile applications on the various mobile platforms.

10.6 User interface libraries

Throughout this research, the focus has been on setting up an open source framework for developing ubiquitous mobile applications. A list of candidate open source libraries that can run on multiple native platforms are described next.

* http://en.wikipedia.org/wiki/Single_page_application.
† http://en.wikipedia.org/wiki/Model_View_ViewModel.

10.6.1 jQuery Mobile

jQuery Mobile is based on jQuery, which is a multibrowser JavaScript library that had been designed to simplify the client-side scripting of HTML.* jQuery is a free, open source software that facilitates the selection of Document Object Model (DOM) elements and their manipulation using by jQuery's selector engine (Sizzle). The jQuery development effort is marshaled by a nonprofit organization the jQuery Foundation.† The jQuery Foundation is also responsible for coordinating the efforts behind the development of jQueryUI and jQuery Mobile.

jQuery Mobile is a unified, HTML5-based user interface system for all popular mobile device platforms, built on the jQuery and foundation.‡ jQuery Mobile supports all major mobile, tablet, e-reader & desktop platforms—iOS, Android, Blackberry, Palm WebOS, Nokia/Symbian, Windows Phone 7, MeeGo, Opera Mobile/Mini, Firefox Mobile, Kindle, Nook, and all modern browsers with graded levels of support.

The jQuery Mobile *page* structure is optimized to support either single pages, or local internal linked *pages* within a page, which satisfies a requirement that was specified earlier. In addition, jQuery Mobile supports building native mobile platform applications using the PhoneGap implementation of Apache Cordova.

Finally, mobile applications can be structured as MVVM with a proper separation between the UI design (View), the code that interacts with the backend (Model), and the interactions between the View and the Model (View–Model).

Table 10.2 shows the open source framework for building ubiquitous mobile web applications.

The last issue remaining is a developer-friendly development environment. While jQuery Mobile does not provide a development environment it does provide a tool called download builder, which lets a developer create

Table 10.2 Revised Framework for Creating Native Agnostic Mobile Applications

	User interface
Development environment	jQuery Mobile Apache Cordova or PhoneGap HTML 5, CSS 3, JavaScript WebKit or Trident Mobile Devices

* http://en.wikipedia.org/wiki/JQuery.
† https://jquery.org/.
‡ http://jquerymobile.com/.

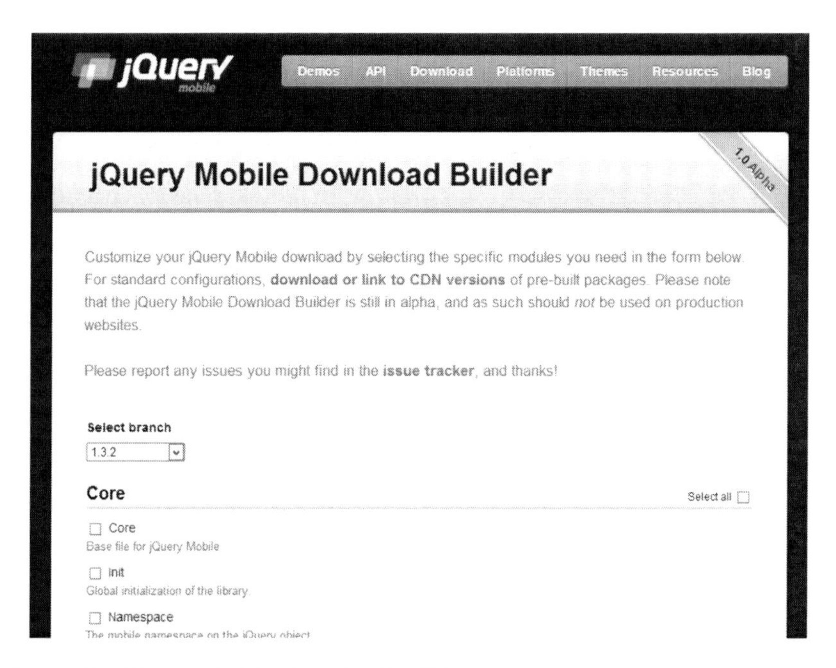

Figure 10.2 jQuery Mobile download builder.

a customized bundle that contains only the components required for the mobile application being designed as shown in Figure 10.2.

jQuery Mobile can be used with the Eclipse Development platform.* Eclipse is an open source multilanguage integrated development environment (IDE) comprising a base workspace and an extensible plug-in system for customizing the environment.† In particular, Eclipse can be customized with the JavaScript Development Tools (JSDT), which provides plug-ins that implement an IDE supporting the development of JavaScript applications.‡ JSDT adds a JavaScript project type and perspective to the Eclipse Workbench as well as a number of views, editors, wizards, and builders. In addition, there is a new plug-in for JDST that adds jQuery auto-completion support to JavaScript project.§ Eclipse, however, does not include emulators for the various mobile platforms. Some emulators, such as Android, can be added by installing the Android software development kit (SDK), but there is no complete emulation capability.

There are two additional notable libraries that offer comparable capabilities, a software development environment, and more complete emulation capabilities (Table 10.3).

* http://eclipse.org/.
† http://en.wikipedia.org/wiki/Eclipse_(software).
‡ http://eclipse.org/webtools/jsdt/.
§ http://marketplace.eclipse.org/content/jsdt-jquery#.Ulcgh1AgcaR.

Table 10.3 Framework for Creating Native Agnostic Mobile Applications with Limited Emulation

	User interface
Eclipse	jQuery Mobile Apache Cordova or PhoneGap HTML 5, CSS 3, JavaScript WebKit or Trident Mobile Devices

10.6.2 *KendoUI*

The KendoUI JavaScript library allows developers to create mobile applications that offer native-like experiences for end users automatically, without any extra coding.* KendoUI supports development for the following mobile platforms: Windows Phone 8, iOS, Android, and Blackberry. In addition, KendoUI integrates with Apache Cordova to provide native application packaging and deployment to the various mobile devices. KendoUI is utilized in the Icenium IDE to facilitate cross-platform mobile application development and emulation.† Icenium provides complete support for Apache Cordova to facilitate the creation of native mobile application packages and deployment. Icenium is available in two versions: Graphite, a stand-alone Windows desktop version, and Mist, a web-based development environment. Figure 10.3 shows the Graphite development environment with a sample project—Tip Calculator.

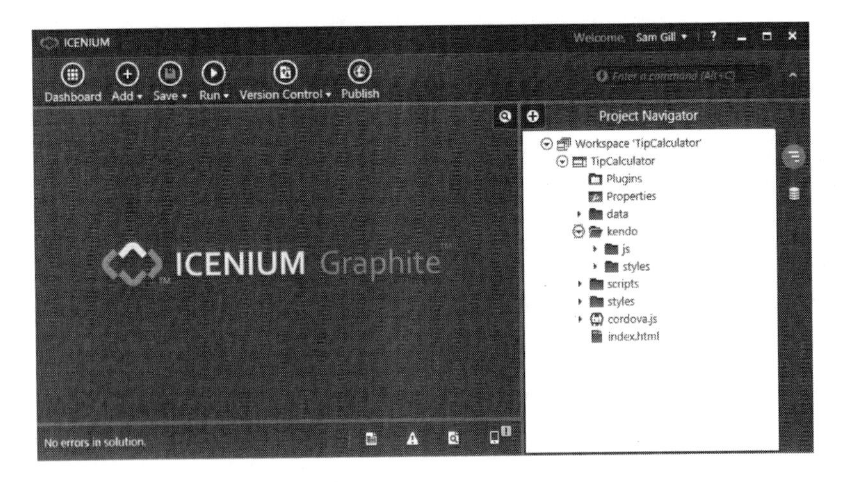

Figure 10.3 Icenium graphite integrated development environment.

* http://www.kendoui.com/mobile.aspx.
† http://www.icenium.com/.

Figure 10.4 Icenium iPhone and Android simulation.

Table 10.4 Icenium/KendoUI Stack

	User interface
Icenium	KendoUI
	Apache Cordova
	HTML 5, CSS 3, JavaScript
	WebKit or Trident
	Mobile Devices

Icenium includes emulators for most of the mobile platforms. Figure 10.4 shows a side-by-side simulation of the Icenium for the iPhone and Android mobile devices, respectively.

Table 10.4 shows the development framework with the Icenium IDE and the KendoUI API.

There are two main drawbacks with the Icenium/KendoUI solution: it costs money and it is not open source that diminishes the likelihood that it will ever become a standard.

10.6.3 Sencha Touch

Sencha Touch is an alternative framework (API) for building mobile web applications.* Sencha Touch integrates with either Apache Cordova or PhoneGap to provide native mobile application packaging and deployment. Eclipse is the IDE used for Sencha Touch and the Eclipse Add-In provides emulators for most mobile device platforms. Table 10.5 shows the development framework with the Eclipse IDE and Sencha Touch.

* http://www.sencha.com/products/touch/.

Table 10.5 Sencha Touch Stack

	User interface
Eclipse	Sencha Touch Apache Cordorva or PhoneGap HTML 5, CSS 3, JavaScript WebKit or Trident Mobile Devices

Like the KendoUI solution, Sencha Touch has the same two drawbacks: price and a proprietary API/framework.

There are many other Mobile UI libraries, but they all have to be integrated into a third-party IDE and they do not provide complete simulation/emulation capabilities.

The choice is clear: a developer can either choose a free open source solution with limited emulation/simulation capability or a proprietary solution with full emulation/simulation capability.

10.7 Additional considerations

Experience shows that developing a ubiquitous native mobile application will generally require a website that users can access with the same capabilities. Since our three proposed solutions are built on the HTML 5, CSS 3, and JavaScript components, all three solutions can be utilized to create a web application with the same capabilities.

Finally, most business mobile applications will require access to backend data. Backend can be accessible either through proprietary web services (private cloud) or through a public cloud implementation. Of the three solutions proposed, the Icenium/KendoUI solution integrates with its own cloud data storage—Everlive.* Figure 10.5 shows the revised Icenium/KendoUI stack.

One last consideration: How difficult is it to set up the software development environment? This is a very subjective measure and after setting up all three solutions, the following ranking from easy to difficult is based on the author's experience: Icenium, Sencha Touch, and jQuery Mobile.

10.8 Concluding remarks

All three solution stacks utilize either the Apache Cordova or the PhoneGap component in their stack, and therefore, all three solutions will yield a native mobile application that can be packaged and deployed through the corresponding vendor stores or enterprise deployments. All three solutions will also yield a similar application look and feel. All three

* http://www.icenium.com/product/everlive.

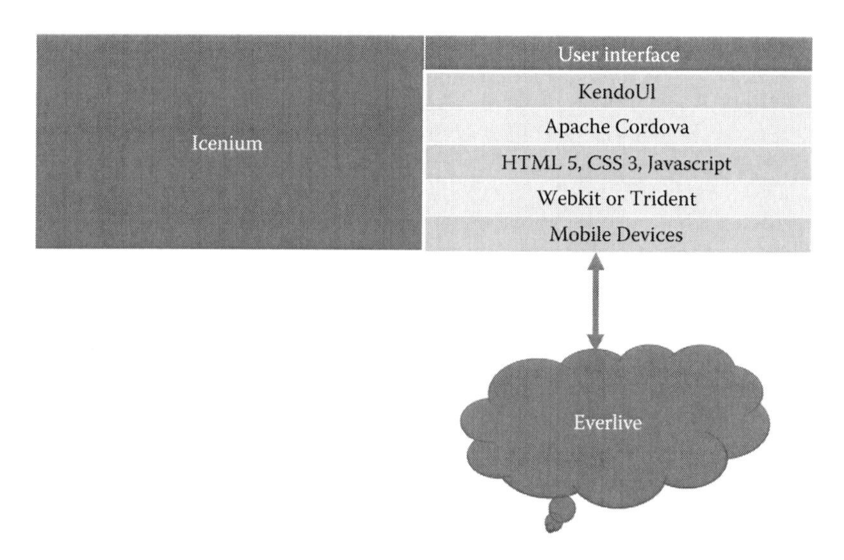

Figure 10.5 Revised Icenium/KendoUI/Everlive stack.

solutions utilize a full featured development environment. The open source solution will have less emulation capabilities than the two proprietary solutions. Having worked with all three solutions they are all somewhat buggy and have not yet sufficiently matured to yield bullet-proof native business mobile applications.

As far as helping my friend, the Icenium/KendoUI/Everlive stack was chosen, since it was the easiest development platform to install and it also provides easy access to cloud storage for storing the application data.

In the future, additional challenges need to be addressed. One of these challenges is how to manage the state of the mobile application, so that when users switch from one mobile device to another device running on a different mobile platform, the users can continue their work seamlessly.

References

http://cordova.apache.org/#about.
http://cordova.apache.org/docs/en/3.1.0/guide_overview_index.md.html#Overview.
http://eclipse.org/.
http://eclipse.org/webtools/jsdt/.
http://jquerymobile.com/.
http://marketplace.eclipse.org/content/jsdt-jquery#.Ulcgh1AgcaR.
http://phonegap.com/.
http://phonegap.com/2012/03/19/phonegap-cordova-and-what%E2%80%99s-in-a-name/.
http://phonegap.com/about/faq/.

http://techcrunch.com/2011/10/03/adobe-acquires-developer-of-html5-mobile-app-framework-phonegap-nitobi/.
http://wiki.apache.org/cordova/PlatformSupport.
http://www.crunchbase.com/company/nitobi-software.
http://www.icenium.com/.
http://www.icenium.com/product/everlive.
http://www.kendoui.com/mobile.aspx.
http://www.sencha.com/products/touch/.
https://jquery.org/.

chapter eleven

MobiCash

Smart mobile payment system

Amila Karunanayake and Kasun De Zoysa

Contents

11.1 Introduction

In this modern world, people try to find more comfortable ways to satisfy their day-to-day needs. Banks have competitively introduced new technologies to make these financial transactions more hassle free and user-friendly. Today, most financial institutions, including banks, provide services to satisfy their customers' needs and continuously introduce new services.

Thus, information and communication technologies play a vital role in the financial and banking industry. During the last two decades, researchers and developers have applied information and communication technology concepts to introduce smart solutions for the banking industry. The e-commerce concept has been introduced as an alternative to traditional methods. Examples of such solutions are automated teller machine (ATM) services, credit card/debit card services, and online payment systems.

With the rapid enhancement and popularity of mobile technologies, banking institutions try to go beyond the e-commerce applications by introducing *mobile commerce*. This opens up a new arena for electronic banking. It is one of the newest approaches to the provision of financial services through the wireless network, which has been made possible by the widespread adoption of mobile phones everywhere. The process involves the use of mobile devices to perform various financial transactions either directly with the recipient (micropayments) or indirectly via a customer's bank account. The functional capabilities of mobile telephony have been rapid and have extended operations well beyond classical applications (telephone calls and short messaging). There is mounting evidence of positive financial, economic, and social impact of those technologies all over the world.

Remarkably, in most countries, the mobile telecommunication sector has achieved rapid expansion [1] in recent years. Therefore, the mobile communication infrastructure can be used as a good deployment platform for the electronic base of banking and financial systems.

Given these capabilities, the most important consideration regarding such transactions is the issue of security. Mobile networks are based on the use of poorly secured wireless protocols. This makes mobile financial applications even more vulnerable to fraud and illegal use than similar transactions performed over open networks. Therefore, one of the main prerequisites for successful, large-scale deployment of mobile financial services applications is their security.

11.1.1 Related works

The first electronic wallet CAFÉ, which contains digital currency, was introduced in 1994 [2]. It was a pocket-sized workstation that lets users undertake cash payments electronically. The CAFÉ [2] electronic wallet prototype transacts via a short-range infrared link either directly with compliant cash registers or over computer networks, such as Internet, to merchants' tills and service points provided by banks. They proposed this as an off-line electronic wallet. The rationale for denoting the CAFÉ as off-line is because mobile devices at that time lack embedded functionalities for mobile online data communication. The system can use electronic money stored in mobile phones to purchase from nominated merchants. In 2003, CAFÉ has been modified as an online wallet [3]. This wallet consists of several modern communication channels to communicate the payment details with merchants.

In the year 2000, SMART (mobile phone–based payment system) [3] has been introduced and SMART Money coupled a customer's mobile phone with his bank account [3]. It facilitates the process of depositing and withdrawing money from the bank account, filling prepaid mobile phone credits, transferring money between other mobile accounts, and bill payment. Although SMART facilitates several mobile banking functions, it uses electronic cash stored in the mobile phone as the transaction medium.

Moreover, near field communication has been used in mobile payment systems [4]. The advantage of this system over other systems is that it does not require high-cost data protection security protocols. In the payment transaction path, a personal mobile phone is viewed as a single, user-trusted touch point. Compared to other solutions available, this approach protects user credentials, provides better user control over the transaction, and supports only proximity transactions [4].

VTBN [5] is another example for m-commerce, which uses existing mobile infrastructure to perform the credit card transactions securely. Users can use their mobile phone as a credit card. The mobile phone syncs with a credit card account maintained by a bank. From a data security perspective, one of the major disadvantages of the traditional credit card is the repetitive use of the same credit card number. The VTBN system is designed to reduce the repetitive use of the fixed credit card number.

Furthermore, we have studied mobile commerce systems similar to the systems we have described. Most of the systems provide similar functionalities. Due to space constraints, we do not discuss those here.

11.1.2 Problem description

Numbers of commercial systems are available for enabling banking services on mobile devices. Most of these systems are based on the Global System for Mobile Communications (GSM) network infrastructure and security features provided by the GSM network [6,7].

The GSM system strives to make a provision for security services, but still has limitations in its security. Previous studies [8] pointed out the lack of data integrity and confidentiality in the GSM network. The GSM service provider has control over the data flow through its network. This is the most critical concern regarding the deployment of mobile banking applications on GSM networks. A highly secured mobile banking application requires end-to-end secured channels for data communication.

Moreover, these m-commerce systems use only electronic financial materials (such as electronic coins) for the transaction [9]. In developing countries, most of the day-to-day transactions still happen by exchanging physical currency [10], not electronic payments. Even in well-developed countries, people still use coins and notes for day-to-day transactions.

Therefore, this research focuses on proposing a low-cost m-commerce solution to withdraw money (physical currency) from the customer's bank account. We propose a money withdrawal/deposit system (ATM system) that would allow people to perform their banking transactions based on mobile technologies with additional security features.

11.1.3 *Why we propose another ATM system?*

ATMs play an increasingly prominent role in the lifestyle of many humans. Although it has several inherent weaknesses, at present there are thousands of ATMs scattered throughout the world. Such growth reveals that the ATM is a successful technology that has been adopted by many people. The ATM is a pay-now payment system and is typically used for microtransactions [11].

Even though the ATM is an extremely popular electronic transaction system, it has several weaknesses.

ATM machines are not evenly scattered throughout different regions. Therefore, some users have to travel a long distance to use ATM facilities. In rural areas, the situation is worse.

The initial cost of installing ATMs is extraordinarily high. Also, ATMs typically connect to the ATM transaction processor securely, via either a dial-up modem over a public telephone line or directly via a leased line, which is expensive. In addition to that, banking organizations need trained staff to maintain ATM devices. As a result, maintenance costs of ATM machines are not economical.

Security, as it relates to ATMs, has several dimensions. There are reports that ATMs have become targets for vandalism. Sometimes thieves attempt to steal entire ATMs. Shoulder attack [12], a situation where a thief will look over the user's shoulder to identify the PIN number, is another famous security threat related to ATMs. In order to protect ATMs from these threats, a security guard needs to be employed for every ATM. Both banks and ATM customers face lots of problems related to the use of ATMs.

The proposed system is designed in such a way that it can solve most of these problems and enhance the security of the transaction. Moreover, the proposed system will reduce some of the barriers (long queues, ATM machines not scattered everywhere) of using ATM network and improve security related to ATM transactions.

11.2 MobiCash

MobiCash is an m-commerce application, which provides the services of ATM. In other words, MobiCash is an alternate to the ATM. The traditional ATM network [11] can be replaced by the MobiCash system. Meanwhile, MobiCash enables the mobile phone to be used as a debit card, which allows access to customers' bank accounts to make purchases.

The next section discusses the key consideration points, which motivate us to design MobiCash system.

11.2.1 Mobile commerce

It is increasingly becoming an understatement to say that the Internet and related technologies are changing the ways people live. One of the most significant changes promises to be in the way business is conducted. M-commerce is one of these relatively new technologies using mobile Internet. M-commerce has promised to revolutionize online financial transactions and be more powerful than anything the ordinary Internet has offered earlier.

Recent research on mobile technologies introduces new services to fulfill the growing demand of mobility. In this context, mobile commerce has a vital role to play. These types of applications/services include buying over the mobile phone, purchasing and redeeming of tickets and reward schemes, accessing travel and weather information, and writing contracts on the move [9].

Today, there is a tendency to provide adequate infrastructure for mobile commerce applications on a mobile network layer. This trend opens up huge business potential for mobile network operators. There is a significant and growing demand for deploying banking and financial services over mobile networks [10].

11.2.2 Strengths and weaknesses

M-commerce applications are extremely useful for mobile users in several ways [13]. Any user with a mobile phone can access m-commerce applications in real time at any place. Also, the mobile devices provide more security to a certain extent over online transaction systems [14,15].

However, there are limitations to mobile devices, as most devices are equipped with limited memory, display, and processing power [16]. In addition to that, the communication through the open air links introduces additional security threats (e.g., eavesdropping).

The security challenge related to mobile commerce is not limited to the mobile devices, but is interconnected with the radio interface, the network operator infrastructure, and the nature of mobile commerce application [17].

In m-commerce applications, each party that participates for a particular transaction does not meet each other physically. However, in financial transactions, trust should somehow be established between each party [16].

General cryptography concepts can be used to accomplish the trust between each participant [10]. These include the following:

Authentication: Authentication is the process of proving user identification. Authentication in the m-commerce environment is performed successfully through public-key cryptographic systems incorporated into public-key infrastructures (PKIs). In fact, the primary goal of authentication in a PKI is to support the remote and unambiguous authentication between entities unknown to each other, using public-key certificates and CA trust hierarchies.

Integrity: Integrity means ensuring that data cannot be modified and transactions cannot be altered. Public-key certificates and digital signature envelopes are good in assurance of integrity. Public-key cryptography is typically used in conjunction with a hashing algorithm such as SHA-1 or MD5 to provide integrity.

Confidentiality: Confidentiality means ensuring the secrecy and privacy of data. With guaranteed confidentiality of data, customers can assure that sensitive data have not been altered by the third party.

Nonrepudiation: Nonrepudiation means ensuring that data cannot be renounced or a transaction denied. This is provided through public-key cryptography by digital signing. Nonrepudiation is a critical security service of any e-commerce application where value exchanges, legal, or contractual obligations are negotiated. Nonrepudiation is a by-product of using public-key cryptography. When data are cryptographically signed using the private key of a key pair, anyone who has access to the public key of that pair can determine that only the owner of the key pair itself could have signed the data in question.

PKI has been used to implement the earlier-described security features. Here, we present brief description of PKI in domain of MobiCash.

11.2.3 Public-key infrastructure (PKI)

Public-key cryptosystem has been designed to securely communicate between two parties without a preagreed secret key. Each party does not need to have a secret key exchange mechanism before starting secure communication.

The setup for a public-key cryptosystem is of a network of users $u_1, \ldots, u_n$ rather than a single pair of users.

Each user u in the network has a pair of keys P_u, S_u associated with him, the public key P_u, which is published under a *public directory* accessible for everyone to read, and the private key S_u, which is known absolutely only to u.

Let A and B be two users where A needs to securely communicate message m to B. To send a secret message "m" to B, A uses the same exact method, which involves looking up P_u, computing $E(P_u, m)$ where E is a public encryption algorithm, and sending the resulting cipher text "c" to B.

Upon receiving cipher text c, user B can decrypt by looking up his private key S_u and computing $D(S_u, c)$ where D is a public decryption algorithm.

The operation of each individual key is a one-way operation: a key cannot be used to reverse its operation. In addition, the algorithms used by both keys are designed so that a key cannot be used to determine the opposite key in the pair. Thus, the private key cannot be determined from the public key.

11.2.4 Digital certificates

Digital certificates allow someone to combine their digital signature with a public key and something that identifies them, an example being their real life name. This certificate is used to allow computer/mobile users to show that they do own the public keys they claim to [18]. In other words, it is a security mechanism for public keys.

As mentioned earlier, a digital signature is required for the PKI certificate. This signature can either be made by an authority figure who assigns the certificates, the person whose identity is being confirmed, or even endorsers of the public key.

As with electronic financial transactions, a digital signature is a way for other parties and people to verify that a person is in fact the owner of the public key they claim is their own. Furthermore, the PKI certificates allow other people to verify that they are indeed communicating with the right person and using the right public key.

In the proposed system, we utilize the properties of PKI certificates to provide the essential security features for electronic financial transactions.

11.2.5 System design

The key components of the anticipated system are bank, customer, and the MobiCash agent. The roles of these components will be discussed later in this chapter. Both the MobiCash agent and the customer should have mobile phones, suitably modified to perform the MobiCash functions. In the case of use as a debit card, the merchant will replace the role of the MobiCash agent. The bank has a MobiCash server as the front end, connected to the bank's back-end transaction management system.

The proposed system is built on PKI cryptography. Therefore, before subscribing to the system, all participants should complete two steps of the user verification process as follows:

1. User registration process
2. Certificate issuing process

All the functions of the MobiCash system work on the information gathered from the aforementioned steps. Details of the two processes are defined next.

11.2.5.1 User registration

The registration process is for the collection and verification of the user's information relevant to MobiCash system, such as bank account details. This registration process should be completed as a face-to-face procedure. Each user should be present at a registration desk for the one-time process.

The registration process should be performed by a trusted registration agent. This trusted registration process can be performed by the bank or other independent identification provider. We suggest an independent identification provider for this task. If all the banks trust the same third-party identification provider, the system can be used for interbanking transactions. Therefore, we recommend a central bank or other responsible government authorities to perform registration process. Figure 11.1 illustrates the registration process.

All information gathered in the registration process is stored securely in a database for user authentication.

11.2.5.2 Certification issuing process

The proposed MobiCash system uses the PKI to enable the security for transactions. Therefore, all the participants who are involved with a particular transaction should have PKI security certificates. The certification authority has the right to issue these certificates, based on the user registration information gathered from the user registration process. The certificate issuing authority collaborates with the registration authority to get the user information. Bank can act as a certification authority. If an external, trusted third-party certification authority is appointed to

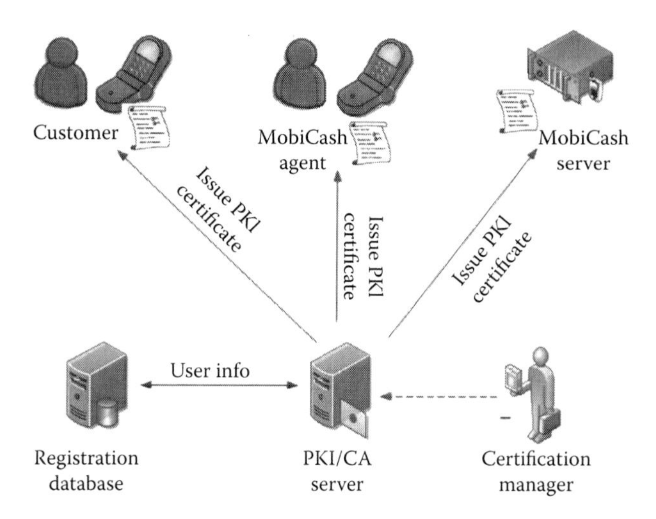

Figure 11.1 User registration process.

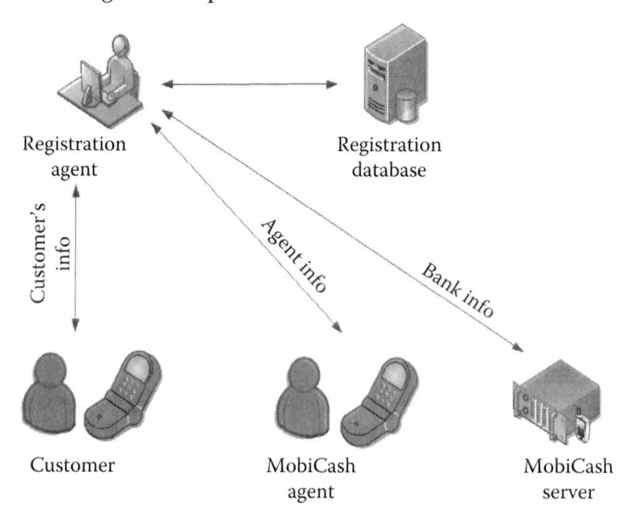

Figure 11.2 Certificate issuing process.

issue certificates, then the system can be extended to interbanking ATM transactions. This certificate includes the user's private key, public key, and digital signature and personal identification information. Figure 11.2 illustrates the certification issuing process.

The certificate provided by the certification authority (CA) server is stored in the user's and MobiCash agent's mobile phone [19]. Thus, the security of all the transactions relies on the public-key cryptography.

After a reliable and verifiable registration and certification issuance, an instance of the system is ready for secure transactions.

11.3 Cash dispensing

One distinguishing feature of the proposed system is the way it dispenses cash among users. The cash dispensing happens through mobile cash dispensing agents. They dispense cash to customers upon receipt of the authorization messages from banks based on customer cash requests. The sequence of steps and exchange of messages for the cash dispensing is described later. We will discuss the security features related to each step in Section 11.4. Figure 11.3 illustrates the overall system architecture.

1. Customer goes to a Mobile-ATM agent's place and sends a secure SMS to the bank (withdrawal request) including MobiCash agent's phone number and requested amount.
2. MobiCash server authenticates the user using PKI certificate chains support of CA server and registration server.
3. MobiCash server verifies the customer's account and sends an authorization SMS to the customer together with a confirmation number (a random number).

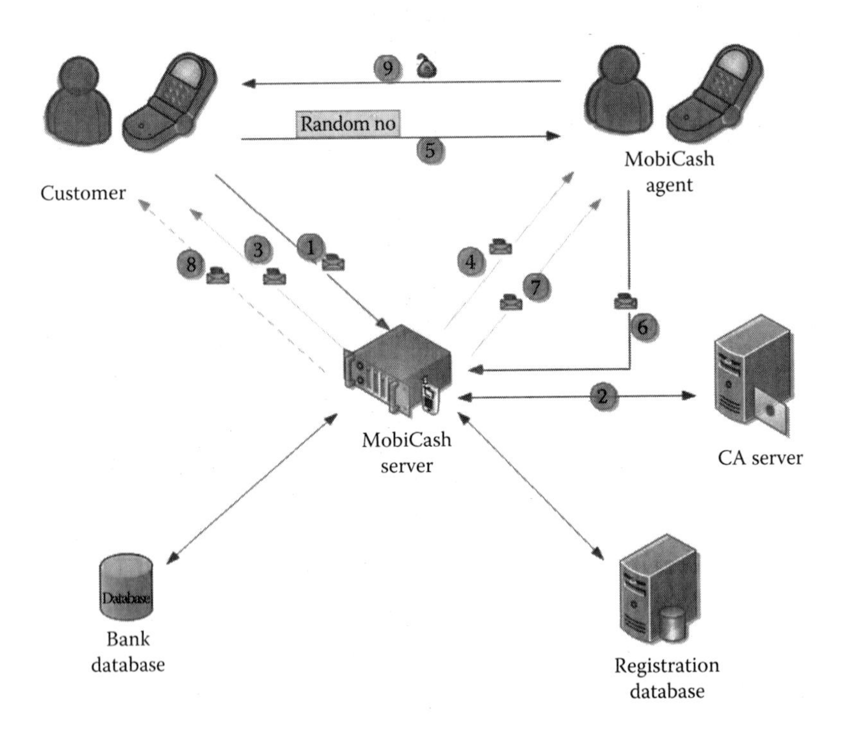

Figure 11.3 System architecture.

4. At the same time, the bank sends a payment authorization SMS to the MobiCash agent together with a transaction number (a random number is different from the confirmation number, but these two numbers are mathematically related to each other).
5. Customer discloses the confirmation number to the MobiCash agent.
6. MobiCash agent sends a confirmation SMS to the bank together with the transaction and the confirmation number.
7. Bank transfers the amount from the customer's account to the MobiCash agent's account and sends a transaction confirmation SMS to the MobiCash agent.
8. Bank also sends a transaction confirmation SMS to the customer.
9. MobiCash agent gives the money to the customer.

11.3.1 System roles

The operations of the system can be divided into three parts based on the actors in the system:

Customer: Customer is a person who needs to perform an ATM transaction. He has a bank account and mobile phone. A special program should be installed in the customer's mobile phone to operate MobiCash functions.

MobiCash Agent: Like the customer, the MobiCash agent (Mobile-ATM agent) must have a bank account and a mobile phone. This mobile phone is also modified to perform functions of MobiCash in a secured manner. He is authorized by the bank to perform MobiCash transactions. The MobiCash agent keeps money on hand to give to the customer when there is a request for such service.

Bank Organization: The bank should have MobiCash servers to deal with transactions between customers and MobiCash agents. These MobiCash servers should directly connect to the bank's databases. In addition, the bank maintains the accounts of both the customer and the MobiCash agent. Furthermore, bank servers have connections with registration servers and CA servers to revoke certificates or verify user information.

11.3.2 System deployment

All the components involved with a transaction must register in advance. After completion of the registration process, each participant receives a PKI X.509 certificate. On the customer side, there is a special mobile application, suitable to operate the MobiCash functions. The customer certificate also has to be stored in the customer's mobile phone.

The application requires the customer's PIN number (which is received in the registration process) for authentication purposes. Moreover, the application requires the MobiCash agent's mobile phone number and the amount of money to be withdrawn. Finally, the customer-side application sends a secure SMS (which is encrypted by both the bank's public key and customer's private key) to the bank with the MobiCash agent's mobile number, the amount of money to withdraw and the customer's account number.

On the MobiCash agent's side, there should be a mobile application that is capable of receiving secured SMS from the bank. A certificate is also stored in the agent's mobile phone, which should be able to securely send a transaction completion notice and the customer's mobile phone number to the bank. This application also requires the agent's PIN number for authentication purposes. Furthermore, the MobiCash agent generates a tracking number for audit purposes, indicates the transaction completion, and sends the information to the MobiCash servers.

The MobiCash server provides an interface between the core banking system and users. Meanwhile, it uses registration information to verify the subscribers. The MobiCash servers are responsible for decrypting customers'/agents' messages and redirecting the messages to the core banking system. MobiCash servers encrypt the outgoing messages with their own private key and the recipient's public key. The MobiCash servers are responsible for generating two random numbers (confirmation number and transaction number) for every transaction. There is an algorithmic relationship between the confirmation number and the transaction number. A random number generation program will not generate the same number for another transaction. After the sixth step, MobiCash servers are able to identify the two numbers that belong to the same transaction.

11.4 Security architecture

We strongly assume here that the bank has concrete security policies on how it internally handles access to its client details like PIN in its database [17]. Here, we discuss the security architecture of the system in customer's perspective and MobiCash agent's perspective.

11.4.1 Customer-side security

Customer-side security is based on asymmetric key cryptography, assuming that the bank is a trusted entity. From our earlier example, recall that user A wants to withdraw money from his account. For money collection, he selects Agent B.

The customer has to enter his PIN number (α) to the customer-side application, and the application itself gets the customer mobile phone number (β) and application ID (γ) to generate a secure Hash Code. Let H be the hash string,

$$H = h(\alpha + \beta + \gamma)$$

where h is the hash function algorithm.

Another hash code (H') is generated using the aforementioned three parameters and is stored in the phone's SIM card at the very first run of the mobile application. When a customer tries to log into the system, it matches these two hash codes. If $H = H'$, the customer can log into the system [20]. The purpose of this hash function matching is to protect the user from losing their individual information via mobile phone thefts.

Let us consider the following notations, which are used in rest of the paper.

Bank public key Pb and private key Sb

Agent public key Pa and private key Sa

Customer public key Pc and private key Sc

The withdrawal request includes the MobiCash agent's phone number, the customer's phone number and the amount to be withdrawn. This message (Wr) has been encrypted first from the bank's public key and next using the customer's private key.

$$EWr = E(Sc,(E(Pb,Wr))$$

Next, the encrypted version of information (EWr) is sent to the relevant bank [21]. According to the assumption mentioned earlier, first the bank tries to decrypt the data using the bank's private key. Let us name this intermediate message as DWr'

$$DWr' = D(Sb,EWr)$$

If DWr' has been successfully retrieved, the bank knows the message is sent by a registered customer and not tampered [19], because (EWr') encrypted with Pb and Pb is known only by the bank. Then the bank retrieves the certificate belonging to the customer and retrieves the customer's public key Sc and decrypts the intermediate message (DWr') using the customer's public key (Pc).

$$Wr = D(Pc,DWr)$$

If it is successful, the decryption process returns. Hence, the bank knows the message comes from the actual customer, who needs to withdraw money from his account. Next, the bank generates a hash code of the customer's PIN number and mobile phone number and matches that against the hash code stored in the registration database. The stored hash code is equal to *H*, hash string. The idea behind this hash code matching is to protect customers from using a fake phone number. The bank can ensure that the correct person is using the mobile application, as only the real user should know his PIN number. Thus, the system has been designed in such a way so as to invalidate the transactions from stolen mobile phones, so that a bank can authenticate the customer. Also, an encrypted version of the customer message provides the integrity and the confidentiality of the customer information.

Bank sends two reply messages to customer: transaction authorization message (*TA*) and transaction completion message (*Tc*). Both the messages are encrypted with bank's private key and customer's public key.

$$ETA = E(Pc,(E(SB,TA)))$$

$$ETc = E(Pc,(E(SB,Tc)))$$

Therefore, the customer can ensure the integrity and the confidentiality of the information [22].

11.4.2 Agent-side security

The MobiCash agent is a person authorized by the bank. The same security architecture described in previous section has been applied here.

At step 6 of Figure 11.3, the MobiCash agent encrypts the confirmation number (*c*) (which has been obtained from the customer) and the transaction number (*t*) received from the bank. The calculated hash code (h_a) from the MobiCash agent's PIN, the agent's phone number, and the application ID are also included in the message.

Let A_r be the request for payment approval from the agent. When the MobiCash agent first registered, a hash code similar to H_a was stored in the bank database, similar to the customer hash code. Then the bank can authenticate the MobiCash agent. The precalculated hash code is matched with the hash code stored in the bank to prevent phone number faking attacks,

$$Ar = h(c + t + ha)$$

where Ar is the payment authorization request from the agent.

$$EAr = E(Sa,(E(Pb,Ar)))$$

This message also provides the integrity and confidentiality of information.

At steps 7 and 8 of Figure 11.3, the bank sends the confirmation note that indicates whether the transaction was completed or rejected. These two messages (*Ma = Message to agent and Mc = Message to customer*) are encrypted by the bank. First, these messages are encrypted by the bank's private key and next by the agent's/customer's public key.

$$EMa = E(Pa,(E(Sb,Ma)))$$

$$EMc = E(Pc,(E(Sb,Mc)))$$

If these messages are successfully decrypted on the receiving sides, the customer and MobiCash agent can verify that the messages came from the bank.

11.4.3 Nonrepudiation

In the proposed protocols, nonrepudiation is ensured through the use of the two random numbers (transaction number and confirmation number) generated by bank back-end servers [23]. The transaction number is sent to the client. He has to disclose the transaction number to the MobiCash agent. Then in step 6 of Figure 11.3, the MobiCash agent should send these numbers to the bank again. If the two numbers are mathematically related to each other, then it confirms to the corresponding client that he has disclosed the correct transaction number to that agent. Thereafter, the client cannot deny participating in the transaction and ask to roll back the transaction [9]. Besides ensuring nonrepudiation, the two random numbers play an important role in preventing replay attacks.

11.4.4 Privacy

Another important consideration is the privacy issues of the financial transactions. Most of the customers do not wish to disclose their financial statements to outsiders. Although the customer does not need to disclose his financial statement, unfortunately, due to the system design, the customer needs to disclose how much money he needs to withdraw. This problem may be a barrier to the popularity of the system, in the case of deployment [24].

11.5 Evaluation

So far, we have discussed the technical aspects/designs and the specific issues of the MobiCash system. This section introduces a detailed technical evaluation and a small user study evaluation. MobiCash is a community-based application. Therefore, community acceptance of the system is an important factor. We believe real-world evaluation has to be done to evaluate the system instead of evaluating the system in a lab environment. For evaluation purposes, we deployed the MobiCash system in a bank, which has a customer community that is both urban and rural.

We have used 40 user samples for the evaluation.

The technical evaluation focuses on evaluating the system performance and the scalability. The performance of the deployed MobiCash system has been measured by the time taken to complete a transaction. The completion time of a transaction consists of the time needed to feed the data to the application, the throughput SMS delivery time, and the time taken by the core banking system to complete the transaction. Thus, the time taken to feed the data to the application also depends on the computer skill of the user. Therefore, we removed that factor to extract precise measurements about the system performances. For a transaction, typically there are two situations where users should feed data to the application: the withdrawal request and the agent's request. Hence, we have to divide time measurements into two parts.

1. Time period starting from the customer sending the withdrawal request and ending with receiving both the confirmation SMS and transaction SMS (T_1)
2. Time period starting from the agent sending the confirmation and transaction numbers and ending with receiving the transaction completion SMS (T_2)

According to Table 11.1, standard deviation values of T_1, T_2, and total are relatively small in comparison to the mean values. That means the spreading of T_1, T_2, and total time is in a short time interval. For example, the bulk of the total completion times for transactions is between 84 and 104 s. From this, we can conclude that the system is stable for most of the transactions.

In addition, we have evaluated the system with a 40-user sample. This user sample was divided into two groups, each with 20 users. Each user in the first group completed one MobiCash transaction at a location of extremely low GSM signal strength (around 7 km away from the base station). The next 20 observations from group 2 were measured at a location of extremely high GSM signal strength (at around 250 m distance from the base station). Given these observations (as shown in Figure 11.4), there was no considerable

Table 11.1 In Network Performance

	T_1 (s)	T_2 (s)	Total (s)
Mean	49.677	44.34	94.021
Standard deviation	8.216	7.311	10.534

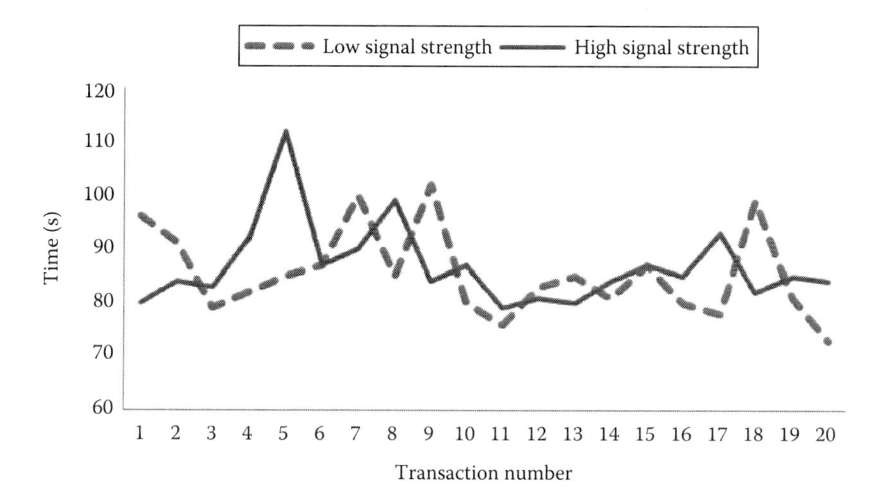

Figure 11.4 Transaction completion time. (First 20 transactions done in where signal strength is low. Last 20 transactions done in where signal strength is high.)

variation of the total time among the first 20 measurements and latter 20 measurements. We observed that GSM signal strength does not affect the system performances to a great extent. Given this finding, the system is suitable for rural areas, where the GSM signal strength is considerably low.

Since there is no effect from GSM signal strength, all 40 users were asked to do a transaction on traditional ATM machine. Time taken to each transaction has been measured. Figure 11.5 illustrates the measurements.

According to Figure 11.5, the traditional ATM takes less time to complete a transaction. However, the traveling time to a traditional ATM machine and waiting time on queue for an ATM machine makes

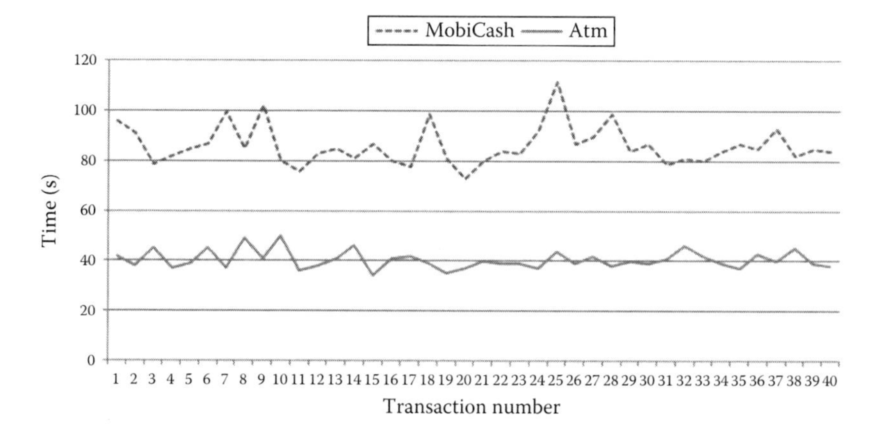

Figure 11.5 Transaction completion time comparison (MobiCash and ATM).

the completion time higher than the transaction completion time for a MobiCash. Even though MobiCash takes more time than traditional ATM to complete a transaction, given the argument related to accessing an ATM machine, MobiCash has the potential to be accepted by the community as a more efficient process for certain financial transactions.

Furthermore, we have performed 120 transactions on MobiCash and all were completed with a successful response from the system. We are confident in the stability of the MobiCash system.

11.6 Discussion

All participants in the MobiCash network are registered through a face-to-face procedure by a trusted third-party registration agent. Therefore, all identities are strongly verified. Identity, financial, and authorization data are stored in databases in an encrypted form. Data cannot be illegally accessed by unauthorized individuals.

All participants in the system have personal security credentials: key pairs, certificates, and other security tokens. For customers, they are safely stored in their mobile phones, encrypted, and accessible only after personal authentication.

Hash codes generate from subscribers' PIN number, application ID, and phone number to match against the hash code stored in registration servers. This feature ensures safety against faking phone numbers and cloning the mobile application. Additionally, it guarantees the identity of the person who uses the application, because of the PIN number included in the hash code.

All messages are digitally signed, encrypted, and enveloped for the targeted recipient. Thus, the system provides integrity and confidentiality to a particular transaction [25]. It is also convenient to have nonreputability for the electronic transactions. As noted in Figure 11.3, the bank generates and sends two random numbers: the confirmation number to the customer and the transaction number to the MobiCash agent. To complete the transaction, the system design requires the customer to disclose this confirmation number to the MobiCash agent. Only after that does the MobiCash agent have authority to give the money to the customer. In the context of the MobiCash agent, the confirmation number received from the customer is strong evidence to verify that the transaction has happened completely. Very similar to the scenario as described earlier, the bank transfers money from the customer's account to the MobiCash agent's account only after receiving the transaction number and the confirmation number from the MobiCash agent. Hence, none of the parties can roll back the transaction in an illegal way. Moreover, the transaction number received from the MobiCash agent is evidence to verify that the MobiCash agent has completed the transaction. So, the proposed system provides the guarantee of nonrepudiation for both customer and MobiCash agent.

The system uses asymmetric key cryptography to provide security as well as a third-party registration agent [26]. Therefore, subscribers have to accept the cost of retrieving the certificate and registration, being the main disadvantage of the system. Thus, the use of trusted third-party certification authority and third-party registration agent enables interbanking money withdrawals after few modifications to the MobiCash servers.

Only five essential SMS messages are used for a transaction, and a commission must be paid to the MobiCash agent. This commission should be paid by the customer. The cost of the SMS messages can be distributed among the customer and the bank. From the bank's point of view, it is economically advantageous to pay for the SMS instead of deploying a traditional ATM.

11.7 Conclusion

The proposed system intended to address difficulties in accessing ATM services. One of the distinguishing features of the MobiCash system that makes it different from any other similar system is its security. The MobiSys system provides confidentiality, integrity, authenticity, and nonrepudability, which are essential in financial transactions. It enables more security features relevant to the ATM transaction. The proposed MobiCash system also provides legally acceptable evidence about transactions.

Without having any additional cost on new infrastructure, existing mobile networks can be used to deploy the system.

The system can scale smoothly to interbanking level and further to the international level. It is compatible with existing security and communication protocols, so further extensions to the system are feasible.

To sum up, MobiCash system addressed a major service gap in banking industry and serve people in rural areas to solve their problems, which was critical to their social and economic development.

For the future study, we will investigate the community acceptance level of the system. Perception of system security and privacy issues are still needed to be identified.

References

1. Garner, P., Mullins, I., Edwards, R., and Coulton, R. Mobile terminated SMS billing exploits and security analysis. *Proceedings of Third International Conference on Information Technology*. IEEE Press, 2014, pp. 294–299.
2. Boly, J., Bosselaers, A., Cramer, R., Michelsen, R., Mjølsnes, S., Muller, F., Pedersen, T. et al. The ESPRIT project CAFE–high security digital payment systems. *Proceedings of Third European Symposium on Research in Computer Security*, 1994, pp. 217–230.
3. Mjølsnes, F.S. and Rong, C. On-line e-wallet system with decentralized credential keepers. *Journal of Mobile Network and Applications*, 8(1), 87–99, 2003.

4. Kadambi, K., Li, J., and Karp, A. Near-field communication-based secure mobile payment service. *Proceedings of 11th International Conference on Electronic Commerce*. ACM Press, 2009, pp. 142–151.
5. Vishwas, P. and Shyamasundar, R.K. An efficient, secure and delegable micropayment system. *IEEE Conference e-Technology, e-Commerce and e-Service*, 2004, pp. 394–404.
6. Lukkari, J., Korhonen, J., and Ojala, T. Smart restaurant: Mobile payments in context-aware environment. *Proceedings of Sixth International Conference on Electronic Commerce*. ACM Press, 2004, pp. 575–582.
7. Bouska, P. and Drahansky, M. Communication security in GSM networks. *IEEE International Conference Security Technology*, 2008, pp. 248–251.
8. Tasneem, G.B. and Paul, C. Mutual authentication, confidentiality, and key management (MACKMAN) system for mobile computing and wireless communication. *Proceedings of 14th Annual Computer Security Application Conference*, 1998, pp. 308–317.
9. Min, Q., Meng, D., and Zhong, Q. An empirical study on trust in mobile commerce adoption. *Proceedings of IEEE International Conference Service Operations and Logistics, and Informatics*, 2008, vol. 1, pp. 659–664.
10. Karunanayake, A., Zoysa, K., and Muftic, S. Mobile ATM for developing countries. *Proceedings of Third International Workshop on Mobility in the Evolving Internet Architecture*. ACM Press, 2008, pp. 25–30.
11. Kumar, P., Shailaja, G., Kavitha, A., and Saxena, A. Mutual authentication and key agreement for GSM. *Proceedings of International Conference on Mobile Business*. IEEE Press, 2006, pp. 25–27.
12. Li, Z., Sun, Q., Lian, Y., and Giusto, D. An association-based graphical password design resistant to shoulder-surfing attack. *Proceedings of IEEE International Conference on Multimedia and Expo*. IEEE Press, 2005, pp. 245–248.
13. Goyal, S. 10 Years of mobile and ubiquitous computing. *Proceedings of IEEE Pervasive Computing*, 4(2), 88–90, 2005.
14. Wang, Z., Guo, Z., and Wang, Y. Security research on J2ME-based mobile payment. *Proceedings of ISECS International Colloquium on Computing, Communication, Control, and Management*. IEEE Press, vol. 2, pp. 644–648.
15. Boudriga, N. Security of mobile communications. *IEEE International Conference Signal Processing and Communications*, 2007, pp. 24–27.
16. Li, X. and Autran, G. Implementing an mobile agent platform for m-commerce. *33rd International Computer Software and Applications Conference*, 2009, vol. 2, pp. 40–45.
17. Thanh, D.V. Security issues in mobile ecommerce. *Proceedings of First IEEE Internationals Conference Electronic Commerce and Web Technologies*, 2000, pp. 467–476.
18. Lin, Y., Studer, A., Chen, Y., Hsiao, H., Kuo, L., Lee, J., McCune, J.M. et al. SPATE: Small-group PKI-less authenticated trust establishment. *IEEE Transactions on Mobile Computing*, 9(12), 1666–1681, 2010.
19. Chen, C. and Jensen, M.A. Secret key establishment using temporally and spatially correlated wireless channel coefficients. *IEEE Transactions on Mobile Computing*, 10(2), 205–215, 2011.
20. He, L.S. and Zhang, N. An asymmetric authentication protocol form-commerce applications. *Proceedings of Eighth IEEE International Symposium on Computers and Communications*. IEEE Press, 2003, pp. 244–250.

21. Islam, S. and Ajmal, F. Developing and implementing encryption algorithm for addressing GSM security issues. *International Conference Emerging Technologies*, 2009, pp. 358–361.
22. Shibli, M., Yousaf, I., Dar, K., and Muftic, S. MagicNET: Secure communication methodology for mobile agents. *Proceedings of 12th International Conference Advanced Communication Technology*. IEEE Press, 2010, vol. 2, pp. 1567–1573.
23. Ahluwalia, P. and Varshney, U. A framework for transaction-level quality of service for m-commerce applications. *IEEE Transactions on Mobile Computing*, 6(7), 848–864, 2007.
24. Morawczynski, O. Examining the usage and impact of transformational m-banking in Kenya. In *Proceedings of Third International Conference on Internationalization, Design and Global Development*, Aykin, N. (Ed.). Heidelberg, Berlin: Springer-Verlag, 2009, pp. 495–504.
25. Li, H. and Wang, Y. Public-key infrastructure. In *Payment Technologies for E-Commerce*, Weidong, K. (Ed.). New York: Springer-Verlag, Inc., 2003, pp. 39–70.
26. Le, V., Matyas, J., Johnson, D., and Wilkins, J. A public key extension to the common cryptographic architecture. *IBM Systems Journal*, 32(3), 461–485, 1993.

chapter twelve

Mobile electronic commerce development

Sathiadev Mahesh

Contents

12.1 Introduction

Mobile device technology is in its growth phase, and there are a wide variety of mobile devices, with new products being introduced at a rapid pace. Competing vendors rush to introduce new and innovative features, with much emphasis placed on user interaction richness. This contrasts with developments in mainframe and desktop systems, where most of the development was focused on benefitting from increasing processing power and storage, with one major shift from the text-based, command line, interface to the window, icon, mouse, pointer (WIMPy)–based graphic user interface in the 1980s. The other major paradigm shift in business computing was the move from in-house computing, with trained employees accessing systems, to customers directly accessing business systems, as is the case in e-commerce. Systems developers, building systems for direct customer access, had to build adequate safeguards to prevent data corruption and design the interface to meet customer expectations. A dominant platform arose quickly for business developers with the MS Windows environment dominating in the case of desktop computing and a combination of hypertext markup language (HTML) and JavaScript becoming the dominant web e-commerce environment. This simplified development and developers could focus on one dominant technology. However, in the case of mobile computing, there are multiple vendors promoting their own platforms; in some cases, a single hardware vendor may support multiple operating systems. Mobile computing places a premium on high interconnectivity, rich multimodal interaction, and user mobility. These features are rare in conventional systems development, and developers need to understand the benefits and limitations of the technologies used in mobile computing. In addition, mobile network providers favor some modes of connectivity over others, making it necessary for developers to alter applications based on communications network providers. The highly fractured marketplace makes it challenging to develop mobile electronic commerce applications that can function effectively on different devices.

Developers may either choose to build and optimize for one platform or build a multiplatform application that provides satisfactory performance on multiple platforms. This impacts the choice of a suitable development tool, since many mobile hardware providers offer toolsets that are designed to develop optimized applications for their hardware. Hence, a major issue for mobile development is whether to develop native applications optimized for one class of hardware or to develop web-based applications where the mobile device is merely the user interface and performs very little data processing. In native code development, a significant part of the application's code is placed on the mobile device. Hence, different versions of this code need to be developed and maintained for each mobile device on which the application is used. The web-based application approach is most beneficial from the developer's perspective since it can be ported from established desktop/laptop browser-based versions, with the primary rework being limited to screen size variation for small mobile devices. However, many of the specialized features of mobile hardware can be accessed only through the code specifically designed for the device, and applications with a rich user experience need to be designed to optimize relatively scarce resources on the mobile device.

Electronic commerce applications benefit from the network effect, where the applications become more valuable, when many other users utilize them. Mobile applications are richly interactive, supporting interaction with the user and between users. Consequently, they benefit even more from the network effect. The development process needs to consider this and develop applications in a manner that facilitates rapid acceptance and adoption by many users to benefit from the network effect. As a result, mobile developers prefer to use modular, prebuilt widgets and rapid applications development (RAD) approaches to quickly put together prototype applications, which can be deployed quickly. A challenge in mobile development is the short half-life of mobile applications, which need ongoing modification to remain competitive. While system maintenance has always been more expensive than new systems development, the problem is much worse in the mobile e-commerce marketplace, due to the constant need to adapt to new market pressures and technology changes. Mobile e-commerce developers need to continually monitor how their applications use device and network resources, such as battery power and net bandwidth, and ensure that they do not hog device resources. A mobile application that is not maintained will rapidly lose market share and be replaced by more nimble competitors (ABI Research, 2013).

The user interface in mobile devices uses a variety of convenient inputs from the user, ranging from touch input common in current hardware to voice, gestures (shake, point), and visual inputs. In addition, the devices can capture data about the user's location, acceleration, and environment automatically from the device's built-in sensors. Mobile electronic commerce applications need to be designed to meet the needs and capabilities of rich input from these sources. The user interface for mobile apps needs to consider the overall user experience in a highly competitive marketplace.

12.2 Development methodologies

Systems development methodologies range from the traditional waterfall approach of the SDLC to agile technologies such as Scrum and Extreme Programming. Mobile applications are designed to respond to a rapidly evolving marketplace, with an emphasis on rich customer interaction, location awareness, and the ability to effectively handle microtransactions. Agile methodologies are used extensively in mobile applications development due to development time pressures and market adaptation requirements. The D-mobile model proposed in 2004 has five phases, and each phase has 3 days of development, one each for planning, working and releasing the product. This provides a quick 15-workday development cycle with close user–developer contact and interaction (Abrahamsson et al., 2004). The process is suited to small mobile projects that deliver data over mobile devices. A proposed spiral model, based on multiple loops that can ensure application interface usability, is more suited to larger and riskier projects that support complex user interaction with the mobile application (Nosseir et al., 2012).

12.2.1 Web applications

Mobile electronic commerce applications can be developed as web-based applications that run in the browser or as native applications that run on the mobile device. The first attempts in mobile computing ported standard browser screen designs to small mobile devices. These web pages, adapted from desktop browser designs, rendered poorly, and the interface elements such as drop-down lists and buttons were inconvenient for mobile users. The wireless access protocol (WAP) promoted by the WAP Forum was developed to adapt e-commerce web pages and scripts designed for larger screens to render effectively on small screens and devices, especially those connected through low-bandwidth connections. The protocol used intermediate servers to store and serve large web pages in a mobile format. The technology failed to gain market acceptance due to a lack of widespread agreement on the WAP standard and the cost of implementing the intermediate servers by different mobile network providers. Modern mobile devices have sufficient bandwidth to deliver the web content for electronic commerce and, in most cases, have browsers that can handle HTML, CSS, and JavaScript, which are the underlying technologies for web-based applications. There have been problems with some specialized web technologies, such as Flash on Apple's devices. Often, these problems have been the result of conflicts between vendors. The small screen size of mobile devices makes it necessary for web applications developers to redesign web page formats. In addition, mobile e-commerce is conducted in shorter sessions and in a wide variety of environmental conditions, such as lighting and street noise, making it necessary for application developers to consider usage conditions in designing the interface.

12.2.2 Native applications

Native applications offer lower latency, deal with loss of connectivity in transit situations, and can be optimized for specific hardware. In web applications, user interactions are processed on remote servers, leading to latency in user feedback. JavaScript code runs on the browser and eliminates the round trip for many simple interactions. While improvements in JavaScript have made it competitive in many e-commerce environments, the language cannot provide, at present, the responsiveness offered by native code that runs on the hardware. When does it make sense to build native applications?

The choice is analogous to the thin client–fat client decision that was common in the client–server application development era. Fat client software performs most of its work on the device and offers fast interactive response, while thin clients perform most of their work on the remote server and are hampered by network latency. System administrators preferred thin clients due to the overall efficiency of resource use and centralized control over the application, while most clients preferred fat clients due to low latency and personal control. Mobile device hardware has become more capable, with memory and processing power exceeding early desktops. This has made it feasible to run fat-client applications on mobile devices. However, mobile devices continue to lag behind current laptop and desktop hardware in both hardware capability and wired network bandwidth and face a major problem in battery power limitation. Mobile applications need to be optimized to run effectively under hardware, bandwidth, and battery power limitations. This suggests the need to build customized hardware-specific versions of mobile applications. Early recommendations for mobile development were focused on the resource limitations (Tarasewich, 2003). A mobile applications development framework proposes a taxonomy of mobile applications based on mobile media delivery (broadcast), real-time location-sensitive data, business transaction completion, and collaboration and describes suitable development and maintenance approaches for each category of application (Unhelkar et al., 2010).

However, if native applications are developed, they also need to be maintained. When a new version of a phone operating system (OS) is released, or when new phone hardware is introduced, the native application needs many more changes than a web application. Unless there is a major change in the underlying hardware, web applications, even those that call hardware features such as a camera image or GPS readings will continue to function. On the other hand, native applications, which run more effectively, will make hardware-specific calls that may need to be updated for new hardware. This adds to development and maintenance complexity, with developers needing to build and maintain multiple versions of the product.

12.2.3 Hybrid

Many mobile applications are developed in a hybrid format with core web applications, enhanced with native features to enrich interaction. Hybrid applications are often built on mobile development toolsets that allow the developer to use standard JavaScript in development, with custom application programming interface (API) calls for specific features. During application compilation, the toolset adds hardware-specific code to the delivered application, customized for each type of hardware. Finally, many mobile applications depend on cloud services for data, notifications, and user interaction. As a result, these applications can be built effectively on cloud-based toolsets, which offer access through convenient APIs to cloud-based services, enabling a plug-and-play approach to custom mobile app development.

12.3 Mobile applications development platforms

12.3.1 Integrated development environment

Integrated development environments (IDEs) provide a platform for developers to build and deploy multitier applications. Eclipse is a well-known open source IDE that supports development in many different languages. Develop toolkits can be plugged into Eclipse to facilitate development in different languages and for different platforms. For example, the Android Developer Tools (ADT) software development kit (SDK) can be plugged into the open source Eclipse IDE to develop android apps, which will run on mobile devices using the Android operating system. Proprietary development platforms are supported by vendors, who promote the platform as a convenient way to develop applications for their devices. Many proprietary platforms have polished interfaces, custom plug-ins that reduce the chores of programmers, and support code optimization for their devices. Developers focusing on a particular platform, say, the iPhone, would prefer to use the Xcode IDE, and program using the objective-C language, to develop apps optimized for Apple's iOS. Microsoft's Visual Studio is a widely used IDE for many languages and is well suited for developing Windows phone applications using the C Sharp programming language. Visual Studio add-ons are available for developing Android apps and apps that work on other platforms. Google provides the Android Studio IDE, a customized version of the IntelliJ IDEA IDE.

Features offered by integrated development environments include (1) layout editors with drag-and-drop features to position and size user interface components, (2) template-based wizards to create common designs, (3) tools to document app performance on hardware, and (4) the ability to quickly build apps to run on different versions of the OS. IDEs provided by vendors have features specific to the mobile device OS and typically can add plug-ins with highly optimized, prebuilt features. Many app development IDEs also support cloud-based app support, such as cloud-based databases, server side components that handle business processes, and other features such as messaging.

Applications for mobile use, like many other web-based applications, use the model-view-controller (MVC) approach. This approach splits the application into three major categories. The first is the model, which is the data model used to store data, that is, a database. The controller refers to the code that manages the business process, the link between the user and the business system. The data are held in cloud storage, and the controller runs largely on the server, though parts of the controller could run on the user device. The view is the part of the application that presents the user interface, displays data, and interacts with the user. It is easy to build models and controllers that are shared across multiple hardware devices and operating systems. The major change occurs in the view component, which needs to be customized and optimized for devices.

12.3.2 Cloud-based development platforms

Many mobile applications are developed for the cloud. While cloud applications can be built from the ground up, using a stand-alone IDE, it is preferable to use development platforms targeted for cloud-based mobile services. The reason is the short half-life of new mobile applications. Many mobile applications fulfill a short-term need and need to be updated in the future. A full-fledged development process will take a long time, and application designers will be hard-pressed to find knowledgeable users for testing a yet-to-be-developed application. Platforms for developing cloud-based mobile applications provide toolsets for the most common features, enabling a quick drag-and-drop development of mobile application prototypes. In addition to beta testing, cloud-based applications can be scaled for growth. Hence, they can be used to grow market share, and full-fledged development can be initiated after market growth patterns are understood and a revenue model is established. Many platforms for cloud-based mobile application development are themselves on the cloud, enabling development teams to quickly access and build prototype systems.

Many mobile applications use common features for mobile computing such as social networking, cloud-based data storage for both customer data and business data, and push notification services to deliver customer-focused updates. Microservices provide very small chunks of content and could come from one source or multiple individual/community sources (Christophe et al., 2011). The mobile Backend-as-a-Service (BaaS) model offers these services as an easily accessible API. Developers using a cloud-based BaaS can quickly access these resources and build their applications. At present, many BaaS providers offer a freemium model to encourage application developers to use their platform for development. The freemium model charges for services only when demand rises above a preset level. Hence, a prototype model can be built and rolled out at low cost, with charges accruing only after business picks up.

12.3.3 *Platform comparison*

There are many frameworks for creating mobile web applications based on HTML5 and JavaScript. These frameworks offer custom script modules to quickly develop robust mobile web applications. Apache Cordova can be used for many devices including Windows Phones (Eberhardt, 2012). A comparison of Rhodes from Motorola, the popular PhoneGap from Adobe, which is based on Apache Cordova, Dragon Rad from Seregon, and MoSync, an open source solution, presents tables summarizing their features for development, architecture, preferred software model, and mobile OS support (Palmieri et al., 2012). Another study compares the performance of two versions of a simple test application developed using different cross platform tools, Phone Gap, and Titanium (Dalmasso et al., 2013).

Various vendors provide toolsets that can facilitate mobile development. Their toolsets run on existing platforms for development, and these could be general-purpose IDEs or specialized mobile development platforms. It should be noted that the tools are all under continued development, and any feature set presented in a study needs to be rapidly revised to current conditions. Table 12.1 contains a list of currently available mobile e-commerce development tools.

12.3.4 *Vendor-supported/vendor-sponsored application stores*

Many mobile device vendors, and mobile telecommunications networks that provide access, maintain control over the applications they permit and support on their platforms. The control is primarily exercised through a licensing and approval process for the installation of the native application on hardware. This contrasts with the situation in desktop- and web-based computing, where applications can be developed and marketed by the developer with little interference, approval needs, or licensing fees from the hardware, operating system, or network provider. In the case of mobile applications using native code, vendors typically take a cut on revenues,

Table 12.1 Mobile E-Commerce Development

Appcelerator/titanium http://www.appcelerator.com/titanium/	Titanium is a JavaScript development tool with many UI widgets for RAD. Apps can be packaged to run in native mode on common mobile platforms. Appcelerator is a cloud-based service that delivers back-end systems such as authentication and databases.
AppMobi http://www.appmobi.com/	Cloud-based development tool enabling push, in-app purchases, ads, and analytics. Build using AppMobi API, develop to multiple platforms.
Appery http://appery.io/	Cloud-based tool for quickly building web apps and hybrid apps, with PhoneGap widgets. It uses drag-and-drop visual tools for building apps.
BuzzTouch http://www.buzztouch.com/	Web-based software to build Android and iOS apps, develop online, and create app code.
Conduit http://mobile.conduit.com/	Web-based software development with prebuilt templates for common mobile apps.
Dragon Rad http://www.seregon.com/overview	Development tool for multiplatform development. Output includes native code modules. Largely drag and drop with optional Lua scripting language add-ons.
Each Scape http://www.eachscape.com/	It builds app on cloud with action blocks for common apps.
NetBiscuits http://www.netbiscuits.com/	SaaS platform for mobile development. Library of 6000 devices and many features for each device. It includes an analytics tool to monitor applications.
PhoneGap http://phonegap.com/	PhoneGap is powered by the ApacheCordova open source engine. Available as an add-on to IDEs. PhoneGap Build is a cloud-based tool to build apps. Toolsets such as AppGyver Steroids wrap HTML5 code developed in PhoneGap to create faster apps that can be customized for hardware.
Pub Nub http://www.pubnub.com/mobile	Development platform for real-time feeds to mobile devices with a focus on good encryption and low power consumption on mobile devices.
Rho Studio http://www.rhomobile.com	Motorola IDE plug-in to Eclipse. It uses Rhodes and RhoElements API libraries with JavaScript modules and Ruby language coding.

(*Continued*)

Table 12.1 (Continued) Mobile E-Commerce Development

Sencha Architect http://www.sencha.com/	Sencha Touch provides touch widgets and a framework for layout and events. Architect is a tool to manage widgets and coding. Output can be made into app by other tools.
Xamarin http://xamarin.com/	Commonly used UI and mobile app tools are provided in native code. The code modules are wrapped to be connected to app. Available as an IDE plug-in. Mono is an open source version.

ranging up to Apple's 30% share. In addition, many platform owners such as Apple, Microsoft, Google, or Blackberry maintain application stores, where the product is offered for sale. These application stores are proprietary malls, and in addition to enforced revenue sharing, the platform owners restrict both the functioning and content of the applications. The focus of limits on code functions is primarily to ensure security and control interaction between applications loaded on the hardware. These limitations screen and stop applications containing the code that violates the rules of the marketplace, as enforced by the platform owner. In addition, applications are tested to ensure they do not hog excessive resources or deter other applications from performing their operations. While most restrictions on code operations are designed to prevent badly behaved applications from creating hardware conflicts and crashing the system, the approval process takes time and delays deployment. Mobile system developers need to consider the delay in deployment when planning the rollout schedule for their product. Often, using the development platform prescribed by the platform owner, following prescribed coding guidelines, and limiting the code to using only authorized features will be adequate to meet approval requirements. Another more variable factor is the content of the application. The application could face delays in approval and rejection if it violates content guidelines laid out by the platform owners as a prerequisite for approval. Note that unapproved applications cannot be installed on devices, and many platforms lock their devices. Violations of device locking rules amount to civil and criminal violations of the Digital Millennium Copyright Act (DCMA).

Another concern for mobile developers is the limits placed on apps operating on mobile networks. Net neutrality on the wired Internet has been debated for a long time. The only certainty regarding net neutrality is that there is much disagreement on what it implies. Mobile network providers have much more control over their broadband traffic than ISPs providing wired access. As a result, mobile providers can choose to shut down applications either through cooperative agreements with mobile device makers or by packet shaping and bandwidth throttling any application they deem unsuitable. This can include applications that conflict with their other business needs, such as the use of VoIP apps over mobile systems that reduce voice communications usage.

12.4 *Mobile application technology*

12.4.1 *Hardware interaction*

Developers prefer to use a programming tool/language that allows them to write code focusing on the business process, while leaving hardware interaction to the OS. In the case of mobile development, there are many different types of hardware and operating systems. These platforms gain or lose market share, and market relevance, rapidly. A true cross-platform development tool should support the development of application source code that can be easily deployed on many devices. In practice, it is necessary to make minor and sometimes major modifications for each device due to ongoing technology changes. In addition to multiple hardware vendors, there is the problem of new interfaces with changes from single to multitouch, accelerometers, and gesture recognition. The task of code maintenance is simplified when common code can be developed for many devices, with only minor variations for special features.

The hardware abstraction layer (HAL) is code that translates instructions from the application to hardware-specific instructions. Each class of device hardware, whether the CPU or an interface device, has its own instruction set, termed the instruction set architecture (ISA). Highly optimized code should send the most suitable instructions for a specific hardware using its ISA. However, this will mean that developers should write and maintain hardware-specific code. The HAL provides a common API for sending instructions to the device, allowing the developer to code using the common API. As a result, the application can work on a variety of devices, as long as the operating system uses an appropriate HAL to interact with the hardware. This simplifies development and, more importantly, maintenance, since code maintenance costs are much more than code development costs over the life cycle of the application. If new hardware allows for unique input types, such as pressure-sensitive input to a tablet screen, a new version of the OS needs to accommodate the enriched API, and the HAL for the new device must translate the new values. In addition, the enhanced application will have features that will not be available on old platforms. This will create a fork in the application development process, with one version supported on old hardware and another on new hardware.

The rapid influx of new hardware features such as cameras with new features such as 3D photography, GPS systems, accelerometers that detect motion speed and direction, and new communications technologies like near field communication (NFC) have made it necessary for developers to maintain multiple versions of applications that adopt the new features. There are also differences between the products offered by different vendors. An Android device typically uses Google Android APIs for the interface, Apple's devices use the CocoaTouch APIs, Windows phones use Silverlight, and a web application using a mobile browser uses the HTML5/JavaScript framework. Incidentally, while HTML5 for web page tags is a W3C standard, there are multiple mobile frameworks for script development. Standards include the open source jQuery standard for scripting and proprietary frameworks that offer advanced features for quickly developing mobile apps such as Sencha Touch and Appcelerator Titanium.

12.4.2 *API intermediary layer*

One solution offered by mobile applications developer tools is to create a layer of tools that emulate device hardware, for example, a screen tool that emulates the screen. The developer will write code to interact with the emulated screen. The developer toolset uses its own tool, which takes the applications requests and translates them into the specific device hardware. While this approach provides benefits to developers, these applications are slower and less responsive than native apps. In addition, the developer toolset needs to be continually updated for new hardware and new OS-specific APIs that are introduced. Developers face the risk of vendor demise, which will leave their applications unsupported. Standards for intermediary APIs have been proposed to enable the creation of device-independent applications. The MODIF framework proposes a user interface (UI) markup language, uniform APIs for the device interface, and transformational models that connect the common APIs to device-specific calls (Cimino and Marcelloni, 2012).

12.4.3 *Modularization and hardware dependency*

A challenge facing mobile application developers is the wide range of hardware and operating systems for mobile devices. In particular, mobile devices have numerous hardware features that vary between devices. In the desktop sphere, business computing adopted the Microsoft OS on the Intel chip instruction set, and the Windows, Icons, Mouse, Pointer (WIMPy) interface grew to be widely accepted, providing business application developers with a common GUI. While other operating systems from Apple and the Linux open source OS were also used, these products became niche markets for developers. In many cases, business applications were first developed for the dominant platform, and versions for other platforms were either unavailable or performed suboptimally. In the case of mobile systems, both hardware and software markets are highly fragmented, and worse, each platform has multiple versions due to rapid technology development. As a result, mobile business system development has more in common with game development, where there are multiple platforms with different hardware, operating systems, and interfaces. This level of hardware dependency has made it challenging to develop one application for many devices. One solution is to design an interface with only those features that are found in all commonly used platforms. This can be narrowed to focus on a web application that limits the interface to a web browser capable of operating on all systems, a least common denominator approach. However, this approach leads to very conservative interfaces that do not exploit new technology and will not satisfy users who have made major investments in leading-edge technology, especially in the case of consumerization of corporate IT, where users bring their own devices for use. Savvy users expect their apps to exploit the high-end mobile device they have acquired with their own money.

Mobile applications need to be built in modules, with user-facing modules developed independently of the rest of the system. This will support the easy replacement of this module to adapt the system to new hardware and operating systems. This is especially useful since mobile device technology is in the growth phase of its life cycle, and operating systems for mobile devices are updated more frequently than for the more stable, late-mature phase desktop environment.

12.4.4 Version control

The wide variety of hardware and OS, as well as the high level of innovation in the field, make it necessary for mobile system developers to build and maintain different versions of their applications. Branching in systems development is a process in which a common system is customized to operate effectively on a particular type of hardware/OS. Since there are multiple final versions, a common system is developed, tested, and archived. There is an archived codebase, which is customized for each platform. These customized versions are called branches. Each branch may be further forked for new versions of the hardware/software. At any time, if a new version of hardware/OS is introduced in the market, developers may need to go back to the archived base and build from that point to get a usable version for the new hardware/OS. A process of merging is used to recombine multiple branches, especially when there are too many branches and code maintenance becomes unsustainable.

12.5 Mobile user experience

12.5.1 Web pages

Web applications are designed to be delivered over the Internet. The MVC design used in web applications uses the view component to deliver the user interface. Conventional web application design is focused on desktop and laptop screens, delivering the content through a browser. When users connect to web applications through mobile devices, they use a much smaller screen for viewing the page. In the past, screen resolutions were much lower for mobile devices than for desktop devices, but that difference has been almost eliminated with newer mobile devices. This results in a subpar user experience. (1) The user may find it difficult to discern screen elements, (2) screens may require multiple interaction steps to be completely rendered, or (3) parts of the screen may be ignored by the user. Since many mobile devices use a touch interface, screen icons need to be sized for reliable touch interaction. An icon size that is suitable for a mouse click is often too small for a fingertip touch. In addition, latency is typically greater in mobile devices, leading to a longer page load for a content-rich site. A responsive web design (Marcotte, 2011) delivers a suitable user experience regardless of screen size, by scaling web page elements and reorganizing content based on the user device's screen capability. Challenges in designing responsive websites exist for many industries such as banking (Crosman, 2013), health services (Beil, 2012), newspapers (Serm et al., 2006), and education (Fisher and Baird, 2006). Over 50 tools to help build responsive websites are presented by Denise and Peter (2012).

Cascading Style Sheets (CSS) are used to change the display format for HTML pages. The changes can range from mundane color and font variations based on user and device preferences to complex formatting changes for page elements. It is the ability of CSS to change how page elements are formatted that enables responsive web design (Frederick, 2013).

While mobile devices' interfaces have rapidly progressed to fully functional touch–based screens, and the primary difference between desktop designs and mobile designs for the UI is the limited screen real-estate mobile devices are used in a wide variety of environments, unlike desktop devices that are used in controlled work environments. The user lacks control over the environment in which mobile devices are used, due to noise, lighting conditions, and traffic hazards. Because of the differences in the context in which mobile devices are used, the mobile UI designer needs to develop resilient designs that can be effective in a range of environmental conditions. A controlled experiment conducted in 2007, when mobile devices had a highly constrained UI, discusses the need to consider the fact that the mobile UI cannot be as immersive as the desktop UI due to environmental conditions under which it is used

(Isomorsu et al., 2008). The study presents many interesting metaphors to contrast mobile and desktop net usage. The mobile UI needs to be optimized for standardized use where the user performs a standard action many times, rather than a nonstandard highly customized, incrementally developed interaction. For example, in a pizza store application, the mobile user may order the same pizza many times and needs an interface that is optimized for this repetitive task.

12.5.2 Device load management

In addition to screen size, the smaller bandwidth and mobile connectivity interruptions impose limitations on the mobile web application design. Applications that impose a heavy load on mobile device resources will impact user experience. An application that drains battery power, the weakest and scarcest resource in a mobile environment, will quickly fall out of favor. It is recommended that web applications reduce the number of HTTP requests, use compressed versions of content, preload necessary interaction components, and plan the use of caches to optimize mobile web application performance (Matsudaira, 2013; Nicolaou, 2013). This is not significantly different from web application guidelines used when net bandwidth is a scarce resource. All this implies is that developers need to return to a focus on parsimony in design, which was the hallmark of early software engineering. Mobile application performance needs to be monitored in real time to ensure that users get the quality of service (QoS) promised by the application.

A cyber foraging approach, in which applications off-load processor-intensive operations onto remote servers based on a decision model that considers mobile device capability and cloud server availability, is described for photo processing and gaming applications (Verbelen et al., 2012). A similar load management approach has been described for data mining of continuous data streams using mobile devices, where resource awareness is handled by variation in analysis detail, rather than off-loading processing (Haghighi et al., 2013). The off-loading should be built into the mobile application for overall effectiveness. Using standard virtual machine images on mobile devices adds too much load on the device to provide significant computational advantage (Shiraz et al., 2013).

12.5.3 Touch and haptic technology

While mobile devices lack screen space, they often incorporate a wide range of user input formats and enhanced outputs to improve user interactivity. Consumerization of information technology in the workplace refers to the use of new technology in mobile devices purchased by the user and utilized in the work environment. In the past, computing devices were expensive, acquired by the business and assigned to employees for work. Corporate IT departments specified and controlled the acquisition of devices for work and, in most cases, managed them with strict rules. The rapid reduction in the cost of mobile devices, coupled with their rich interface and connectivity, has created a generation of users who prefer to use personal devices at work. These users have rapidly enhanced their skills in using preferred personal devices and invested significant human capital in learning to use the tool effectively. Mobile devices support high connectivity and offer far richer interfaces than office computers provided by corporate IT. In addition, since mobile devices are typically held by the user and in close physical contact, they utilize creative approaches to alerting and interacting with the user. Mobile application developers need to incorporate touch interaction in their application view component. While this has generally meant a separate fork in UI development for mobile devices, the influx of touch-sensitive tablets and introduction of touch-capable laptops points to a future where all e-commerce applications will use a touch-capable interface.

Touch is a major component of the interaction, and vendors are moving toward pressure-sensitive touch (Greene, 2009) that will allow users to vary their input by varying pressure. Though pressure-sensitive touch has been available through dedicated sketch pads (Wacom, 2013), it has not been available for on-screen use. Workarounds have been developed using a pressure-sensitive stylus (Steele, 2013) that however impact device mobility.

Haptic technologies provide a feedback approach that uses tactile responses, such as vibrations. Haptic technology has developed from early iterations where the entire device vibrated to vibrations focused on a small part of the surface area, thereby providing feedback customized to an area of the device. Improvements in this technology will enable effective Braille responses on mobile devices, leading to improvements in communications with individuals with disabilities. An electro-vibration technology under development uses electrical charges to simulate vibration (Liszewski, 2013). Though being experimental at present, some of these technologies are likely to become mainstream in the future.

12.5.4 Voice input and output

Voice input is useful in mobile devices used in a highly variable environment, where complete user immersion in the device is neither feasible nor recommended. Speech recognition (and text to speech) is typically performed by a separate application that transfers the data to the mobile application. The HTML Speech Incubator Group of the W3C has developed a report on how HTML can grow to accept speech inputs (W3C, 2011). At present, vendors use different approaches to coding and delivering speech output to applications. Nuance's SpeechAnywhere, iSpeech, and ATT Watson are cloud-based services that can be called by client-side scripts to decipher and return text. Software development kits for these services allow the developer to call the cloud service in their application code. Applications that need speech recognition for a few words, typically commands, can use a client-side code module from OpenEars. This follows a freemium pricing model, with a paid version offering better recognition, including real-time recognition. Other client-side speech recognition modules include CMU Sphinx, an open source product that can link to web applications written in the python language, and SpeakEasy, based on the Nuance engine.

Voice output for mobile applications is often essential for handsfree mobile computing in industrial workspaces and on the road. An option for mobile apps that delivers text to speech is VoiceOver, a screen reader app, which reads screen elements. When developing web apps for mobile e-commerce, developers must ensure they follow screen reader guidelines to ensure mobile device usable pages (Rangin and Hahn, 2011). Guidelines include ensuring that interactive icons' actions are clearly stated. For example, if clicking on an icon will open a new window, the screen reader will, by default, merely state that it is a clickable icon, and read the text caption provided for the icon. The mobile-enabled application should clearly explain the action that will take place if the icon is clicked. Another guideline relates to long drop-down lists that are easily visually searchable, such as a list of 50 states, but turn out to be extremely inconvenient for a screen reader. In addition, multimedia links must be accessible; for example, audio tracks should have transcripts.

12.5.5 *Gestures and movement*

Gesture input recognizes human gestures mostly from the hands or from facial gestures. Gesture recognition can extend to body posture and movement. Gesture recognition and feedback can be based on intrusive devices such as sensor gloves and body suits, and minimally invasive devices such as multiple cameras capable of stereoscopic vision and motion sensors embedded in mobile devices. True gesture recognition should not involve touch, and the location and movement in a three-dimensional space should be recognized by the device. Swype technology used to recognize text using continuous touch motions is not true gesture recognition. Current technology for hand, face, or body gesture recognition requires large sensors and is not yet suitable for small mobile devices. However, motion sensors embedded in mobile devices can be used to specifically recognize hand and body movements, such as a shake for near field data transfer. A set of user-defined motion gestures for mobile interaction is presented by Ruiz et al. (2011). Advances in this technology using multiple motion sensors and dedicated hardware for data analysis offer the promise of effective gesture recognition (Zhu et al., 2013). Accelerometers and compasses provide movement and direction information. The gyroscopes present in some smartphones have been demonstrated to provide accurate orientation information that can be used in applications providing accurate walking directions in enclosed areas (Barthold et al., 2011).

12.5.6 *Visual and augmented reality*

Most mobile devices have a camera, often two (one facing the user and the other rear-facing). Camera quality has improved, both in pixel resolution and in digital signal processing of images, which enhances image quality. While the physical size limits of mobile devices limits the optical quality of built-in cameras, rapid advances in optical sensors and digital signal processing have enabled the production of higher quality images. Mobile applications routinely file and post camera images to user social network sites. Advanced features include using face recognition capability to identify individuals and enhance interaction.

Augmented reality adds features to a real worldview, enhancing the user experience. This can range from adding descriptive data to a camera view (building address, name, and features) to showing the product specs when a product code is scanned by a phone camera. Mobile applications need to sense GPS location data and movement from accelerometers and combine it with cloud-based information to create augmented reality views on the screen or through audio output.

12.5.7 *User attention spans*

Mobile application developers need to consider user attention spans on mobile devices. In addition to the small screen size in mobile devices, and multiple distractions in the mobile environment, which we discussed in the earlier section, users tend to interact in multiple short sessions with their mobile devices. Internet traffic is typically composed of asymmetric bursts, with short bursts from users followed by longer bursts from the service, and is conducted over an extended session. In contrast, mobile device interactions are typically a series of short bursts (Oulasvirta et al., 2005), interspersed with other unrelated activities. Texting and tweeting interaction captures this dynamic effectively. Hence, mobile applications need to be developed for this manner of user interaction. Long web pages need to be split into shorter segments, with effective search tools to provide the necessary information.

12.5.8 Security

Passwords have been used as an authentication tool since the mainframe era. The approach is now well past its prime and needs to be updated to suit current needs. Single-factor authentication with passwords is considered unsuitable even for in-house access from desktops, and business has moved toward two-factor authentication, with access cards and other technology. The password approach to authentication assumes a secure workplace, and a log-on at the start of a session, which continues uninterrupted for a few hours. In the case of mobile devices, user interaction modes are very different, with repeated access in short microbursts from many locations. The concept of a work session, with clearly defined start and stop authentication events, is replaced by an authenticated device that is authorized for multiple, independent interactions. In many mobile applications, the password approach is ported to the mobile version, with a password entered on the first access from the device, which is henceforth authenticated for a prolonged period. This creates a security hole, since the authenticated device can be used by another individual. Many mobile devices have a second layer of security, a device access code, which is required after a short break in use, when the device shifts to a dormant state. This code is trivial, typically a few digits or a pattern on the screen.

Multifactor authentication on mobile devices should use biometrics, as in recently introduced fingerprint scans, or proximity to other items such as smart cards. In any event, the mobile application developer faces a clear challenge in authentication due to the manner of access. Users connect in multiple short bursts and access from a wide range of nonsecure environments. Device sensors need to detect multiple biometric patterns, preferably automatically, and force a reentry of complete authentication parameters, once the separation between the user and device exceeds a threshold, before authentication becomes truly reliable.

System and data security is ensured through a combination of authentication, authorization, and security monitoring. After a user is authenticated, computer systems typically use access control lists to authorize these users to access permitted resources. Again, this approach made sense in the traditional workplace environment, where users accessed the network from an office machine. When users access resources from a mobile device, it is important to introduce the concept of location-aware authorization. For example, if the access to corporate data is from the business office, a salesperson can have authorization to access all sales data. However, if the access is from a customer site, then the authorization should automatically extend to only records about that customer and related products. In the absence of a specified reason, and additional approvals, automatic access should not be provided to all customer records. Even tighter control should be placed for access from undesirable locations. A tighter access control process will prevent a stolen or misplaced device from being used to access a treasure trove of valuable corporate data.

Many e-commerce apps store data on the mobile device to support customer use when connectivity is not available. Strict limits must be placed on data stored on devices, and retention rules should be enforced in mobile applications, especially native applications that can store data internally. In addition, mobile applications share hardware space with many other applications, often downloaded by the user on a whim. It is important that applications secure themselves and their data from other authorized applications on the device (Floyd, 2006).

12.6 Testing mobile applications

Mobile applications have characteristics that make them susceptible to a multitude of field problems. First, native mobile applications are designed to take advantage of special hardware features. If these features are newly introduced and nonstandard, the application will face problems when deployed in the field. Second, while mobile hardware capabilities have increased tremendously in the past few years, they continue to have limited battery power, depend on mobile networks offering widely varying levels of connectivity, and often share resources on mobile devices with other poorly behaved applications. Third, many mobile applications are cloud based, and problems in configuring cloud resources can impact user experience. Some of the issues that make mobile testing different from conventional software testing include the difficulty of setting up test benches for touch interfaces and the need to test mobile apps in varied mobile network and security environments (Muccini et al., 2012).

Hardware testing: A test for mobile applications should be conducted on the designated class of hardware, including both current and old versions of the hardware. It is possible that the app will work on the version of the hardware used in the lab, which could be quite different from the wide variety of hardware configurations used in the real world. We also need to consider user workarounds to hardware, including "jailbroken" hardware, which uses nonstandard operating systems, and how the app functions with modified hardware. In particular, developers need to ensure that the app remains secure when used on a modified device. The mobile app tester needs to determine the hardware configurations used by the potential market, and the test bed should include hardware in all these configurations. Since it is expensive to acquire all versions of the hardware, most initial testing is done on hardware emulators, software that emulates the hardware. While these emulators can identify system bugs, process steps, data transfer, and process delays involved, they lack the actual feel of the device and user interface impact on events such as the chances of hitting the wrong button or the visibility of the screen display in a brightly lit public space (Keynote, 2013). In addition, most test benches use programs that automatically enter user responses to test the application. Such programs were perfect for web-based forms entry, where users typically enter alphanumeric data in fields, and follow up with a return keystroke or press a button to submit data. Mobile devices use touch and voice input and allow for gestures and camera scans. The test bench must be able to simulate all these forms of data entry effectively.

Installation testing: Mobile apps are typically downloaded, installed, and executed by users in one integrated process. The test must include the download process time, installation success rate on devices hosting other common applications, and the quality of the first interaction with the applications. Due to the circumstances under which mobile apps are installed, users do not have the time to resolve conflicts and do not go through a long training and learning process to use the app.

Connectivity testing: Mobile app performance is dependent on network connectivity. A good test should simulate the real connectivity available to your app in the market environment. In addition, the test should use the bandwidth, latency, jitter, and packet loss rates for your application on the real-world network. Another critical issue in testing mobile applications is the impact of firewalls and packet shaping between the real-world user and your service.

Load testing: Since mobile apps are accessed from portable devices, they will be used by many users from different locations. Connectivity testing should include the impact of log-ins from a geographic and access location distribution that matches the expected customer profile. In addition, since connections from individuals will follow a queuing pattern with highly varying loads, simulation studies should be conducted to determine the impact of peak and sustained demand for the service.

Security testing: Mobile apps face extensive threats from focused hackers and scattershot attacks. Many mobile apps are accessed from personal devices, and the design standard is to require the user enter authentication information once and retain these data for future use. Stolen devices have been used to access secure applications, and until biometric tools are incorporated in these devices, mobile app developers need to consider alternate means of security. A security test for mobile apps should consider the robustness of the application to compromise by social engineering, intrusive hacking, and scattershot attacks.

Crowdsourced testing: One model for testing mobile apps is based on a crowdsourcing-type approach and is promoted by a new venture, Testlio.

The business model offers a team of validated testers who have passed a screening process and completed training. These testers will follow a standard protocol to access from a globally distributed base and subject a mobile app to a range of tests. The company claims access to nearly 10,000 types of mobile hardware configurations. The business submits a report to the developer on the functionality and robustness of the app.

12.7 Mobile web application technologies

12.7.1 HTML5

The hyper-text markup language (HTML) is used to tag web pages and enable them to be rendered on browsers. Tim Berners Lee, who worked as an independent contractor at the Center for Nuclear Research (CERN) in Geneva, Switzerland, proposed an interlinked system of documents based on hypertext and connected using net technology in 1989. Hypertext is text that is linked to other text files and was proposed in 1945 by Vannevar Bush, and versions of hypertext were demonstrated by Douglas Engelbart in 1968. Berners-Lee put these concepts together and combined them with the basic approach outlined for a standard generalized markup language (SGML), to create a system where text files could be created independently, and linked together by hyperlinks to other documents, creating the web concept of "surfing" for information and at the same time eliminating the need to centrally manage a hierarchical document structure. HTML was first proposed in 1991, with the Internet Engineering Task Force (IETF) standard, version 2.0 released in 1995. The most recent version of HTML is version 5, which includes many features for highly interactive websites. When considering the value of the HTML5 standard, we should understand that the first version of HTML did not even have a standard way of incorporating images in a document, and even today, images are separate from a HTML tagged page. HTML 5 includes well-designed and integral support for a document object model (DOM) that segments the document presented in the browser in separate units for convenient processing within the browser sandbox. As a result, code in the browser sandbox, for example, JavaScript, can reliably interact with page elements, based on user actions. While this could be designed in earlier versions, there was no established standard, and developers could never be certain that their code would work on all browsers, or even if a browser update would break the code, and render the website useless. Another feature standardized in HTML5 is drag and drop, the ability of the user to take an object displayed on the screen and move and drop it on another part of the screen. This feature is especially useful in touch-enabled interfaces.

12.7.2 jQuery

JavaScript is used to code processes that can be executed within a browser. They support dynamic web pages that are capable of rich user interaction. Pages containing JavaScript can complete business data processing steps within the user's browser, cleaning and streamlining data transfer to the application server. User experience is enhanced, with a fast acting, richly interactive interface on a web browser. Since data transfers are conducted only after a sequence of interactive steps is completed, the inherent latency of web applications is reduced. JavaScript was first introduced by Netscape in 1995, as a feature available on the Netscape browser. It is a scripting language, having almost nothing in common with the multiplatform programming language Java. JavaScript code was interpreted, that is, translated to machine code sequentially, on the user's browser. As a result, early versions were slow and could be not be used for much beyond adding a few interactive features, such as drop-down lists, to a static web page. Over time, JavaScript processing has improved and rich interactive interfaces have been developed. These interfaces function within the web browser's JavaScript sandbox. JavaScript is also used on the web server to code simple business processes using server-side scripting.

Web applications are accessed through the Internet, and users preferred applications that offered rich interactions, without the latency of web data transfers. Enhancements to the language, in particular, improvements in the speed of JavaScript processing on browsers, made it possible to develop web applications that presented a highly interactive interface on the user's devices. JavaScript coding was popular with web developers who worked in a highly dynamic e-commerce environment and developed systems that were continually fine-tuned. While traditional coders viewed the language as poorly designed, and unreliable, the language became the dominant means of providing rich interactivity for websites. It became common in the industry to use script modules to automate common website features. An example is the Spry Library, which initially provided chunks of reusable code to add special effects to web pages. Its initial use was limited to web page enhancements, and JavaScript faced some criticism for its security weaknesses, especially because of exploits where it could be used to reach outside the browser sandbox and infect the host machine.

It returned to prominence with the popularity of Asynchronous JavaScript and XML (AJAX), where the segments of the web page could be processed locally on the user's browser. AJAX created the rich web interfaces of web applications like the Google Maps application. New libraries were developed to add additional features to websites by adding prepackaged code modules.

12.7.3 *JavaScript modules*

jQuery is a popular JavaScript library for document navigation and event handling. One feature, traversal, makes it easy to write code that works through the elements in a page. One use of this feature is to work through all the fields in a form and ensure that data entered in the fields meet standards. Since the code works on the browser, the test is performed when data are entered, with no net latency or traffic, and only clean data are sent to the business server. Another related feature, manipulation, allows the code to replace or alter data in different elements of the page. The jQuery architecture also supports the development of plug-ins for new features, which can be easily added to a site. There is a large library of jQuery plug-ins, and this makes it easy to develop a website with rich elements, such as interactive calendars, calculators, or image galleries. Coding productivity and reliability is enhanced, and coding maintenance cost is reduced by writing once and reusing the code. In jQuery, the approach is to reuse rather than to write, that is, to use prebuilt components and patch them together into a working site, rather than to write code. Since this approach uses tried and tested code modules, not only does it reduce the time to develop the code but also turn out to be more reliable than home-grown code. An interesting study comparing the education effectiveness of mobile in-class interaction showed significant improvements from a mobile app, which incidentally was built using the jQuery framework (Kert, 2013).

12.7.4 *Courses*

MIT App Inventor is a web-based platform that provides a drag-and-drop interface to design a mobile device application. It is designed to teach the basics of app development to students working on their first programming course, especially high school students. The interface is limited, but easy to learn. The Stanford iPhone App Development course is a full-fledged undergraduate course in iPhone Application Programming (CS193P) using Objective C, the language used for Mac OS/X and iOS development, and focuses on developing mobile applications for the Apple iPhone. It is an intense course and requires prior knowledge of computer programming using a procedural programming language. Stanford has a couple of prerequisite courses that are also offered through iTunes University. The course has been offered online, in the semester format, since winter 2009. While the course is offered for credit to Stanford students, lectures have been posted on iTunes University for public access. Another course recognizes the multidisciplinary nature of mobile app use and focuses on bringing together students from diverse backgrounds to build mobile apps (Massey et al., 2006).

12.8 Summary

Developing applications for mobile e-commerce is in many ways similar to developing other applications. E-commerce applications tend to have many cloud-based components and depend on effective interaction with other businesses. Mobile applications need to consider usage context differences in location, interaction, and interface. As a rule, due to the short half-life of mobile applications and the need to satisfy rapidly changing consumer tastes, mobile applications are built using rapid application development techniques, with many prebuilt plug-in modules and cloud-based services. Another challenge for mobile e-commerce developers is the need to enhance app performance on different types of hardware, using custom routines.

A mobile application developer can choose to build a web-based application, customized for mobile device screens, and limit customization to screen size and touch input. Since many laptops are being touch enabled, and large size tablets are becoming popular for web browsing, the resulting web application could become truly cross-platform. The other approach is to build a native app utilizing hardware-specific features and gain richly interactive performance. This approach is required for gaming and immersive applications and economical for widely used business applications. In many e-business situations, a hybrid approach can be used where web applications are developed, with small native apps that enrich the view part of the model, using custom hardware features to improve interaction richness.

References

ABI Research. (2013). Enterprises and mobile apps—How to get it right, October 15, 2013, https://www.abiresearch.com/webinars/enterprises-and-mobile-apps-how-get-it-right/.

Abrahamsson, P., Hanhineva, A., Hulkko, H., Ihme, T., Jäälinoja, J., Korkala, M., Koskela, J., Kyllönen, P., and Salo, O. (2004). Mobile-D: An agile approach for mobile application development. *ACM, 19th Annual ACM Conference on Object-Oriented Programming, Systems, Languages, and Applications (OOPSLA)*, Vancouver, British Columbia, Canada, October 2004.

Barthold, C., Subbu, K.P., and Dantu, R. (2011). Evaluation of gyroscope-embedded mobile phones. *2011 IEEE International Conference on Systems, Man, and Cybernetics (SMC)*, Anchorage, AK, October 9–12, 2011, pp. 1632–1638. doi: 10.1109/ICSMC.2011.6083905.

Beil, G. (2012). Technology speaks: Health and human service agencies must embrace the mobile revolution. *Policy & Practice*, *70*(4), 30–31.

Christophe, B., Narganes, M., Antila, V., and Maknavicius, L. (2011). Mobile execution environment for non-intermediated content distribution. *Bell Labs Technical Journal*, *15*(4), 117–134.

Cimino, M.A. and Marcelloni, F. (2012). An efficient model-based methodology for developing device-independent mobile applications. *Journal of Systems Architecture*, *58*(8), 286–304. doi:10.1016/j.sysarc.2012.06.001.

Crosman, P. (2013). The debate around responsive design in mobile banking. *American Banker*, *178*(318), 18.

Dalmasso, I., Datta, S.K., Bonnet, C., and Nikaein, N. (2013). Survey, comparison and evaluation of cross platform mobile application development tools. *Ninth International Wireless Communications and Mobile Computing Conference* (*IWCMC*), Sardinia, Italy, July 1–5, 2013, pp. 323–328. doi: 10.1109/IWCMC.2013.6583580.

Denise, J. and Peter, G. (2012). 50 Fantastic tools for responsive web design, Creative Bloq, April 24, 2012, http://www.creativebloq.com/css3/tools-responsive-web-design-5132770.

Eberhardt, C. (2012). Develop HTML5 windows phone apps with apache cordova. *MSDN*, *27*(5), 28–42.

Fisher, M. and Baird, D.E. (2006). Making mLearning work: Utilizing mobile technology for active exploration, collaboration, assessment, and reflection in higher education. *Journal of Educational Technology Systems*, *35*(1), 3–30.

Floyd, D. (2006). Mobile application security system (MASS). *Bell Labs Technical Journal*, *11*(3), 191–198.

Frederick, K. (2013). Responsive web design 101. *Computers in Libraries*, *33*(6), 11–14.

Greene, K. (2009). A touch of ingenuity. *Technology Review*, *112*(5), 112–114.

Haghighi, P., Krishnaswamy, S., Zaslavsky, A., Gaber, M., Sinha, A., and Gillick, B. (2013). Open mobile miner: A toolkit for building situation-aware data mining applications. *Journal of Organizational Computing & Electronic Commerce*, *23*(3), 224–248. doi:10.1080/10919392.2013.807713.

Isomursu, P., Hinman, R., Isomursu, M., and Spasojevic, M. (2008). Metaphors for the mobile internet. *Knowledge, Technology & Policy*, *20*(4), 259–268. doi:10.1007/s12130-007-9033-5.

Kert, S. (2013). Using J-Query mobile technology to support a pedagogical proficiency course. *Journal of Educational Computing Research*, *48*(4), 431–445. doi:10.2190/EC.48.4.b.

Keynote. (2013a). Testing strategies and tactics for mobile applications, white paper, October 15, 2013, http://www.keynote.com/docs/whitepapers/WP_Testing_Strategies.pdf.

Keynote. (2013b). November 1, 2013, http://mite.keynote.com/support/use-cases.php.

Liszewski, A. (2013). Disney lets you feel features on a touchscreen by zapping your fingers, *Gizmodo*, http://gizmodo.com/disney-lets-you-feel-textures-on-a-touchscreen-by-zappi-1442569940. Accessed October 15, 2013.

Marcotte, E. (2011). *Responsive Web Design*, A Book Apart, NY.

Massey, A.P., Ramesh, V., and Khatri, V. (May 2006). Design, development, and assessment of mobile applications: The case for problem-based learning, *IEEE Transactions on Education*, *49*(2), 183–192. doi: 10.1109/TE.2006.875700.

Matsudaira, K. (2013). Making the mobile web faster. *Communications of the ACM*, *56*(3), 56–61. doi:10.1145/2428556.2428572.

Muccini, H., Di Francesco, A., and Esposito, P. (2012). Software testing of mobile applications: Challenges and future research directions. *Seventh International Workshop on Automation of Software Test* (*AST*), Zurich, Switzerland, June 2–3, 2012, pp. 29–35. doi: 10.1109/IWAST.2012.6228987.

Nicolaou, A. (2013). Best practices on the move: Building web apps for mobile devices. *Communications of the ACM*, *56*(8), 45–51. doi:10.1145/2492007.2492023.

Nosseir, A., Flood, D., Harrison, R., and Ibrahim, O. (2012). Mobile development process spiral. *Seventh International Conference on Computer Engineering & Systems (ICCES)*, Cairo, Egypt, November 27–29, 2012, pp. 281–286

Oulasvirta, A., Tamminen, S., Roto, V., and Kuorelahti, J. (2005). Interaction in 4-second bursts: The fragmented nature of attentional resources in mobile HCI. *Proceedings of the SIGCHI Conference on Human Factors in Computing Systems*, Portland, OR, April 2, 2005, pp. 919–928.

Palmieri, M., Singh, I., and Cicchetti, A. (2012). Comparison of cross-platform mobile development tools. *16th International Conference on Intelligence in Next Generation Networks (ICIN)*, Berlin, Germany, October 8–11, 2012, pp. 179–186. doi: 10.1109/ICIN.2012.6376023.

Rangin, H. and Hahn, J. (2011). Mobile accessibility, challenges & best practices. *14th Annual Accessing Higher Ground Accessible Media, Web and Technology Conference*, Boulder, CO, November 14–18, 2011.

Ruiz, J., Li, Y., and Lank, E. (2011). User-defined motion gestures for mobile interaction. *CHI 2011 Proceedings*, Vancouver, British Columbia, Canada, May 7–11, 2011, pp. 197–206.

Serm, T.C., Blanchfield, P., and Su, K.S.D. (2006). Mobile newspaper development framework: Guidelines for newspaper companies for creating usable mobile news portals. *ICOCI '06 International Conference on Computing & Informatics*, Kuala Lumpur, Malaysia, June 6–8, 2006, pp. 1–8. doi: 10.1109/ICOCI.2006.5276443.

Shiraz, M., Abolfazli, S., Sanaei, Z., and Gani, A. (2013). A study on virtual machine deployment for application outsourcing in mobile cloud computing. *Journal of Supercomputing*, *63*(3), 946–964. doi:10.1007/s11227-012-0846-y.

Steele, B. (2013). Sketch it out, engadget. http://www.engadget.com/2013/08/26/wacom-intuos-creative-stylus-hands-on/. Accessed October 15, 2013.

Tarasewich, P. (2003). Designing mobile commerce applications. *Communications of the ACM*, *46*(12), 57–60.

Unhelkar, B. and Murugesan, S. (May–June 2010). The enterprise mobile applications development framework. *IT Professional*, *12*(3), 33–39. doi: 10.1109/MITP.2010.45.

Verbelen, T., Simoens, P., De Turck, F., and Dhoedt, B. (2012). AIOLOS: Middleware for improving mobile application performance through cyber foraging. *Journal of Systems & Software*, *85*(11), 2629–2639. doi:10.1016/j.jss.2012.06.011.

Wacom. (2013). August 20, 2013, http://www.wacom.com.

W3C. (2011). HTML speech incubator group final report, December 6, 2011, http://www.w3.org/2005/Incubator/htmlspeech/XGR-htmlspeech-20111206/.

Zhu, W., Liu, L., Yin, S., Hu, S., Tang, E., and Wei, S. (2013). Motion-sensor fusion-based gesture recognition and its VLSI architecture design for mobile devices. *International Journal of Electronics*, *101*(5), 621–635. doi: 10.1080/00207217.2013.794482.

section five

Mobile electronic commerce applications and mobile business

chapter thirteen

Mobile advertising*

The Indian perspective

Pradeep Nair and Harsh Mishra

Contents

13.1 Introduction

At present, 60% mobile advertising in India takes the form of text messages and is evolving significantly in terms of creativity. Rich media advertisements delivered on mobile allow consumers to interact with the advertisement in a nonintrusive manner (Barnes, 2002: pp. 399–419). For example, users could tap to watch a video, get a 360° view of a product, locate a store, call a dealer, and return gracefully to the content. Higher Internet speeds make the user experience better, but rich media technology has evolved to an extent where immersive user experience is possible even at lower data speeds. Also, multimedia messaging service (MMS) facility available in majority of handsets provides a good opportunity for the modern advertisers to reach to the consumers as these ads are stored in mobile phone's memory and are not dependent on data download speeds. Corporate brands in India are using the capabilities of the mobile phone to deliver engaging experience to consumers (Sharma et al., 2008).

The number of mobile phones in use is also growing much faster than the number of computers. According to the report of Telecom Regulatory Authority of India (TRAI) on the Indian Telecom Services Performance Indicators, January–March (2013), released in August 2013, there are more than 867.80 million mobile phone subscribers with an urban subscriber share of 60.11% and a rural share of 39.89%. This is more than 70% penetration rate for the country. The Internet and Mobile Association of India (IAMAI) reported (2013) that there are 46 million active mobile Internet users in India, a statistic that suggests that mobile Internet users have already surpassed the desktop Internet users by a huge margin. According to a report published by Associated Chambers of Commerce and Industry of India (ASSOCHAM) in December 2012, there were only 23.1 million PC Internet users in India at the end of June 2012 (Broadband for India, 2012). Although there is a contradictory report from research firm Juxt that says that there are only 23.8 million active mobile Internet users in India (Juxt, 2013).

Currently, there are around 13 network operators providing mobile communication services in India including the two public service undertakings (PSU), BSNL and MTNL. Private operators hold around 87.78% of the wireless market share (based on subscriber base), whereas the PSU operators hold only 12.22% market share. At present, prepaid subscribers account for 96% of the total mobile subscribers in India. Out of the total 150 million Internet users in India, there were around 87.1 million mobile Internet users till December 2012, and this number is expected to reach 130.6 million by March 2014 and 164.8 million by March 2015 (Indian Media and Entertainment Industry Report, 2012).

The average monthly consumption of a normal mobile user who uses her wireless device to access Internet is approximately Rs. 198. According to a Gartner Report titled "Forecast: Mobile Advertising, Worldwide, 2009–2016" released in November 2012, India is slated to develop into the biggest smartphone market by 2017 next only to China and the United States. Also, at present, about 6% of the total mobile phone users (67 million) in India use smartphones to access wireless communication services, and the number is slated to grow at an astounding rate of 52% in the coming years (Gartner, 2012).

Mobile users carry their devices to almost everywhere they go and this gives this medium an edge over traditional devices such as televisions or personal computers. Mobile advertisements develop their selling points based on the relevance of the advertisement to the target audiences. Park asserts that most advertisers believe that the *advertiser's waste* in case of traditional advertising is more than half the amount spent as it does not reach to the desired audiences (Park, 2005: pp. 63–80).

The issue with online advertising regarding the efforts and money is that almost half of all the online advertisements are sold on a *pay-per-click* basis, which means that advertisers pay only when consumers click on an advertisement; however, it is not an effective format as its measurability is minuscule. The advertising through text messages using mobile platform is more focused and if the marketers use mobile firms' profiles of the subscribers intelligently, they can tailor their advertisements to match the habits of individual subscribers (Hackley, 2010: pp. 29–33). Furthermore, when users consume content on the mobile Internet, through either mobile websites or mobile applications, they provide an opportunity for advertisers to reach out to their target audience (Tahtinen, 2006: pp. 152–164).

As a communication medium, mobile is fundamentally different and more interactive than other media on all counts that matter to advertisers—availability, targeting, engagement, and measurement (Bulander et al., 2005).

The mobile is an extremely personal device that most of us tend to access every moment we are awake. This gives advertisers access to their target audience nearly 24 × 7. Mobile offers better targeting than other communication media. Advertisers can target their audience based on the type of mobile device (basic phone/feature phone/smartphone) they use, operator they use, and operating system, in addition to the usual targeting and segmentation variables that exist for other media (Harte, 2008: pp. 43–49).

Mobiles also have immense computing power, 4.3″ screens, touchscreen capability, and sensors such as accelerometers, gyros, in addition to technologies such as GPS. All these elements allow for creative advertisement formats and immersive engagement. Unlike other media, the friction between engagement and call to action is nearly nonexistent for mobile, given it is a communication device at the core. Mobile enables tracking and measurement at a far greater level compared to conventional media. Advertisers can easily track the number of seconds a consumer spent on viewing a video, or the percentage of consumers that chose a particular product or service (Leppaniemi and Karjaluoto, 2005: pp. 197–213).

13.2 Mobile advertising—a conceptual framework

Mobile refers to any portable wireless communication device capable of delivering multifarious messages to the end users. The communication may be one-way or two-way and involves two or more than two persons. The messages may range from basic text and voice messages to rich multimedia messages. In the past few decades, mobile phones have achieved phenomenal success in the field of human communication owing to their inherent characteristic of portability and increasing affordability resulting from the introduction of cheaper technologies.

Technological advancements witnessed in the last few years have catapulted mobile phones into a different league altogether (Noonam, 2001). Modern mobile phones have evolved from simple communication devices to advanced gadgets employed for accessing the Internet, reading books, teleconferencing, video chatting, and a lot more. Worldwide, the number of mobile phone users is growing at lightning pace outpacing even the growth in the number of personal computer users. According to Internet Advertising Bureau's (IAB) (2013) Internet Advertising Outlook Report 2013, in 2012 the number of personal computer users globally declined by 4% than in 2011. The same report also proclaims that in 2012 about 15% of the total web traffic was on mobile phones. Mobile phones have become the order of the day in developing countries like India where these devices have been successful in filling the communication gaps and are proving helpful in bridging the digital divide. Mobile phones have now become items of necessity for the common public (Clay and Zainubhai, 2013: pp. 84–87). According to a report published on an Indian news website ibnlive.in.com on September 8, 2013, the number of actual mobile users in India at present is 554.8 million (Juxt, 2013). This means that nearly half of the Indian populace has access to the mobile phones making it the most preferred communication device for the masses. An interesting datum mentioned by the website is that there are more rural mobile subscribers than the urban ones. That is, approximately 54% or 298 million active subscribers belong to rural India. The combination of factors discussed earlier provides an opportunity to businesses to reach to a wide assortment of customers through mobile advertising in a cost-effective manner.

Mobile advertising refers to any legitimate paid personal or nonpersonal promotional message/s sent by an identifiable advertiser to the end user. The concept of legitimacy is attached with the concept of mobile advertising because mobile phones are distinctly personal devices and unwanted promotional messages are perceived as an infringement of privacy by the customers. It is the dissemination of multifarious text, voice, or multimedia messages to one or more existing or potential customers through mobile phones and devices. Basically, mobile advertising refers to the advertising and marketing activities that deliver advertisements to mobile devices using wireless networks, the advertising solutions to promote the sales of goods and services, and a marketing strategy to build brand awareness. Mobile advertising creates a new media channel to conduct product and business advertising and marketing through mobile devices such as mobile phones, personal digital assistant (PDA), and tablet/pocket PC (Lawrence, 2008). Mobile advertising includes advertising through mobile Internet, mobile banner ads, mobile search engine advertising, mobile portals, mobile in-apps advertising, rich media mobile advertisements, multimedia advertising messages, advertising through mobile radio/television broadcasts, and spillover Internet/print advertising, that is, the advertisements meant originally for the desktops or print media but accessed on mobile. Mobile advertising provides the businesses with an opportunity to promote augment or reinforce their corporate as well as product brand in a cost-effective manner with great degree of personalization and customization (Denk and Hackl, 2004: pp. 460–470).

Mobile advertising has already become one of the fastest growing forms of advertising in the developed world. According to IAB's Internet Advertising Revenue Report 2010–2012, the total mobile ad spend in the United States by the mid-year 2012 was $1200 million representing a growth of 95% on a year-on-year basis. Mobile advertising has several distinguishing features that provide a significant edge over other media used for advertising. The most important features that it offers to the advertiser are that it is portable and is *always on* unlike other advertising media such as television or radio (Tsang et al., 2004: pp. 65–78). The chances of messages being lost due to the device being switched off are considerably less in comparison with other media. Mobile phones or devices are characteristically personal in nature and carry the user's identity making it possible to track the individual characteristics of the user. This means that customized marketing messages may be delivered to the existing or potential customers (Dholakia and Dholakia, 2004: pp. 1391–1396). This also provides an opportunity for promotion of taboo goods such as contraceptives, products related to personal hygiene, and so on in a culturally sensitive society like India (Terence et al., 2013: pp. 46–49). Modern tracking technologies have paved way for real-time tracking of the physical location of the user making user centric advertising and customized advertising messages an effective reality. For example, if a user uses her mobile's GPS service to search for a book store, then she may be targeted with advertisements for book stores and since she is searching for one, the advertisement may generate an actual sale. Another important attribute of the mobile advertisements is that a number of different mobile advertisements can be comfortably stored in the memory of the device and can be accessed by the user at an appropriate moment even after considerable lapse of time. For instance, a user may save a number of text messages pertaining to tour and travels services provided by different service providers and may use them while planning for her vacation (Kalakota and Robinson, 2002).

However, despite having a long list of advantages attached to it, mobile advertising does pose some challenges in front of the advertisers. The first and foremost is the difference in the quality of mobile phones being used by different customers. That is, depending on the purchasing capacity, a customer may purchase a basic feature phone, while another customer may have a gadget overflowing with apps. The factors such as different screen sizes, variations in the speed of the processors, and different connectivity plans render it difficult for the advertisers to produce universal advertisements suitable for different kinds of devices. The scope for creativity is delimited by comparatively smaller display areas than traditional advertising media such as newspapers and television. Also, for the same reason the information provided is also not as comprehensive as in other media. Nevertheless, for brand augmentation and reinforcement and customer retention, it is an excellent medium as a large number of messages may be sent to a large number of customers superseding the geographical boundaries, and a database of the loyal customers may be maintained and retained by businesses by continually providing the existing loyal customers pertinent information regarding the changes taking place in the realm of their preferred brands and various promotional schemes and offers (Yuan and Tsao, 2003: pp. 399–414).

13.3 *Indian market and industry dynamics*

According to a report jointly published by PricewaterhouseCoopers (PWC) and Confederation of Indian Industry (CII) in 2012, the mobile advertising market is at a significantly evolved stage in developed markets as compared to India. Countries such as the United States, the United Kingdom, Japan, and China are considerably ahead of India in mobile advertising. In fact, in many developed countries, the size of mobile advertising market is comparable to the television broadcast advertising market. This is due to high mobile Internet penetration and usage across almost all sections of the society.

Nearly all the key markets such as the United States, the United Kingdom, Japan, and China are expected to exhibit strong growth rates as consumers continue to spend significant amounts of their time online on mobile. As far as India is concerned, it has a fledgling market at the moment with a current spend at 33 million US$. But with smartphones increasingly becoming the norm and adoption of 3G and 4G expected to increase further, the mobile advertising market is expected to continue to grow at a rapid pace of 40% by the end of 2014. Despite having more than 120 million mobile Internet users in the country, a majority of advertisers are not yet exploring the medium at all bringing the overall digital spends in India at 5%–7% of overall advertising pie, which is pegged at 5 billion US$ to 5.1 US$ (PWC and CII, 2012).

Mobile advertising is an exciting, fast-growing category, but its share in the overall advertising pie is exiguous. Presently, it is growing at a rapid pace and is expected to grow faster than many of its peers over the next few years. The driving factors include increasing adoption and usage of mobile Internet from a low base and rising interest level of advertisers. India's mobile Internet advertising market is expected to grow even more rapidly as the global trend of increasing adoption of smartphones and high-speed mobile data access is being witnessed in India as well and this will lead advertisers to follow the target audiences on this platform.

The experts of mobile advertising often argue that one of the first challenges that mobile advertising faces in becoming a marketing medium is the absence of a widely accepted metric that can gauge how the medium has delivered for those who have invested in it and become a currency. Value-added services, promotions on mobile, new mobile applications, display, search, and SMS/MMS are the key drivers contributing to mobile advertising spends in India, and are expected to help boost the market from 50 million US$ to 75–80 million US$ by the end of 2013 (Mobile Advertising Spend Report for India, 2012).

Marking a notable growth of 12.6% in 2012, the Indian media and entertainment industry grew from US$ 11.90 billion in 2011 to US$ 13.42 billion in 2012. With an increase of 9%, the total advertising expenditure across media stood at US$ 5.36 billion in 2012. The print advertising leads with 46% of the total advertising pie at US$ 2.45 billion. Covering nearly 80% of advertising market in India, the conventional advertising media newspapers and television were the preferred media by the advertisers, even though Internet advertising showed substantial increase with a 7% share in the total advertising spend in 2012 (PWC and CII, 2012).

Out of the total online advertising spend in 2012, search advertising alone accounted for about 38%, that is, US$ 139.10 million while display advertising formed a considerable 29% (US$ 108.34). The market for advertisements delivered on mobile phones and tablets has grown to 10% amounting to spends of around US$ 37.64 million of the Indian online advertising market in FY 2012–2013 with a growth of 3% from the previous financial year 2011–2012. Other advertising formats like social media witnessed a 13% share in the online advertising market, while email advertising accounted for 3%. Video conferencing advertising witnessed a 7% share, respectively (Global Mobile Advertising Spends, 2012).

The reasons behind the mobile advertising market's growth in India are external to the advertising environment. Mobile handsets are increasingly becoming cheaper leading to greater accessibility for the masses. Another reason for growth is the introduction of new mobile data plans, which has allowed access to the Internet on-the-go at comparatively affordable rates starting as low as Rs. 10 per day. Mobile, as a consequence of the aforementioned factors, has now evolved from just being a communication device into a much more advanced medium capable of delivering an assortment of rich multimedia messages (Okazaki, 2005: pp. 160–180).

In addition to the external changes occurring on the consumer's side, the advertisers have also realized the utility of mobile as an effective advertising platform. Five years ago, although there was excitement about the medium, there were not as many avenues as there are now. The reasons at the consumer and the advertiser end coupled together have made mobile a lethal combination for both the advertisers and the consumers. With the introduction of 3G in 2012 and now 4G, the market grew even more significantly as the users are now being introduced to better services with better connectivity.

The mobile advertising market has evolved from a stage of communication to a stage of entertainment. Now it has become the preferable tool to get instant access to the content. The difference between India and other countries is that it is a mobile-first market. In India, for many people, the first Internet experience is on the mobile, making it the primary mode of content consumption. This is in contrast with other developed countries where the first Internet experience is on the computer (Yunos and Jerry, 2008).

The game plan for the mobile advertising companies in India is simple. They create money in two ways—one by creating the technology and products around the mobile advertising market and another by having partnership with publishers to place advertisements. Thus, the companies provide the advertisers and brands with an end-to-end holistic mobile ad campaign, by creating larger supply at one end and large demand at the other (Pura, 2002: pp. 62–71).

One of the largest sectors in mobile advertising is the automobile industry—with brands such as Maruti, Honda, and Yamaha engaging in mobile advertising. FMCG brands such as Unilever, PepsiCo, and Coca Cola, and telecommunication advertisers such as Nokia, Samsung, Karbonn, and Micromax, follow the automobile industry. The top five spending brands in India in mobile advertising industry in 2012 were Maruti Suzuki, Nokia, PepsiCo, Coca Cola, and Unilever (On Device and Decision Fuel, 2012).

13.4 *Key trends in mobile advertising in India*

Mobile advertising is slated for a phenomenal rise across the globe in the next few decades. India, a country with the second largest mobile phone subscriber base next only to China, and way ahead of most developed countries and inherent characteristics such as democracy, humongous consumer market, and an economy growing at a considerable pace, is going to be one of the most crucial and exciting playgrounds for mobile advertisers (Lee, 2013: pp. 231–247). India has certain unique features that distinguish her from other developing countries. It is a complex market with attributes of developed nations interspersed with those of developing and sometimes even of underdeveloped countries. An example of this phenomenon is that, according to a Reuters report, a number of India's dollar millionaires will cross the 4 million mark by 2015, while on the other hand, the number of people suffering from poverty in India is also very high and if the recent planning commission of India data is believed nearly 27% (approximately 320 million) of the population is living below the poverty line (Reuters, 2012). However, according to TRAI data, almost 75% (approximately 900 million) Indians have access to mobile phones. Even if the conservative estimates given by a report published on an Indian news website ibnlive.com quoting PTI are considered, the number of active mobile subscribers in India is almost 600 million. Out of these, approximately 23 million consumers access the Internet from their mobile phones. This is a big number considering that most countries do not even have this much population. The fact that Indian market is huge cannot be contested; however, it is still nascent and to fully explore the market and realize its true potential we need to understand the key trends that are going to shape the future of Indian mobile advertising.

Subscriber base—and more significantly active subscriber base—in India is going to increase rapidly throughout the next decade and is not going to witness a downward spiral in the near future. This is due to a combination of factors such as availability of cheaper feature mobile phones as well as smartphones, declining access costs, lack of alternate modes of communication, and the impetus provided by the central government as well as the state governments to increase the penetration of mobile subscribers in India. Mobile phones have the advantage of portability and affordability in comparison to personal computers and even laptops (Basu, 2000: pp. 83–99). Also, almost all the basic feature phones besides the smartphones are Internet ready and provide an opportunity to anyone with an investment capacity of as low as Rs. 2000 to buy a feature mobile phone and a monthly spending capacity of around Rs. 100 to access the world of information through the Internet. Smartphone pricing wars have commenced with companies like Apple, which used to focus more on product than affordability, coming up with cheaper products for mass markets like India. Apple's Ipad 5C is priced as low as $99. Smartphones from Indian companies like Micromax, Karbonn, etc., start at prices as low as Rs. 3000, and even multinationals like Samsung, LG, and HTC have considerably affordable smartphone range. These low prices coupled with comparatively cheaper accessibility costs translate into a lucrative market for mobile advertisers.

Telecom giants like Airtel, Vodafone, Reliance, and Tata have already invested thousands of crores of rupees to acquire 3G licences, and Airtel has even launched its 4G services in metros like Mumbai. They will have to monetize their investment by popularizing mobile broadband data plans and in a price-sensitive Indian market overflowing with cutthroat competition. This means that consumers will be the winner and have data plans to access the Internet that are within their reach. At present, according to a recent report published by an Indian news website quoting PTI, the number of active Internet users in India is nearly 144 million and only a very small portion to these users, that is, only approximately 23 million of them use the Internet through mobile. Given the affordability of

Internet-ready smartphones in comparison with the laptops or personal computers, the number is bound to increase and even smaller efforts on the part of the data plan providers may catapult the number into a different orbit altogether (Juxt, 2013).

A distinctive feature of the Indian Internet users is that the number of those who have accessed the Internet for the first time through mobile phones is greater than those who have done it through personal computers. This is due to triple factors of affordability, accessibility, and availability. India has the second largest mobile subscriber base next only to China, and this is going to increase further, as in order to bring about transparency in the administrative system, the Indian government is pushing for e-governance initiatives and innovations, and once its ambitious Adhaar Project is completed, most of the government services will be shifted to the online platform. This will make the Internet access a necessity instead of a luxury, and affordability of mobile Internet will be a big factor in deciding the medium through which the masses access the Internet. This will give a boost to the rural market. The advent and maturing of rural markets will force the development of content in regional languages.

Regional language content would provide an opportunity for the advertisers to reach to the masses. Area-specific targeting would also become possible as there is little possibility that an ad designed in Malayalam would be accessed by the Marathi speaking people. This along with the fact that the advertisement will be sent only to the mobile numbers belonging to consumers in Kerala state (whose native tongue is Malayalam) would minimize the wastage. Indian consumers are language sensitive and content in their language would be more emotionally appealing to them than in any other language.

Mobile-based e-commerce will also gain momentum in the coming decades. According to Master Card Mobile Payment Readiness Index (MPRI), India ranked 21st among 34 countries with a score of 31.5 on a scale of 100. This index also points out that only 14% of Indian consumers are familiar with P2P and m-commerce transactions and only a meager 10% are familiar with POS transactions. However, these percentages are bound to increase owing to growing mobile penetration, ease of transactions, increasing Internet literacy, and impetus provided by the government. The Finance Ministry of India has directed the Indian Public Sector Banks to focus on mobile and Internet banking to minimize cash transactions in order to tackle the problem of fake currency, minimize black market transactions, and bring about transparency in the financial transactions (MPRI, 2012). Indian e-commerce websites like flipkart.com, jabong.com, etc., are witnessing a growth in their revenues on a daily basis. Mobile recharges, DTH service recharges, and travel bookings are gaining popularity day by day and are among the fastest growing e-commerce transactions in India (ERNST and YOUNG, 2012).

Price sensitivity is also a key factor in the context of mobile advertising and the fact that even the slightest of variations in price may drive the consumers away means that value-added services and free applications will be the order of the day. A new trend will be the blurring of the boundaries between the mobile advertising and the mobile applications. For example, if an advertiser of automobiles wants to establish and augment its brand among the youngsters, that advertiser may device games that target the audience that the advertiser wants to focus upon. The advantage would be that if the game gets popularized then it will not be intrusive upon the privacy of the customer and this will lead to the creation of favorable brand image of the company in the minds of the existing as well as the prospective customers (Tao et al., 2013: pp. 2536–2544).

Privacy is a big concern for the consumers, and advertising messages that infringe upon their privacy may drive them away and may even antagonize them toward the brand. It will be incumbent upon the mobile service providers as well as the advertisers to respect the privacy of the consumers and take prior consent of the consumers before advertising to them (Hyunsook, 2012). In India, the TRAI has already formed strict rules regarding unwanted messages, and companies neglecting them may entail heavy fines and may have to pay damages to the customers. This may be disastrous for the company as not only will it incur financial losses but its reputation will also suffer greatly. One way to address the privacy concern of the consumers is to follow companies like Life Insurance Corporation of India, which asks prior permission of the consumers to contact them. This not only generates goodwill among the consumers, but also establishes the brand as someone who shares the concerns of the consumers regarding privacy (Barwise and Strong, 2002: pp. 14–24).

These are the key trends that are going to shape the future of mobile advertising in India.

13.5 *Mobile advertising strategies*

India is a tremendous market for mobile advertising. One of the key reasons for the mobile advertising wave in India is that almost 87% of Internet users access the Internet through mobile. The mobile Internet user base in India is increasing day-by-day and will fast approach the 200 million mark by the end of 2016 (Page, 2013). Around 40% of all mobile Internet users in India have mobile as their only Internet access point. Their first Internet experience is on mobile as the mobile Internet accounts for 59.36% of the total Internet usage in India, as of December 2012, while Desktop Internet accounted for 40.64% of the total Internet usage in India. This revolution is being driven by the proliferation of mobile applications that are driving the mobile ecosystem, similar to the manner in which websites powered desktop Internet (KPCB, 2012).

The market trends that drive mobile advertising in India concern with smart mobile advertising. Advertisers are devoting more time to evaluate the mobile usage practices and psychographics of audiences to produce nifty advertising campaigns that could effortlessly and effectively establish a bond with the customers, irrespective of the mobile platform the consumer use rather than focusing only on smartphones, with an increase in the demand of mobile applications coming from emerging markets like India and China with a 36% market share in terms of downloads in 2010. It is the second largest market share in the whooping 6.8 billion US$ market (World Mobile Application Market Report 2010–2015, 2012). Thus, developers are initiating strategies to generate revenues from their applications' user base by integrating them with telecommunication billing for microtransactions and are candidly using potent mobile advertising solutions.

Consumer's mobile lifestyle is allowing the brands to converge advertising, distribution, and transactions over multiple delivery platforms (Grunewald, 2013). Today, brands have the opportunity to leverage the mobile for all the 4Ps of marketing. Combined with the microtransaction capability of mobile, mobile coupons (m-coupons) are already altering the pricing paradigm with hyper-segmented offers. m-coupon is a mobile solution that enables customers to receive, save, and redeem coupons on their mobile phones. The m-coupon applications are mostly integrated with other mobile applications and social media websites like foursquare to extract, save, and redeem coupons at the point of sale, anywhere and anytime. m-coupons is a loyalty solution that helps to reach out to a wide consumer base by reducing the delivery cost of coupons and improve redemption efficiency. m-coupons can be quickly accessed and can enhance new revenue flow through brand marketing and promotion. This is blurring the lines between advertising, distribution, and transaction.

The mobile advertising strategy most preferably in practice these days is to deliver personal messages based on the consumer's local time, location, and preferences. Location independence and ubiquity are the two key arguments for mobile advertising. Properly applied, location-based advertising services can create or reinforce virtual communities. A personalized SMS campaign relies upon databases with enough active and potential clients to reach the target group profitably. A database containing information about the clients such as leisure activities, holidays, music and media interests, type of Internet access, occupation, marital status, car ownership, and income will help the advertisers to engage clients in an interesting way (Zanton, 1981: pp. 141–146).

While strategizing mobile advertising campaigns, advertisers have to take care of consumer control, permission, and privacy. There is a hitch that the mobile operators have lot of personal information about their clients that would be of great interest to advertisers, but privacy laws may prevent them from sharing it. There is a further trade-off between personalization and consumer control. Gathering data required for tailoring messages raises privacy concerns. The corporate mobile advertising strategies and policies should consider legalities such as electronic signatures, electronic contracts, and conditions for sending SMS/MMS. According to the experts of mobile advertising, advertisers should take permission and should convince the consumers to *opt-in* before sending advertisements. A simple registration ensures sending relevant messages to an interested audience. The advertisers, mobile operators, and middlemen have not agreed on a common format for the sharing of this information, neither they have worked out how to share the revenue it might yield.

Unsolicited advertising messages, commonly known as spam, stifle user acceptance, particularly as mobile phones cannot distinguish between spam and genuine communication automatically. An important problem with mobile advertising is that consumers are used to ads on television and radio but they consider their mobile phone a more personal device. A flood of advertising might offend its audience, and thus undermine its own value. Tolerance of advertising also differs from one market to another (De Moij, 1998). In the Middle East, unsolicited text messages are quite common, and do not prompt many complaints, but subscribers might not prove to be so open-minded in Europe or America.

The cultural and personal mindset of the consumer is also an area of concern to strategize mobile advertising. Consumers in India, especially the rural consumers, are still hesitant in using mobile banking, mobile payments, and m-commerce. A positive shift is required to make them comfortable with mobile-based e-transactions. For this, a consumer-focused solution is required. The transactions should be maintained with the highest levels of performance and security. The payment gateways should be incorporated with specially designed functions enabling various parties involved in the issuance of mobile applications to mobile phones to securely provision those applications. Payment options like contactless payment card applications using NFC, peer-to-peer payments applications should be designed in such a way that they are compliant with emerging industry mandates and standards of due care in order to infuse a sense of security in the minds of the consumers.

Mobile advertising should be designed innovatively with an exceptional use of audio and video to make it as a part of a rich media campaign. World Bank data reveal that in middle- to low-income economies, there are as many as three times the numbers of mobile subscribers per 100 people as those with Internet access (ICT Facts and Figures by ITU & World Bank, 2012, The World in 2013). Moreover, studies conducted by two digital marketing research firms Insight Express and Dynamic Logic in 2012 also showed that mobile campaigns were found to be four to five times more effective than PC Internet campaigns against standard brand metrics such as unaided awareness, message association, and even purchase intent (Insight Express and Dynamic Logic, 2012).

An effective mobile advertising strategy for reaching consumers in developing markets like India needs to target smartphones and tablets with engaging rich media content. It further requires a fine balance between rich media content and the available bandwidth so that the advertisements would load fast. In the same way, to reach the masses, it is important to engage the consumers on mobile sites, but advertisers can engage the consumers more actively only when the ads will load quickly on sites. The important point to be noted here is that without better connectivity at affordable rates mobile advertising will not be able to realize its true potential in India.

The key to any mobile advertising strategy is to make sure that the advertising units are designed for mobile users rather than designing it for some other media like television or print and then adopting it for mobile platforms. The strategies should be well supported by innovative ways to target the audience, and care should also be taken to constantly analyze and optimize the advertising campaigns delivered on mobile. Refreshing the contents frequently will also help the advertisers to combat ad fatigue (Bansal et al., 2009).

A recent consumer research study conducted by InMobi, a mobile customer engagement platform, revealed that mobile web users in key Asian markets such as Singapore, Indonesia, and Malaysia are increasingly becoming more comfortable with accessing the Internet using their mobile gadgets over desktop computers or laptops. This growth is driven in large part by the use of social media on mobile. On an average, 21% of mobile web usage time of consumers in these key Asian markets is spent on social media. Social media usage is also the only mobile activity that over 50% of mobile web users across these key markets expect to increase their time spend on over the next 12 months. Interesting point to note here is that social media consumption is largely categorized as entertainment and is therefore associated with a more relaxed and receptive consumer mindset. Users are, therefore, more open to explore content that is relevant and attractive to them (InMobi, 2012).

Thus, exploring and exploiting social media and its vital elements can give a mobile advertising campaign an edge. Using Facebook or Twitter buttons to share great promotions and interesting sites adds tremendous value by reaching consumers at a time when they are most receptive to such content. Moreover, if a user likes the campaign, there is a further scope that he or she will recommend it to their friends who might in turn pass it on to their contacts (Tuten, 2008).

Tracking the advertising campaign is also very important point to consider while designing the strategies for mobile advertising. With the advances taking place in mobile communications, mobile advertising campaigns can also be tracked, measured, and optimized (Martin, 2011). The key effort should be to clearly define the objectives of the advertising campaign and then to identify relevant metrics to measure that how far these objectives have been achieved after delivering the campaign. The metrics will also be useful to analyze performance or awareness-related clicks, lead generation, and applications downloads or ad impressions, to track the time spent watching videos and other engagement measurements. A strategy for mobile advertising also requires a clear and strong campaign message and well-designed mobile platform-oriented creatives to ensure that consumers are engaged well with the campaign (Feldmann, 2005).

Various mobile advertising firms are developing new standard measurement guidelines to bring mobile first on the mindset for the advertisers and to figure out how effective a mobile advertising campaign could be for the consumers. With 3G and 3.5G facilities, advertisers can use video, animation, photo galleries, and interactive elements, which can make mobile advertising more akin to a television commercial or a slick magazine. Opera software, the mobile browser company, reported in July 2012 that users who clicked on a rich media ad spent an average of 52 s viewing a video and 1 min and 25 s interacting with photos, respectively (State of Mobile Web, 2012). Opera report also points out that advertisers have started using rich media and video advertisements more frequently in 2012 than the traditional banner advertisements.

Another promising part of mobile advertising is location-based product and service information delivered via geo-fencing or proximity. These advertisements mostly catch a person at a time when he or she wants to act and do not factor in a person's preferences. The advertising strategies are now designed mostly in and around location-based and search advertisements (Trimponias et al., 2013). Advertising based on native content is also gaining popularity among the advertisers. The advertisers are now taking more interest in creating messages and content that work within the flow of their delivery platforms. By using the social networking platforms like Facebook and Twitter, advertisers are linking their campaigns with the existing units of content, like a tweet or update. This is an organic and unique way to advertise through mobile that is harder to ignore.

Facebook in its financial statement for the year 2012–2013 said that it gets 14% of all of its revenue via mobile-sponsored stories and install advertisements, which appears right in the news stream of its mobile applications and website. E-Marketer, a digital media firm offering insights essential to navigate the changing digital environment across the globe in its report published in 2013, says that Twitter has made US$ 129.7 million in mobile advertising in the year 2012 (The Global Media Intelligence Report: Asia-Pacific, 2013).

The experts of mobile advertising argue that mobile advertising strategy does not mean only placing the advertisements in an application or website appropriately but also to place advertisements alongside creative activities inside the applications and websites. Advertisers like Pontiflex and Tapjoy help people to earn in-application rewards for watching videos, installing applications, or subscribing to services. Advertisers are creating lock screens for Android devices that can be branded and potentially carry advertising. Smartphone users collectively swipe to unlock their Android phones millions of times each day. Lock screen advertising promises to make the act of unlocking more rewarding by paying users to look at an ad before getting to their phone. When the phone is turned on, a user can swipe left to see a photo ad or watch a movie trailer. The user can swipe right if he or she has no interest in seeing the advertisements.

These are some effective mobile advertising strategies for reaching consumers in emerging markets like India.

13.6 *Cost-effective connectivity solutions for mobile Internet*

A population in excess of 3.7 billion coupled with highest Internet user base in the world makes Asia arguably the most lucrative destination for online advertisers including mobile advertisers. According to Internet world statistics, Asia had more than 650 million broadband users in 2012, a number that has grown 469% since 2000. Major driver for this exponential growth is increasing the availability of low cost, low bandwidth wireless solutions, especially in business sector. According to latest TRAI Report, August 2013, there were approximately 900 million mobile subscribers in India. That is, almost 75% of Indian populace has access to mobile phones. However, only about 145 million Indians use mobile to access the Internet; the number in itself is huge, but when put besides the total mobile subscriber base in the country, it is miniscule. For mobile advertising to become a full-fledged industry in India, it is imperative to popularize mobile Internet among users. This is not an easy task in a price-sensitive market like India, and therefore, low-cost connectivity solutions will have to be developed and made accessible to the users.

There is little doubt that the Internet Age has transformed the way the world does the business. The e-Commerce market in India grew to US$ 9.5 billion in 2012 and is expected to reach US$ 12.6 billion—a 34% y-o-y growth since 2009 (IAMAI, 2012). These data give perspective to the potential in the Indian mobile advertising industry. The opportunities and tangible benefits it accords are now being quantified. Recent studies conducted by World Bank revealed that each 10% increase in broadband penetration would result in 1.21% increase in per capita GDP growth in developed countries and 1.38% increase in developing countries. This suggests that the governments world over will also have to look at providing low-cost connectivity solutions to their citizens and online marketing and advertising industry will have to work in tandem with governments to arrive at affordable connectivity solutions.

Asia is at the forefront of broadband innovation, and India is also witnessing a notable broadband penetration in both urban and rural areas. New wireless technology offered by telecom operators has brought the power of Internet to an increasing number of enterprises and government agencies looking to implement cost-effective connectivity solutions. These technologies have brought next generation broadband network to the rural areas of developing countries that would not be served otherwise (Ahonen, 2008). The low-cost wireless broadband solutions provide an excellent way to expand network connections to rural underserved areas of India, thereby boosting the mobile advertising to a large extent.

Technology like Worldwide Interoperability for Microwave Access (WiMAX) is now expanding very quickly in India and is providing mature 3G and 4G mobile technology to the people at affordable cost. A single WiMAX platform can support fixed, portable, and mobile broadband services in scalable channel bandwidths from 5 to 20 MHz and can support security features through mutual device/user authentication, flexible key management protocols, strong traffic encryption, and security protocol optimization for fast handovers (The State of Broadband, 2012). The deployment cost for WiMAX per megabit is less than other cellular technologies. This is a key to dense urban, suburban, rural, and remote area deployments. The technology also enables higher network capacity to support new value-added mobile advertising services for increased operator revenues.

Because of these broadband revolutions, emerging economies like India are going wireless—with a higher speed and wider range. This is fueling the convergence and is transforming the advertising and communication industry. Wireless communication technologies like Wi-Fi, WiMAX, 3G, 4G, and ultra-wideband are working well for all kind of platforms such as PC, laptops, mobile, and tablets. These technologies are well optimized for various mobile advertising applications, thus taking the advertising solutions and services to the last mile. All these new wireless communication technologies are providing various connectivity solutions to the people of both urban and rural India, thus allowing them to access advertising and promotional campaigns whenever and wherever they want (Online Advertising Performance Outlook, 2012).

13.7 Conclusion

The *mobile advertising* phenomenon is here to stay and for good. The growing importance of wireless communication technologies in a developing country like India cannot be overstated. With nearly 900 million mobile subscribers, an Internet user base of approximately 144 million, and about 23 million mobile Internet users, mobile phones are slated to become the biggest personal, group as well as mass communication medium in a not very distant future. Portability and accessibility coupled with reducing hardware costs, cheaper data, and voice access plans have provided mobile phones an edge over other communication media. Easy accessibility and affordability have enabled the mobile phones to penetrate even the farthest corners of India rendering it a near ubiquitous medium. This makes them the medium of choice for the advertisers who want to reach not only to the urban audiences but rural audiences as well. With the ever-increasing pace of life, most of the working Indian populace does not get sufficient spare time to be able to access traditional media such as television and radio. Mobile phones, however, have become almost indispensable for this portion of the populace and these are the people who have the power to purchase. Therefore, mobile becomes the best medium for the advertisers to reach to such audiences.

Mobile phones in the last decade have evolved from being just a communication device to an advanced tool used for multifarious purposes such as accessing the Internet, making monetary e-transactions, reading newspapers, social networking, and so on. Technological advancements in the field of mobile and computing hardware have led to the evolution of high-end smartphones that are as capable as personal computers and much more convenient to use. Also, these technological advancements coupled with increasing competition among the mobile hardware manufacturing companies have ensured that now smartphones and feature phones are now available at extremely low costs making them acceptable in a highly price-sensitive market like India. Another important factor that is acting as a positive catalyst in the growth of mobile phones in India is the availability of low-cost data and voice access plans. Cheaper data plans supported by inexpensive smartphones and Internet ready feature phones have acted as a positive catalyst in rise in the phenomenon of mobile phones. The number of Internet users in India is increasing day by day, and an interesting fact associated with Indian Internet users is that most of them have experienced Internet first through mobile devices than through personal computers or laptops.

The aforementioned features make mobile devices an appropriate platform for advertisers. Mobile advertising is a phenomenon that is bound to give a hard time to other advertising media such as television, radio, newspapers, and so on. Mobile advertising in simplest terms is any advertisement that is delivered to audiences through wireless mobile devices such as mobile phones, feature phones, smartphones, or any other such device. Mobile advertising offers a plethora of advantages over traditional advertising such as personalization, universality, better audience tracking mechanisms, improved feedback mechanisms, audience segmentation based on factors such as geographical location, data usage, online purchase preferences, and so on, and a relatively nonintrusive way to reach to the desired target audiences. These advantages of mobile advertising translate into minimization of the proverbial *advertisers waste*, that is, the expenditure on advertising that goes down the drain as the messages do not reach to the desired target audiences. Also, mobile advertising is cheaper in comparison to traditional advertising and provides much better tractability of the advertising messages delivered through it. Despite these tangible advantages, mobile advertising is still at its nascent stage and will have to deal with a number of issues in order to be able to realize its true potential in the Indian market.

One of the major issues involved in mobile advertising is of privacy. Mobile devices owing to their very nature are personal. Any advertising message or mechanism that breaches the privacy of the users may get negative response from the audiences. Issues like geographical location tracking are quite complex and need to be handled sensitively by the advertisers. It would be expedient from the advertisers' perspective to take prior consent from the users before delivering them advertising messages (Hyunsock, 2012). They will have to develop mechanisms to make mobile advertising less intrusive and more rewarding phenomenon. Another issue that the advertisers will have to deal with is the development of innovative advertising ideas in a restrictive environment (small screen size, huge variations in screen sizes, different connectivity speeds, etc.) of mobile devices. They will also have to deal with connectivity issues as lower connectivity speed would mean that they will not be able to deliver rich media messages and would have to contend with less appealing text-based advertising messages. These are the issues that need to be resolved as soon as possible in order to give a boost to the mobile advertising industry in India.

There appears to be a very exciting future for the mobile advertisers in India. The opportunities in front of the mobile advertisers are endless, and there are a few threats as well. Resolution of threats such as issues related to privacy, ethical issues such as geographical location tracking, lower connectivity speeds, and so on would lead to proper utilization of unique opportunities such as better audience segmentation, quicker feedback, and access to the most distant markets within India. A proper understanding and handling of the aforementioned issues would ensure that mobile advertising finds its roots in India and move ahead nimbly to reach its true destination.

References

Ahonen, T. (2008). *Mobile as 7th of the Mass Media: Cellphone, Cameraphone, iPhone, Smartphone*. London, U.K.: Futuretext.

Bansal, A., Phatak, Y., Gupta, I.C., and Jain, R. (Eds.) (2009). *Transcending Horizons through Innovative Global Practices*. New Delhi, India: Excel Book.

Barnes, S.J. (2002). Wireless digital advertising: Nature and implications. *International Journal of Advertising*, 21(3), 399–419.

Barwise, P. and Strong, P. (2002). Permission based mobile advertising. *Journal of Interactive Marketing*, 16(1), 14–24.

Basu, S. (2000). Advertising trend in India. *Scandinavian Journal of Management*, 3(4), 83–99.

Broadband for India. (2012). A report by associated chambers of commerce and industry of India (ASSOCHAM) released in December, 2012. Available at http://www.ccaoi.in/UI/links/fwresearch/BROADBAND%20FOR%20INDIA.pdf. Accessed on October 1, 2013.

Bulander, R., Decker, M., Schiefer, G., and Kolmel, B. (2005). Comparison of different approaches for mobile advertising. *Proceedings of the Second IEEE International Workshop on Mobile Commerce and Services*, July 19, 2005, Munich, Germany.

Clay, C. and Zainulbhai, A. (2013). *Reimagining India: Unlocking the Potential of Asia's Next Superpower*. New York: Simon & Schuster, pp. 84–87.

De Moij, M. (1998). *Global Marketing and Advertising: Understanding Cultural Paradoxes*. New York: Sage, pp. 39–42.

Denk, M. and Hackl, M. (2004). Where does mobile business go? *International Journal of Electronic Business*, 2(5), 460–470.

Dholakia, R.R. and Dholakia, N. (2004). Mobility and markets: Emerging outlines of m-commerce. *Journal of Business Research*, 57, 1391–1396.

Ernst and Young Report. (2012). Rebirth of e-Commerce in India. Available at http://www.ey.com/Publication/vwLUAssets/Rebirth of e-Commerce in India/$File/EY RE-BIRTH OF ECOMMERCE.pdf. Accessed on September 17, 2013.

Feldmann, V. (2005). *Leveraging Mobile Media: Cross-Media Strategy and Innovation Policy for Mobile Media Communication*. Heidelberg, Germany: Springer Physica-Verlag.

Gartner Report. (2012). Forecast: Mobile advertising, worldwide, 2009–2016. Available at http://www.gartner.com/resId = 2247015. Accessed on September 20, 2013.

Global Advertising Spend Report. (2012). Report published by eMarketer. Available at http://www.emarketer.com/corporate/reports. Accessed on September 21, 2013.

Grunewald, A. (2013). How effective are mobile ads at driving traffic. Available at http://www.google.co.in/think/articles/mobile-ads-experiment.html. Accessed on September 13, 2013.

Hackley, C. (2010). *Advertising and Promotion: An Integrated Marketing Communication Approach*. New York: Sage, pp. 29–33.

Harte, L. (2008). *Introduction to Mobile Advertising*. New York: Althos Publications, pp. 43–49.

Hyunsook, K. (2012). *Privacy Protection in Mobile Advertising*. Baltimore, MD: University of Maryland Press.

Indian Media and Entertainment Industry Report. (2012). Digital dawn—The metamorphosis begins. A report published by Federation of Indian Chambers of Commerce and Industry (FICCI) and Klynveld Peat Marwick Goerdele (KPMG). Available at https://www.in.kpmg.com/securedata/ficci/Reports/FICCI-PMG_Report_2012.pdf. Accessed on September 17, 2013.

InMobi's Consumer Research Study on Media Consumption. (2012). A study conducted by decision fuel and on device research. Available at http://www.inmobi.com/hstar/consumer-research/. Accessed on September 25, 2013.

Insight Express and Dynamic Logic Report. (2012). Available at http://www.iab.net/insights research/1672/1360. Accessed on September 22, 2013.

Internet Advertising Bureau. (2013). Internet advertising revenue report 2010–2012 and internet advertising outlook report 2013. Available at http://www.iab.net/media/file/IAB Internet Advertising Revenue Report FY 2012.pdf. Accessed on September 29, 2013.

Internet and Mobile Association of India Report. (2012). Available at http://www.iamai.org.in. Accessed on September 25, 2013.

Juxt Report. (2013). Sourced from Press Trust of India (PTI) and posted on ibnlive.in.com on September 8, 2013. Available at http://www.ibnlive.in.com/news/india-has-5548-crore-mobile-owners-1432-crore-internet-users/420444-11.html. Accessed on September 20, 2013.

Kalakota, R. and Robinson, M. (2002). *M-Business: The Race to Mobility*. New York: McGraw-Hill.

Kleiner Perkins Caufield and Byers (KBCB) Report. 2012. Internet year-end update 2012. Available at http://www.medianama.com/2012/12/223-india-has-137-million-internet-users-44-million-smartphone-subscribers-report/. Accessed on September 20, 2013.

Lawrence, H. (2008). *Introduction to Mobile Advertising, How to Setup, Create and Manage Ads for Mobile Telephones*. New York: Althos Publishing.

Lee, In. (2013). *Mobile Applications and Knowledge Advancements in E-Business*. Dallas TX: IGI Global, pp. 231–247.

Leppaniemi, M. and Karjaluoto, H. (2005). Factors influencing consumers' willingness to accept mobile advertising: A conceptual model. *International Journal of Mobile Communication*, 3(3), 197–213.

Martin, C. (2011). *The Third Screen: Marketing to Your Customers in a World Gone Mobile*. London, U.K.: Nicholas Brealey Publishing.

Master Card Mobile Payments Readiness Index (MPRI) Report. (2012). Available at http://www.nextbigwhat.com/is-india-ready-for-mobile-payment-297/. Accessed on September 19, 2013.

Mobile Advertising Spend Report for India. (2012). A study conducted by mobile marketing association and exchange4media. Available at http://www.mmaglobal.com/news/mma-reports-mobile-ad-spend-india. Accessed on September 28, 2013.

Noonam, M.K. (2001). Wireless advertising: Where are we today? Available at http://www.waaglobal.org.WAAWirelessAdvPres.ppt. Accessed on September 19, 2013.

Okazaki, S. (2005). Mobile advertising adoption by multinationals. *Internet Research*, 15 15(2), 160–180.

Online Advertising Performance 2012 Outlook. (2012). A study by vizu, a Nielsen company in collaboration with CMO council. Available at http://brand-lift.vizu.com/knowledge-resources/research/2012-industry-outlook/. Accessed on September 21, 2013.

Page, M. (2013). The mobile economy 2013 report. A.T. KeArney. Available at http://www.atkearney.com/documents/10192/760890/The_Mobile_Economy_2013.pdf/6ac11770-5a26-4fef-80bd-870ab83222f0. Accessed on September 22, 2013.

Park, J.S. (2005). Towards a model of opportunity recognition and development. In W. During, R. Oakey, and S. Kausar (Eds.), *New Technology Based Firms in the New Millennium* (pp. 63–80). Oxford, U.K.: Elsevier.

Price Waterhouse Coopers and Confederation on Indian Industry Report. (2012). Indian entertainment and media outlook 2012. Available at http://www.cii.in/WebCMS/Upload/em%2Oversion%202low%2Ores.PDF http://www.pwc.com/india. Accessed on September 20, 2013.

Pura, M. (2002). The role of mobile advertising in building a brand. In B.E. Mennecke and T.J. Stader (Eds.), *Mobile Commerce: Technology, Theory and Applications* (pp. 62–71). Pennsylvania, PA: Idea Group Publishing.

Reuters Report. (June 19, 2012). Number of India's dollar millionaires will cross the 4 million mark by 2015. Available at http://in.reuters.com/article/2012/06/19/wealth-report-capgemini-idINDEE85I0AJ20120619. Accessed on August 14, 2013.

Sharma, C., Herzog, J., and Melfi, V. (2008). *Mobile Advertising: Supercharge Your Brand in the Exploding Wireless Market*. Hoboken, NJ: Wiley.

State of Mobile Web. (2012). A report by opera software. Available at http://media.opera.com/media/smw/2012/smw072012.pdf. Accessed on September 14, 2013.

Tahtinen, J. (2006). Mobile advertising or mobile marketing: A need for a new concept. In *Frontiers of e-Business Research 2005, Conference Proceedings of eBRF 2005*, pp. 152–164. Available at http://www.academia.edu/2490132/Mobile_Advertising_or_Mobile_Marketing._A_Need_for_a_New_Concept. Accessed on September 15, 2013.

Tao, G., Rohm, A.J., Sultan, F., and Pagani, M. (2013). Consumers un-tethered: A three-market empirical study of consumers' mobile marketing acceptance. *Journal of Business Research*, 66(12), 2536–2544.

Telecom Regulatory Authority of India. (2013). Report on Indian telecom services performance indicators, January-March, 2013 released on August 2013. Available at http://www.trai.gov.in. Accessed on August 21, 2013.

Terence, A., Shimp, J., and Andrews, C. (2013). *Advertising Promotion and Other Aspects of Integrated Marketing Communications*. Mason, OH: South-Western Cengage Learning, pp. 46–49.

The Global Media Intelligence Report: Asia-Pacific. (2013). A Report published by eMarketer. Available at http://www.emarketer.com/corporate/reports. Accessed on September 20, 2013.

The State of Broadband—2012 Achieving Digital Inclusion for All. (September 2012). A Report by the Broadband Commission and published by International Telecommunication Union (ITU) and UNESCO. Available at http://www.broadbandcommission.org/documents/bb-annualreport2012.pdf. Accessed on September 26, 2013.

The World in 2013, ICT Facts and Figures. (2012). A Report by International Telecommunication Union and World Bank. Available at http://www.itu.int/en/Pages/default.aspx http://www.gmanetwork.com/news/story/329766/scitech/technology/5-1-billion-mobile-phone-users-by-2017-mostly-in-asia-report. Accessed on September 11, 2013.

Trimponias, G., Bartolini, I., and Papadias, D. (August 21–23, 2013). Location-based sponsored search advertising. In M.A. Sellis, T. Cheng, R. Sander, J. Zheng, Y. Kriegel, H.P. Renz, and C. Sengstock (Eds.), *Advances in Spatial and Temporal Advertising. Proceedings of 13th International Symposium, SSTD 2013*, Munich, Germany.

Tsang, M., Ho, S.C., and Liang, T.P. (2004). Consumer attitudes towards mobile advertising: An empirical study. *International Journal of Electronic Commerce*, 8(3), 65–78.

Tuten, L.T. (2008). *Advertising 2.0: Social Media Marketing in a Web 2.0 World*. Oxford, U.K.: Greenwood Publishing Group.

World Mobile Application Market Report 2010–2015. (2012). Published by Markets and Markets. Available at http://www.prweb.com/releases/mobile-applications/market/prweb11197451.htm. Accessed on August 29, 2013.

Yuan, S.T. and Tsao, Y.W. (2003). A recommendation mechanism for contextualized mobile advertising. *Expert Systems with Applications*, 24, 399–414.

Yunos, H.M. and Jerry, G. (2008). Wireless advertising. Available at http://www.engr.sjsu.edu/gaojerry/report/wireless-add-paper2.pdf. Accessed on September 27, 2013.

Zanton, E. (1981). Public attitudes towards advertising. In K. Hunt (Ed.), *Advertising in New Age* (pp. 141–146). Provo, UT: Brigham Young University Press.

chapter fourteen

e-CRM, m-CRM, and ICTs adoption in the e-tourism and m-tourism industries

Te Fu Chen

Contents

14.1 *Introduction*

14.1.1 *Backgrounds and motivations of the study*

Customer relationship management (CRM) indicates a comprehensive strategy and an interactive process intended to achieve an optimum balance between corporate investment and the satisfaction of customer needs to generate the maximum profit. Electronic customer relationship management (e-CRM) and mobile customer relationship management (m-CRM) refer to CRM using the Internet and wireless technology plus a database, OLAP, data warehouse, data mining, etc. In order to gain an understanding of the efficiency of implementing an e-CRM and m-CRM system within the business context, there is a need to develop theoretically and empirically an evaluation process for the e-CRM and m-CRM system and survey its impact on service quality.

e-CRM can boost customers' satisfaction and patronage in the tourism and travel industry, as Internet business models have empowered customers with a great amount of information, which, in turn, makes them more price-sensitive, less brand loyal, and more sophisticated. e-CRM is one of the primary strategic initiatives in the industry today with coexisting criticism and overenthusiasm. There are a rapidly increasing number of reports on e-CRM implementation failure and success; these reports, however, are all in piecemeal form.

In recession, being able to create accountable marketing campaigns is crucial for commercial survival—and personal advancement. e-CRM has become a critical element of any marketing strategy, providing the means to generate millions in new revenues. e-CRM strategy is about developing customer relationships through meaningful communication that is targeted and relevant, ultimately building business and improving the bottom line. Both the existing academic literature and practical applications of CRM strategies do not provide a clear indication of specifically what constitutes e-CRM implementation. This research will argue that the reformulation of e-CRM problems within this new framework can result in more powerful analytical approaches.

Evolving from a simple face-to-face handshake, e-CRM has become a science, often requiring large expenditures in manpower and technology. In a people-oriented service industry, e-CRM has become a necessity for any travel and hotel company. More often than not, the results of e-CRM program implementations have been disappointing, according to a published study by IBM Business Consulting Services, called "Doing e-CRM right: What it takes to be successful with CRM." The study finds that just 15% of the companies it surveyed—both small and large—felt fully successful with their CRM programs. Only 15%! In reality, e-CRM means different things to different people especially in the travel and travel verticals. From a working definition, CRM and its online application, e-CRM is a business strategy aiming to engage the customer in a mutually beneficial relationship. Within this context here is the best description that describes e-CRM that is universal for any travel supplier or intermediary: e-CRM in the context of Internet distribution and marketing in the travel and travel verticals is a business strategy supported by web technologies, allowing travel services suppliers to engage customers in strong, personalized, and mutually beneficial interactive relationships, increase conversions, and sell more efficiently.

With the rapid development of information and Internet technology, the industries of tourism, hotel, and entertainment are constantly introducing the content of experience economy (Cooper, 2003). The content of travel channel of Travelocity, Expedia, Yahoo Online, and other websites becomes much richer. A vast virtual travel market based on e-commerce and online travel purchase is promoting the industries of tourism and travel in the United States with an increase of economic scale from 180 billion US dollars in 2002 to 64 billion US dollars in 2007 (Law and Cheung, 2005).

e-CRM and m-CRM can be defined as the translation of existing techniques for finding customers in the electronic environment. It provides products and services customized to the needs of the customers. It helps to retain customer's loyalty and attend the needs for information and support in the use of the tourism products. Many e-CRM and m-CRM techniques are already employed by businesses using nonelectronic methods. Intelligent agent technologies, the linking of call centers to websites, and the use of data warehousing techniques to perform detailed analysis of customer needs are among the new opportunities offered by the Internet and other advanced ICTs. e-CRM and m-CRM are not just customer service, self-service web applications, sales force automation tools, or the analysis of customers' purchasing behaviors on the Internet and wireless world. e-CRM and m-CRM are all of these initiatives working together to enable an organization to more effectively respond to its customers' needs and to market to them on a one-to-one basis.

Tourism can be viewed as very different from most other sectors of e-commerce and m-commerce as the consumer goes and collects the product at the point of production, which is the destination. Consequently, the tourism avoids the need to deliver products around the world. The factors described earlier result in the taking of a larger and larger share of e-commerce and m-commerce globally. As a result, the Internet and wireless technology can be considered as the last revolution in the distribution of tourism information and sales. Internet and m-commerce is even becoming the primary channel for business-to-business communication. It offers the suppliers the potential to by-pass intermediaries in the value chain and thus increases their revenue base.

The travel and tourism sector is one of the fastest-growing sectors in the world. In order to become more efficient and effective in delivering products and services to customers via the use of ICT, travel and tourism organizations have to rethink the ways in which they build relationships with their customers by initiating e-CRM and m-CRM projects. Inappropriate e-CRM and m-CRM decision making and implementation can result in multimillion dollar losses, which can translate into a loss of competitiveness. Therefore, the study conducts to (1) identify critical success factors (CSFs) involved in e-CRM and m-CRM initiatives in general and (2) identify and examine key issues in the implementation of e-CRM and m-CRM in the travel and tourism sector.

e-Tourism and m-tourism offer the potential to make information and booking facilities available to large numbers of consumers at a relatively lower cost. It enables the tourism sector to make large-scale savings on the production and distribution of print and other traditional activities such as call centers and information centers. It also provides a tool for communication and relationship development with the end-consumers as well as tourism suppliers and market intermediaries.

14.1.2 Objectives of the study

This chapter deals with the issues related to the e-CRM, m-CRM, and ICTs adoption in the e-tourism and m-tourism industry. This study aims to systematically explore the current status of e-CRM and m-CRM implementation status in the tourism and travel industry and the CSFs that affect the e-CRM and m-CRM survivability, that is, continued use of an e-CRM and m-CRM system reaching intended profitability or growth. Hence, this study is to investigate the tourism and travel industry organizations with regard to the implementation of e-CRM and m-CRM. The study will seek to bridge the gap between the theories, e-CRM, m-CRM, and its application within the tourism and travel industry. In order to achieve that goal, the study will examine the benefits of e-CRM and m-CRM, its potential and the barriers identified, find out CSFs of e-CRM and m-CRM strategies, and construct an integrated e-CRM and m-CRM model through literature review and relative theories. Finally, the study will examine the model via some case studies.

14.2 Literature review

14.2.1 Adoption of e-CRM solutions

e-CRM (Electronic Customer Relationship Management) can be defined as the translation of existing techniques for finding customers in the electronic environment. It provides products and services customized to the needs of the customers. It helps to retain customers' loyalty and attend the needs for information and support in the use of the tourism products (Lidija, 2008). Many e-CRM techniques are already employed by businesses using non-electronic methods. Intelligent agent technologies, the linking of call centers to websites, and the use of data warehousing techniques to perform a detailed analysis of customer needs are among the new opportunities offered by the Internet and other advanced ICTs (Constantelou, 2002).

According to Constantelou, the most important barriers to the adoption of e-CRM solutions are the maturity of the market, the financial resources required, and the prevailing structures and modes of practices within organizations. Considering that the tourism industry operates with low margins, the important financial investments in the introduction of e-CRM solutions is a barrier to the adoption of these systems. For some companies, the investments must be justified to shareholders and thus compensated in a reasonable period of time. Tourism organizations' culture and organizational structure are other barriers to the adoption of e-CRM solutions. The organizations are not always ready to adopt these solutions as some departments show a resistance to change and slowdown the adoption. That is the reason why a common vision must be developed in order to eliminate conflicts between new and traditional sale methods (Lidija, 2008).

The complexity of travel products and the periodic changes in consumer tastes and behaviors are considered as additional barriers to the adoption of e-business methods in CRM. As a matter of fact, some destinations become fashionable for a certain period of time while travelers' tastes and demands evolve over time. Consequently, the *industrialization* of tourism products is a difficult task to handle (Lidija, 2008). KPMG's survey shows that CRM adoption within the travel sector is primarily driven by three key factors: (1) The industry's developing understanding of the business benefits that can be derived from CRM. (2) Travel companies' increasing inability to compete in terms of price, as the profit margins within the industry are already low. A company's key differentiators, or unique selling points, will then stem from customer rather than price-focused strategies in the near to mid-term. (3) The prioritization of customer's retention within travel company's agendas (KPMG, 2001).

14.2.2 *Step-by-step implementation of e-business methods*

Considering all the barriers described earlier, many tourism organizations remain cautious in the implementation of the e-business methods. That step-by-step approach aims to take into account the company's identity and image with customers (Constantelou, 2002). Even if e-business implies cost savings and the deepening of relationships with regular customers, some companies consider that their adoption of e-CRM solutions must follow the market and the industry circumstances that are highly impacted by unforeseeable events (i.e., terrorist attacks, tsunamis, SARS).

Nevertheless, the vast majority of the tourism organizations are willing to extend e-business in their organization. They believe that as customers will get more used and confident in the Internet, they will extend the implementation of their e-business solutions. Some companies consider that the expansion of e-business is the strategic choice, which would help them becoming the leading players in the online market. Obviously, the increasing maturity of the market and the sufficient number of experienced e-customers are necessary to make the Internet channel a meaningful investment for all the industry players (Lidija, 2008).

That is the very reason why CRM strategies are considered as *peripheral activities* to the companies, which are operating in the less mature and small markets. Investment in e-CRM solutions is not their top priority as they do not expect a large number of users and thus a very scary return on investment. In spite of the fact that consumers are interested in surfing the Internet, the vast majority of e-customers still prefer the personalized service provided by travel agents. As they are overloaded with information, the customers mainly rely upon human interaction with travel agents. If we consider business customers, it seems that human interaction is not going to be replaced by online services as their travel arrangements are complex and often requires the *human touch* of an experienced travel agent (Constantelou, 2002). However, a new *web-savvy* customer emerges. This *web-savvy* customer uses the web as a rich information medium. He seeks travel information, compares prices, and then proceeds to purchase complex travel products online. That is the reason why the leading players, including airlines and tour operators, have increasingly started to offer a variety of customer services online and they are particularly successful in attracting this *web savvy* customer group indeed (Lidija, 2008).

We can assess that e-business technologies have facilitated the shift in the focus of companies from supply to demand while customer retention and satisfaction are shown to have improved as a result of companies' online presence. In order to realize productivity gains, tourism organizations have to increase the efficiency level of their back office operations. As a matter of fact, the value of CRM initiatives depends on back office processes and the flow of information containing important customer data between the front and back offices of a company. This strategy is let as it allows the companies to focus on the key customer groups that can generate additional profits. For the tourism industry, the e-business methods help to transpose the existing techniques onto cheaper distribution channels and thus developing sales potential. However, the e-CRM solutions and ITCs must bring benefits to a supply-driven market, meaning that it must manage the growth and the maintaining of a stable employee numbers, automate back office and fulfillment operations, and standardize products and operations (Lidija, 2008).

Proponents of e-CRM will recognize that a comprehensive understanding of customers' activities, personalization, relevance, permits, and timeliness metric are means for the end of optimization (Milovic, 2011). In the context of Internet distribution and marketing in the hospitality industry, e-CRM is a business strategy supported by web technologies that allows hotels to engage guests in strong, personalized, and mutually beneficial interactive relationships, thereby increasing profitability and sales effectiveness. e-CRM is the latest technique that companies use to increase and improve their marketing skills and capabilities. Integrating technological and marketing elements, e-CRM covers all aspects of online user experience throughout the transaction cycle: prepurchase, purchase, and postpurchase (Alhaiou et al., 2009).

ICT trends in the hotel industry are improving on a daily basis. The development of information communication technology has dramatically changed the way customers interact and seek information, as well as the way of purchasing services (Ip et al., 2010). This represents a video that shows the hotel through the lobby, hallways, and rooms. ICT provides a platform for hoteliers to collect information on guests. Many hotel websites invite customers to register and identify their interests, from which hotel managers can create personalized services and products and increase customer satisfaction (Ip et al., 2010). Personalization increases customers' emotional involvement with the experience, often improving their opinions about it as well as the service provider. As it is important to inform the guests, the responsibility of hoteliers is also to keep information about the guests safe (Luck and Lancaster, 2003).

14.2.3 e-CRM models

e-CRM implies an additional means of communication and level of interaction with the customer where there is a real difference in the technology and its architecture, which allows for ease and self-service to customers. The study summarized the e-CRM models from various scholars as Table 14.1.

Table 14.1 e-CRM Models

Scholars	e-CRM models
Rowley (2002)	Identified the following functions of e-CRM practices, but these practices are also very broad and do not focus on particular organizational processes or customer values: e-commerce, channel automation software, collaborative commerce software, online storefront, multichannel customer management, e-service, e-mail response management, guided selling and buying, product configuration, order management, electronic agents, catalogue management, content management, e-customer, fulfillment software, self-service.
Anton and Postmus (1999); Feinberg and Kadam (2002)	Developed a more customer-centric e-CRM model by identifying a long list of e-CRM features providing customer value.
Feinberg and Kadam (2002)	Recognized that their 42 e-CRM items may not really define e-CRM, as e-CRM is a dynamically changing process. Apart from being criticized for its inclusiveness, their list of e-CRM features also lacks a systematic approach for further developing e-CRM features that add customer value.
Sigala (2005a)	Developed a more holistic and business operations' integrated model of CRM implementation, but its value for this study is limited as it focuses on a strategic rather than on an operational level.
Kotorov (2002); Singh (2002)	e-CRM is also defined as the application of ICT to increase the scale and scope of customer service. Building and maintaining customer relationships online depends on maintaining effective customer service. Thus, an e-CRM customer-centric implementation and model should be viewed in close connection with e-service provision.
Riel et al. (2001)	e-service as "an interactive, content-centered and Internet based customer service, driven by the customer and integrated with related organizational customer support processes and technologies with the goal of strengthening the customer-service provider relationship." Indeed, unless service is maintained, customer loss may result in great inefficiencies and costs. Moreover, relationships in e-commerce heavily depend on information exchange during all functions of the online purchase process. In turn, there are numerous opportunities to gather the information in each function and use it to improve the possibility of continuing profitable customer relationships by improving service, save customer's time and easy frustration.

(*Continued*)

Table 14.1 (Continued) e-CRM Models

Scholars	e-CRM models
Otto and Chung (2000); Voss (2000); Riel et al. (2001)	e-Service functions supporting online shoppers are widely classified into the following consumer purchase behavior stages: problem recognition, search for information, evaluation of alternatives, choice, transactions, postsale services.
Voss (2000); Sigala (2004)	Their study used e-service categorization for modeling e-CRM practices for the following reasons: (1) website design and functionality developed in line with these categories are proved to significantly affect online consumer behavior and patronage, (2) it is a systematic customer value-oriented approach focusing on supporting consumers' use of webstores, (3) by using this framework, one can easily identify and further develop additional e-CRM features that can add customer value, (4) failure to develop e-CRM features in one or more steps may defeat the customer provider relationship, and (5) the categorization allows to relate e-CRM features with organizational functions and processes, which in turn enables managers to better allocate resources and develop more appropriate metrics for monitoring and measuring successes and failures.
Feinberg and Kadam (2002)	Their study illustrates how the consumer purchase behavior stages were used for modeling and measuring e-CRM practices by classifying and further extending e-CRM features.

14.2.4 Strengthening customer service of e-CRM in travel and hotel

The Internet is the best interactive marketing channel ever invented. Therefore, it is the ideal medium for reaching out to your customers and establishing interactive relationships with them. It is important to understand that customer service is only one aspect of e-CRM and is primarily a reactive function aiming to improve performance and efficiency, while e-CRM as a whole is a proactive long-term strategy. On the other hand, e-CRM is more than just a tool to achieve and enhance customer satisfaction. The traditional CRM focus in travel and tourism has always been *customer satisfaction*. The presumption is very simple: Customers will appreciate good service so much they would not go to your competitor. In other words: customer satisfaction + quality of services = customer loyalty.

The truth is that customer satisfaction does not always equal customer loyalty: 40% of satisfied customers switch suppliers without hesitation

(Forum Corp), 65%–85% of customers who choose a new supplier claim to be satisfied and very satisfied with the former supplier (*Harvard Business Review*), 85% of customers claim to be satisfied, yet willing to switch to other suppliers (University of Texas).

A study by Cornell University also calls into question the widely held belief that guest satisfaction means repeat business. The results of this study challenged the theory that satisfied guests generate repeat business in the lodging industry. Analysis showed only a weak connection between satisfaction and loyalty, which is a precursor to repeat business. In travel the Internet has provided hoteliers with unprecedented capabilities to interact with their customers: eNewsletters, online promotions, and sweepstakes, reservation confirmation emails, prearrival emails with value adds and up-wells, poststay "Thank you" emails, and comment cards.

Interactive Web 2.0 applications on the hotel website (experience and photo sharing, customer generated Top 10 lists of coolest bars, museums, things to do in the destination, etc.) Providing innovative customer service on your website and to your customers via latest technological applications is a must. The airlines have perfected some very neat tools in this respect. But once again, this is the customer service side of e-CRM. Establishing mutually beneficial interactive relationships with your customers is the ultimate goal of any e-CRM initiative.

14.2.5 *e-CRM in an integrated e-marketing strategy for tourism organizations*

CRM has been a buzz word in the tourism industry for several years. However, looking at the diffusion among companies and even more at the commitment of the marketing and sales managers. According to E-Business W@tch, Report (2006), CRM does not seem to be a big issue for tourism players. While 36% of the tourism companies in Europe (38% in the Czech Republic) offer online booking facilities, only 11% (8% in the Czech Republic) make use of a CRM system. Certainly this might originate in the structure of the tourism industry itself, which is dominated by small- and medium-sized enterprises. Other reasons may be that the executive management is not fully convinced of the potential of CRM since the benefits are not always visible immediately. Concerns about the change caused by the implementation of CRM within an organization might also be an obstacle to CRM implementation. However, Maurer (2007) indicated, "e-CRM makes use of the Internet and its services in order to support and enhance the communication with customers, and both elicit and refine information about the customers. Knowledge about the markets, the customer segments, and their demands has become a decisive competitive advantage in the tourism industry with tougher and more numerous competitors and an increasing number of products and services on offer.

The shift from customer-oriented to customer-driven business processes, accompanied by increasing consumer maturity, has led to new challenges in the fulfillment of customer requirements. Tourists have matured to prosumers (i.e., professional consumers) who want personalized products and services for the best rates available."

The changes mentioned call for an interactive and integrated marketing strategy in which the Internet acts as an intermediary network for the essence of one-to-one marketing (the use of information from the consumers rather than about the consumers) and the distribution of tourism products. Therefore, successful eMarketing requires the integration of all business units and business processes. The development and implementation of a sustainable eMarketing strategy are based on information about the target groups, competitors, and the market situation (Maurer, 2007).

14.2.6 Integrated e-marketing strategy development

According to Maurer (2007), the eMarketing strategy aims to attract visitors to the company website, generate revenue (i.e., through online bookings) and increase the image and brand awareness of a supplier. This can be achieved by the continuous enhancement of the three strategic processes—customer analysis, competitor analysis, and business processes efficiency—applied to the operational processes for maintaining a website that acts as an electronic market place for both B2C and B2B. Besides an appealing design and usability, the basis for a good tourism website is interesting and relevant content geared to the needs and wants of the targeted customer segments. Visitors to the website who leave electronic footprints (i.e., log files and profiles) behind increase the image and brand awareness of a supplier. Besides an appealing design and usability, the basis for a good tourism website is interesting and relevant content geared to the needs and wants of the targeted customer segments. Visitors to the website leave electronic footprints (i.e., logfiles and profiles) behind that can be used to refine the content and even to develop new products. These data can be merged and enriched with customer information from other contact points such as log-in data, online surveys, online competitions, brochure order, feedback forms, etc. However, publishing attractive content is only one side of the story.

To be found in the virtual globality and hence the promotion of the website and its content have become important issues in particular for suppliers who cannot profit from a well-established brand. According to various market researchers (i.e., Forrester, Nielsen Net Ratings), about 80% of tourism website visitors use a search engine when looking for information about tourism products and approximately 60% of all searches are carried out through Google. It is essential for an efficient and effective website promotion strategy to analyze which information sources the target groups use, as well as which keywords they use to search for information. Moreover, one's website promotion activities can be enhanced by keeping an eye on the marketing activities of the competitors in order to find out what makes them successful. It might be even beneficial to seek strategic alliances and establish ePartnerships. What is important is that all endeavors are monitored and evaluated thoroughly in order to receive essential feedback, which, in turn, can be used to boost the performance.

A successful and sustainable eMarketing strategy can only be achieved by involving and coordinating all relevant business units in order to ensure the bundling of information both from and about the customers and the competitors and eventually add value to it. The Internet has changed the business model in tourism significantly. e-CRM can support tourism suppliers in anticipating tourism trends and consumer wishes to facilitate customer acquisition and retention, to provide personalized products and services, to identify demands, and to link up with partners to meet these demands proactively (Maurer, 2007).

14.2.7 Usages of the Internet and e-tourism

According to Leiper (2000), Tribe (1997), and Tremblay (1999), tourism does not follow the usual rules of economic theory, or of any other theory; besides, many authors are used to make reference about the *indiscipline of tourism*. Indeed, tourism gathers all the activities dedicated to the satisfaction of the needs of the tourists, and borrows to a multitude of other activities. The tourist products are complex and heterogeneous products, combination of elements separated in time and space (Caccomo and Solonandrasana, 2001), often predefined packages assembling interrelated products and services (transport, accommodation services, leisure services…). This notion of packaging, of bundling, is the core of the activity. Contrary to the traditional goods sectors, where resources are transformed to be delivered to the customers, the tourists have to go to the resources: Whatever their intrinsic qualities, the resources acquire an economic value only with the organization of the traveling of the tourists and development of the activity (Spizzichino, 1991). Tourist products and services are often experience goods, the quality or utility are not known ex ante by the consumers; a system of advices and critics is thus necessary to the formalization of choices (Gensollen, 2003).

14.2.8 m-CRM strategy for customer satisfaction

According to Newell and Newell Lemon (2001), m-CRM means using the mobile channel for customer relationship marketing or management. A key to any brand-building effort is gaining knowledge of the relationship that exists between the customer and the brand. Customers have gained more power to choose how they wish to interact with companies; therefore, new tools for customer communication are demanded. Customers want to be treated individually and receive personalized value. Marketers need to develop alternative ways of reaching customers with timely, relevant, and highly personal information of high value, and they also need to provide customers with a means for timely feedback (Newell and Newell Lemon, 2001).

The mobile phone can be used for this purpose. If m-CRM is to have an effect on brand loyalty, per se, marketers need to engage in a dialog with customers and personalize their brand communication messages. For example, the Belgian home improvement store, Brico, collects vital information about customers by inviting "its Discount Club members to join a continued SMS dialogue" (The Forrester Report, 2001). Those who sign in receive questions on home improvement, and by answering the questions, they receive discount coupons or invitations to equipment demonstrations. An eventual goal of m-CRM is to empower customers and let them be in charge of the brand communication and interaction (Newell and Newell Lemon, 2001).

Marketers should consider using m-CRM when a highly personalized and timely dialog with customers is desired and when they want to differentiate the customer service or product offering with an innovative method of interaction. Marketers should be cautious, however, since a long-lasting dialog with customers is challenging to obtain. It will most likely engage only those customers who are highly involved in the product or company. Furthermore, marketers should pay close attention to consumer privacy issues and to handling consumers' concerns regarding how their personal information is used (Phelps et al., 2000).

In recent years, customers using mobile phones have presented a very fast growing on value-added services, such as games, ringtones, stocks, Global Positioning System (GPS), MSN mobile, and information services (Hsu and Lin, 2008). It shows a chance for enterprises that mobile channel is a new opportunity to deliver more complete services to consumers, to increase corporate brand image and to supplement company products (Sideco, 2011). There are little researches in the literature about m-CRM but few articles have been deeply discussing customer satisfaction in m-CRM (Facchetti et al., 2005; Hsu and Lin, 2008). Through the mobile medium access to manage customer relationship, it brings not only new transactional possibilities but also new challenges. To operate a successful m-CRM, a stable technological infrastructure is necessary. Therefore, m-CRM is integrated to the existing customers and CRM activities so it is supported by the technological infrastructure of the mobile medium system (Hsu and Lin, 2008).

A structured method by which mobile business can be introduced to the CRM field with more and more mobile phone users. So that m-CRM has been concentrated seriously (Schierholz et al., 2006; Hsu and Lin, 2008). The m-CRM integrates wireless technologies into existing operations and strategies on customers. Consequently, m-CRM is a corporate strategy and systematic approach realized by a mobile device (such as a mobile phone or PDA). Through mobile value addition, the perception of customer satisfaction is promoted on strengthening a long-term customer relationship (Hsu and Lin, 2008).

While the concept of customer satisfaction is deeply inset with technology as is the case with m-CRM, customer satisfaction helps enterprises in creating new customer relationships, acquiring or maintaining retained customers by the mobile medium, and enterprises delivering their information to customers by wireless networks. Moreover, m-CRM also supports enterprises on marketing strategies through understanding customers' needs and improves the service process through the feedback from satisfied customers (Hsu and Lin, 2008). The m-CRM is not only the critical role of communication in establishing but also aims at maintaining profitable and satisfied customer relationships through mobile communication medium. Therefore, mobile communication sits at a very important position to increase the degree of customer satisfaction in m-CRM (Hsu and Lin, 2008). The m-CRM promotes satisfaction to customers through the mobile medium on communication, either one-way or interactive, which is related to sales, marketing, and customer service activities conducted (Palen et al., 2001; Camponovo et al., 2005). Therefore, the mobile medium provides essentially the same function as any other channel within CRM system. The m-CRM has the most important difference that enterprises and customers are connected through the mobile medium and customer satisfaction is also promoted to customers through the mobile medium. Thus, the personalized communication strengthens the relationship between enterprises and customers (Hsu and Lin, 2008). This also shows that personalization plays a very important role in promoting satisfaction and push up customer relationship. In order to provide a personalized service through the mobile medium, enterprises send a relevant communication to targeted customers who always result in a significantly higher satisfaction than inaccurately targeted customers (Camponovo et al., 2005; Nysveen et al., 2005).

The challenge in m-CRM is to ensure that the right messages are sent in the right way to the right people who come with a positive perception on customer satisfaction. However, message may be missed or re-sent with the wrong medium at the wrong time by traditional CRM (PocketGear Mobile Application Store, 2011; Thomas and Sullivan, 2005). Customers give permission to receive messages by the mobile medium of their preferences and the utilization of the mobile medium has to overcome these problems. According to the case study, this paper indicates that m-CRM is based on several technology- and marketing strategy-related issues to promote their customer satisfaction (Hsu and Lin, 2008). With the rising consumer consciousness, customers have more and more choices that channel the use for communicating with companies. The channel is the new and big opportunity for enterprises to create business and retain customers. Hence, enterprises approach m-CRM to increase customer satisfaction by utilizing the mobile medium effectively (Hsu and Lin, 2008).

14.2.9 *Mobile challenges for travel and tourism*

Scholars and industry representatives are turning their attention toward the promise of electronic wireless media, envisaging that the next—or the real phase of e-commerce growth will be in the area of mobile commerce (Hampe et al., 2000). Keen and Macintosh (2001) stress that the issue is that mobile commerce (m-commerce)is marking the start of another era of innovation in business and that m-commerce will continue to extend the way organizations conduct business—and change the relationships between companies, customers, suppliers, and partners (Keen and Mackintosh, 2001). Mobility means freedom—say Keen and Macintosh—and freedom creates choice and value, something much more than convenience as it may revolutionize the way companies work, buy, sell, and collaborate.

Although the mobile Internet appears to have much to offer as an instrument of commerce, little is known about the consumers' willingness to adopt wireless electronic media, and the factors that influence their adoption decisions and value perceptions relating to m-commerce (Pedersen et al., 2002). Just as we are gradually starting to gain an understanding of the unique characteristics of the fixed Internet, a new medium has emerged, the wireless Internet, which raises many of the same questions in a new context (Guerley, 2000). Building successful strategies for the mobile marketplace begins, no doubt, by recognizing the distinctive forces driving the emergence of m-commerce (Senn, 2000). On the Internet, firms can create value for customers in a manner that is different from that which has been achieved in conventional business (Han and Han, 2001). Correspondingly, m-commerce possesses unique characteristics—Keen and Macintosh (2001) call them the mobilization of knowledge—when compared with traditional (i.e., fixed) e-commerce, and many statements on an impending m-revolution have, in fact, been triggered by the assumption that the potential of m-commerce will involve (1) lower barriers and (2) greater benefits in comparison to both fixed e-commerce and traditional commerce. In view of that, the key question from commerce is to find some way to assess the value of mobile applications to prospective users (Carlsson and Walden, 2002), and to gain an understanding of the factors that may delay the penetration of the mobile Internet on a larger scale (Lee et al., 2001).

There are a number of ideas of what is going to constitute the key success factors for the actors in the global m-commerce arena. This arena is already growing diversified with a number of application areas, which are growing in different directions and at different paces. Travel and tourism, being one of the largest and most rapidly expanding industries in the world and one of the significant users of ICT in its operations will without doubt be one of the trailblazers in the global m-commerce arena with value-added mobile applications. It can be assumed that travelers' and tourists' lives will be enhanced by smart services, accessible via mobile devices anywhere and anytime. Intelligent software technologies will allow mobile services to be personalized and context-aware to improve travelers' and tourists' experiences. Context-aware mobile services will make a difference as the services and contents adapt to both the environment and to personal interests.

Thus, it is not too unrealistic to assume that the competitiveness of tourism organizations and destinations will increasingly depend on their ability to use innovative mobile technologies to promote location-aware services and to serve customers on the move. The key questions to be addressed in the panel are:

1. Which are the key value-adding mobile services (each panelist is expected to propose and motivate one service)?
2. What do we know about consumers' willingness to adopt m-services (each speaker will identify one to two factors that will influence the adoption process of the proposed m-service)?
3. How will these services make the day for the m-traveler and m-tourist?
4. How is the mobile technology going to change the travel and tourism industry?

14.2.10 *Mobile industry and the travel and tourism industry*

Modern technology is constantly changing the way of communication and interaction in different areas of human life. Miniaturization of electronic components and enhancements in transfer speeds of wireless transmissions has pushed the use of mobile devices to almost every area of human life (Kumar and Zahn, 2003). Web technologies based on user interaction in social networks, known as Web 2.0 are created around the idea that the people who consume media should not passively absorb what is available, and rather they should be active contributors and codevelopers (O'Reilly, 2011). Using a new generation of mobile devices, it is possible to consume and change content available on the web from different locations and devices. Hardware components are becoming more powerful, but less power demanding and in a few years the mobile devices could have the prime usage for everyday applications like reading, web browsing, multimedia, and general information manipulation (Palen et al., 2001).

By popularizing technologies like social interaction and location awareness, a relatively new era in the modern business is becoming more and more evident. Those areas are based around using customer's location with all other information that could be gathered, for analysis and service providing (Abowd, 1997; Maervoet et al., 2008; Chiu et al., 2009). These tactics are already being used by big advertisement companies like Google (2010), which provide location-based advertisements to customers, targeting their location and interests (Camponovo et al., 2005).With the radical impact of the mobile networks and mobile devices, this kind of service providing will be even more evident in the future. The emerging technology can be used in all kinds of different areas like finance, marketing, sales, transit, and tourism. The focus of this paper will be on the last two areas—transit and tourism, because the potential use of the technology is huge and promises opening of the new possibilities in the sector that is currently spiraling down in the profit and market share.

Recent studies have shown that, while the economy is in the state of recession, potential customers rather spend their earnings on home entertainment systems and other electronic devices like laptops and mobile phones, than on vacation planning and traveling (Facchetti et al., 2005). This behavior generates huge losses for the industry based on traveling or tourism in general. The potential solution for the problem is connecting the mobile industry with the travel and tourism industry in a way that will animate the customers to travel more and enjoy their time by using interactive or helpful content. That could be of benefit for both sides—the customer and the industry.

14.2.11 *Use of mobile technology services in the travel and tourism industry*

14.2.11.1 *Finding the customers*

In the first place, the technology must animate the customer to think about traveling (Buhalis and Law, 2008). Nowadays it is not enough to motivate potential customers solely with pictures, souvenirs, tales of natural and cultural beauties, and delicious culinary recipes. One of the things the Internet has brought is the outstanding multimedia and user interaction, which enables users to experience almost anything from the comfort of their home. Online activities have advantages compared with traditional offline activities. For many consumers, online shopping represents a more entertaining activity than the same activity done offline. They have a positive attitude toward Internet advertising, which they perceive as being informative and up-to-date (Dobre and Constantin, 2010).

Interesting facts, questionnaires, games, and viral advertising are a good starting point for marketing campaigns. In the last few years, games and viral advertising—specially designed commercial concept relaying on fast, exponential information sharing through social networks (hence the name Viral), have been the most successful tools in animating potential customers into spending their earnings on the advertised products or services (Helm, 2000; Larsen et al., 2007; Porter and Golan, 2011).

Usually, questionnaires are more informative than games or viral advertising, but they can cause the customer to lose interest very fast, so this sort of advertising must be well-thought-out, short, compact, and targeted on specific customers (Kotler and Armstrong, 2010). This can be achieved by incorporating surveys within games or between multimedia contents, so the customers will not be able to ignore the content of the survey. The results of these questionnaires should not be considered completely reliable during the analysis and evaluation, although they could provide some input from the potential customer and subsequently be used as a measure of motivation for certain product or service.

14.2.11.2 Make them an offer they cannot refuse

The mobile devices are the best gateways to the customer these days, because they are the most popular means of communication (The Nielsen Company, 2011; Sideco, 2011) and make customers available at all times (Nysveen et al., 2005). According to Gartner market research in 2009 (Stevens and Pettey, 2011) and 2011 (Pettey and Goasduff, 2011), worldwide smartphones sales to end-users has been continuously growing. Smartphone sales to end-users were up 72.1% from 2009 and accounted for 19% of total mobile communications device sales in 2010.

Different operating systems changed their market position during that time, but in total, recorded sales have been growing steadily. The winner in the market of smartphones the previous year were mobile phones based on Android (2011) operating system, which grew 888.8% in 2010 and moved to the second position in the top list of smartphones sold (Pettey and Goasduff, 2011). Many mobile applications and games are free and accessible to users through web stores and markets. Currently the most popular mobile application stores are Apple's App Store (2011), BlackBerry App World BlackBerry (App World, 2011), Google's Android Market (2011), Microsoft's Store (2011), and Pocket Gear (Pocket Gear Mobile Application Store, 2011). They are open to the developers, hence creating a bridge between customers and content providers, and in this way help spread the information about a potential product.

Customers can also be motivated by obtaining access to knowledge. Interesting literature provided in electronic compact form, specially designed for mobile devices could raise interest within certain user groups and spread the information to many potential customers. Connecting with the mainstream media like the Internet news portals and blogs can be valuable for the success of the advertised product. Investing in applications specifically designed for the social interaction and content sharing is also essential for the potential success of the product or service, allowing it to be discussed and rated, which is beneficial for the future marketing campaigns and product placements (Riegner, 2007).

Customer age groups and gender must also be considered. Misplaced or complex applications could have a great impact on a smaller group of people, but fail to have the desired effect on a larger group of customers (Tarasewich, 2003). Applications or games must be relatively simple, considering that the elderly age group could potentially have problems with their usage. If the probability of reaching a larger group of elderly customers is considered important, then games and complex application should be avoided (Conci et al., 2009). Violence, inappropriate language, and other questionable content should also be avoided. Ergonomics of the mobile application must be considered as important as the content. Small screens can be very difficult to handle, so the design must be as ergonomic and clear as possible Bruner and Kumar (2003), (Cyr et al., 2006).

14.2.11.3 Present options to the potential customers

When the interest among customers is awoken, the time is right to start the second stage of the concept, which is to take all the necessary preparations to present user with the lucrative options for spending their free time or vacation. Since the potential customer has already spent some time with the aforementioned content, it is possible to take action and analyze the data gathered from the user's actions. One of the solutions is to create the necessary backbone system based on business intelligence and data mining concepts (Cech and Bures, 2009; Liu et al., 2010), which will generate enough information to make a decision on what categories of the service should be offered to the customer.

If there is not enough data gathered from the potential customer, common data like user's origin, language, proximate location or patterns of web usage could be used to narrow the list of possible services that could be offered. Some of that data could be obtained from large Internet search engines or other services that offer that kind of information. Customer's proximate preferences can also be obtained by analyzing data from other customers living in the nearby area (Finn et al., 2000).

The last two options offer a part of reference data for the potential customer, but are much less effective and lack the individuality, and should be used only if there is no other way to gather information directly from the targeted person. Also, these solutions are less cost-efficient because every piece of information obtained from large advertisement agencies comes at a price that could grow over time and create significant problems in the budget (Antonio et al., 2011).

After the information is processed, it must be delivered and presented to the potential client in a way that will raise the chances of successful product reception by the potentially interested receiving side. That could also be achieved using games, multimedia, or some other interactive medium like social networks. Customer wants more value for his or her money, and that must be taken into consideration. Because of the nature of today's way of life where almost one billion people own a mobile device that can connect to the Internet (ITU, 2011), tighter and more transparent link between Internet providers, media services, and travel agencies must be established.

Internet and media providers possess the content that is very attractive to the customers traveling via travel agencies. First and the most important part of the chain is the Internet connection that has to be affordable to the average customer. If there is no such thing as a broadband connection, no other processes can be considered. When traveling abroad, most users have the choice of using roaming Internet connection, but that can be extremely expensive and unaffordable for a great majority of potential users. Other choices include using wide area WI-FI connections like ones in the big metropolitan cities like Paris, France or Berlin, Germany. Also a contract between travel agencies and mobile operators could be negotiated, so passengers traveling with the agency could have some benefits while connecting to mobile operators' network, without roaming, or with roaming but with special, lowered prices per data unit (Antonio et al., 2011).

14.2.11.4 Keep track of customers

If a customer chooses to purchase the presented offer, interaction between the tour operator central server and customer's mobile devices could be used to help or entertain the customer in a number of different ways. If the customer buys complete travel package including airplane transport, transfer from the destined airport to the hotel or apartment and sightseeing tours, it is possible to provide them in real time the latest weather forecast, information about airplane delays, bus stops, and other important information. The customer can be led from the moment when he or she leaves home to the moment of arriving back after a trip or vacation (Antonio et al., 2011).

In the urban areas, location-based data could be used to expand the mobile device services. Location-based services make use of a device's integrated GPS components to get the user's current location. The latitude and longitude of the customer/device is usually determined using some sort of positioning service like satellite-based GPS—currently, the only GPS in commercial function (Virrantaus et al., 2001). In the future, devices could be using another type of GPS like the Russian GLObal Navigation Satellite System (GLONASS)—ex-military system, but currently open to public (Russian Institute of Space Device Engineering, 2011), Chinese Compass navigation system (Gibbons, 2011; Grelier, 2011), and the European Union's Galileo positioning system (Issler et al., 2011). Some microchip manufacturers like Broadcom (Broadcom, 2011), or Qualcomm (2011), implemented the support for multiple position systems in their chip architecture to achieve more flexibility in global markets, especially in Russia where the government confirmed a plan to introduce 25% import duty for all GPS devices without support for GLONASS by 2012, as part of the efforts to encourage worldwide adoption of that technology (Bachman, 2011).

Current GPS modules integrated in modern mobile processors provide enough accuracy to be used with the mobile applications interacting in a way that makes it possible for the customer to be leading in almost every situation. GPS positioning is less suitable for use in urban areas because of the significant signal loss, which is attributed to signal bouncing of the walls of tall buildings block (multipath propagation), which can cause inaccuracy in positioning results. Bad weather conditions also contribute to signal degradation. Weak signal contributes to significant delays in precise signal acquisition. Better solution for calculating the mobile device's position is with the help of the network-positioning approach (Antonio et al., 2011). This concept, also called Assisted-GPS, or A-GPS (Jarvinen et al., 2011), is integrated into most mobile devices today and uses the help of mobile network tower cells to calculate geographical position faster and more accurate. Cellular network towers can provide user's device with additional helping parameters as

1. Orbital data with current satellite positions
2. Position recalculation from the gathered GPS data snapshot received from the client
3. Approximate location of the customer using the network multilateration, which narrows the GPS position to a smaller geographical location

If there is no GPS signal to accurately determine users' position, most common solution is based on the network multilateration based on signal strength received on three or more mobile base stations (Wong and Choi, 2007). This concept can narrow the radius of the customer's approximate position up to 50–70 m, depending on the number of network towers and terrain architecture.

14.2.11.5 Take care of customers' needs

If there is a travel delay, the customer could be presented with a choice to purchase some sort of interactive or multimedia entertainment like mobile IP TV or video-on-demand. But if the flight is to be canceled or delayed for an uncertain period of time, the customer should be alerted and presented with the list of choices like booking a hotel/apartment or contacting the taxi service (Antonio et al., 2011). On some flights, the customer is offered with an option to purchase WI-FI Internet connection during the time of the flight so travel agencies could team up with airline companies and create special offers that could also be a part of the travel package (Jahn et al., 2003). If the customer is traveling by airplane, transportation from destined airport sometimes is not included in the purchased package. In that case, customer using the mobile application can contact the taxi service, or he or she can be informed in detail on how to reach the destination. Timetables for buses, trains, or metro transportation could also be provided, as well as prices and categories of tickets. If the customer has a device with location awareness, such as GPS-enabled device, detailed instructions could be provided leading them across the points of interest. The relatively new concept in the development of mobile applications called augmented reality is similar to fighter jet pilots head-up display (Antonio et al., 2011). In this concept, the mobile device camera, electronic compass, and GPS device are connected together via software platform creating advanced visual experience (Feiner et al., 1993). On mobile device screen, the actual image captured with the device's camera, data about current direction and position of the device analyzed via electronic compass and GPS coordinates are mixed with the information about the environment and surroundings gathered from the Internet.

All of this is presented on the device's screen in a compact form allowing the customer to have the synthesis of right information about the surroundings in real time. There are many visually or hearing-impaired people who are willing to travel and can be potential customers. Many of them have difficulty finding their way, frequently feeling disoriented and even isolated. Despite their impairments, they also are potential customers for travel and tourism companies. There is a number of navigation systems developed for visually impaired people, and some of them can provide dynamic interaction and adaptability to change in the environment. Outdoor navigation systems are generally based on GPS (Antonio et al., 2011). They can locate and track the user and give them up-to-date information about routes and rerouting according to changes in the environment. That system should also be able to give information about nearby signs (Virtanen, 2001; Virtanen and Koskinen, 2004).

In an indoor environment, such as airport building or museum, orientation is more difficult because of the confined space with many obstacles like stairs, doors, and furniture (Rajamaky et al., 2007). The next logical step is to adapt hardware and software of the mobile devices to the level that is comfortable for people with special needs. For hearing-impaired people, design should be based around visual interaction, and for those visually impaired, design must have more audio and tactile stimuli, such as screens capable of displaying Braille alphabet (Antonio et al., 2011). The application should be at customers' hand at all times during the trip, providing current information about historical monuments, museums, bars, clubs, and other attractions (Dìez-Dìaz et al., 2007; Paelke and Sester, 2010). Customer's gender, age, and budget should be considered when offering points of interest. This information could be gathered right on spot through the application, or could be extracted from the data collected from the customer before he or she decided to book the trip. The first option is a better choice for customer and travel agency, because it enables more accurate results for places worth visiting for the customer, and more detailed information about wishes and patterns of behavior, for the travel agency. These data also allows better choice to future travelers so they have more options presented even before they arrive at the destined location (Antonio et al., 2011).

The application must have an interface with the most popular social networks like Facebook (2010) or Twitter (2010), so the user can share the experience with friends or followers. Users rating should also be implemented, and he or she should have the possibility to evaluate/rate service, accommodation, visited places, etc. That information could be shared between other users and travel agencies making the service better and more advanced. The trend today, especially with the younger population is avoiding travel agencies and turning directly to web enabled company services for finding flights and accommodation. In that way younger people feel more free and unbound by agency-created itinerary and can travel with significantly smaller budget than with the common options presented by the agency (Antonio et al., 2011). This group of potential customers should not be neglected since it is a great target for offering a mobile application that could help them plan the travel their way. For a small amount of money, they could purchase an application that would offer the best way to plan a trip to the wanted location. The quantity of purchased applications by users could largely surpass the funds spent for creating it (Antonio et al., 2011).

14.2.12 Conceptual model of mobile device integration

Conceptual model of integrated mobile services consists of three essential parts (Antonio et al., 2011):

1. Mobile application
2. Enterprise architecture consisting of cloud-based SaaS (Software-As-A-Service), which is the business-oriented part of the application
3. Decision support system for business intelligence and information gathering. Information is generated via multiple sources, mainly related to the Internet

Goal of the presented architecture is to provide synthesized information from different sources, protocols, and technologies that can be found on the global network. The center of the architecture is a user carrying a mobile device able to connect to a wireless network. Using application developed for that device, the user can communicate with the cloud-based architecture owned or leased by a network operator, travel agency, or some other enterprise industry subject. The user can also be a source of information providing feedback via mobile application itself or through blogs, webcasts, reviews, social networks, or other Web 2.0 communication tools. Described system would not be based only on feedback from one user, but on massive amount of data gathered from the Web (Antonio et al., 2011). This data could come from blogs, podcasts, forums, reviews, frequent web searches, news portals, social networks, etc. Because the number of users connected to the Internet grows linearly every year (Internet World Stats, 2011), today almost two billion people have Internet access (ITU, 2011). The amount of data on the web increases by approximated 10% yearly (Wilhelm, 2011) nearing 50 billion unique web pages in 2011, and it is extremely hard to extract useful and precise information from the Web content. These statistics include only data gathered from the number of pages existing online, but the Web consists of many information sources that are constantly changing, or is in a format that could not be described as a web page, for example, social content and multimedia, both containing valuable information about the behavior of Internet users (Antonio et al., 2011).

To be able to analyze so much data, nontraditional approach must be considered. Analyzing data in detail, one at a time would be time-consuming, resource demanding, and virtually impossible. The technology that could provide synthesis of content is based around fuzzy linguistic agents, semantic clients, and neural networks (Bigus, 1996; Herrera-Viedma et al., 2005). The process during which data of interest are gathered from the Internet is called content acquisition. To be able to distinguish between small amounts of interesting and large amounts of potentially uninteresting data, a selection must be applied. Data selection is enforced with the help of the customer profile database. This database can be created via customer's interests gathered from aforementioned questionnaires or with the help of the data gathered from the social networking sites (Antonio et al., 2011).

This process leads to knowledge creation from which the set of relatively strict rules of content filtering can be created and inserted into the customer profile database for filtering the content in content acquisition phase (Srivastava and Cooley, 2003). Implementation of this technology would gather precise information based on a set of rules, and produce approximate values in synthetic form, compressing large amounts of data to only crucial information that could be used by the client or the enterprise software owned by the service provider (Antonio et al., 2011).

Synthesized information can be further processed by the process called data cleansing and preprocessing (Bigus, 1996) in which incorrect data is ignored and correct data is saved to the data warehouse for further analysis. Data obtained in this step could be connected with the offerings from advertisers in partnership with service or application provider, so the end-user gets the detailed information with helpful hints combined with the latest offerings from the business partners. Ranking of the most interesting results must also be considered to be an important issue. Although the final results are synthesized, there can still be a significant amount of information that complies with the rules defined from the user's interests. Location could be used to help ranking the results, with the offers closer to the user's location positioned on the top of the presented result. The user can be interactively forced to make the selection between two or more additional options and the result of the chosen option(s) can be used to narrow the search and display of the results (Antonio et al., 2011).

For example, if the user wants to book a hotel or private accommodation, he or she could be faced with the big set of results, although the data has already been cleaned and synthesized. To narrow the search to a smaller set of data, the user can be presented with the set of choices like accommodation special offers, category and type, relative distance from the user's current location, view etc.... Commercial side of the result ranking should be considered if significant amount of financial resources is gathered from the advertising. The paid advertisements could be placed on the top of the presented results or marked for greater visibility. Google presented paid ad ranking concept in which the commercial companies go through bidding and the competing process to reserve the better ranking in search results (Antonio et al., 2011).

The information received by the user can be divided into two types:

1. General helpful information about places, tourist attractions, means of transportation, etc., which is essential for the user to enjoy the trip.
2. Special offers, such as current business and service offerings by the partners and advertisers, making the user's experience potentially even more enjoyable.

14.2.13 From m-Tourism to m-Tourism 2.0

What is m-Tourism? m-Tourism stands for mobile tourism. Year after year, the daily use of mobile phones has constantly evolved. m-Tourism takes into account all kind of touristic use of mobile phones. Tourism destinations, as well tourism organizations or any company linked to the field of

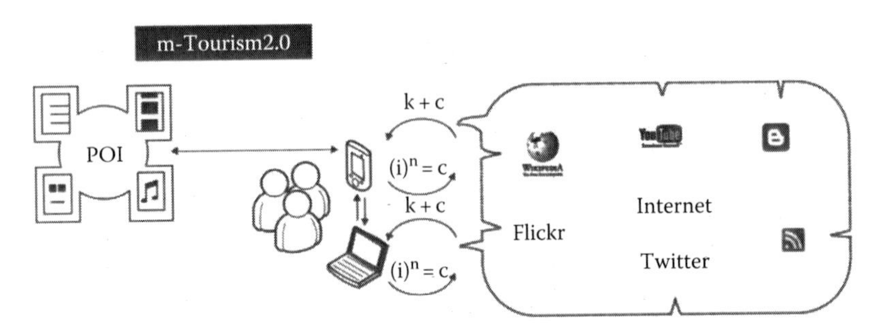

Figure 14.1 m-Tourism 2.0. (From Beça, P. and Raposo, R., m-Tourism 2.0: A concept where mobile tourism meets participatory culture, Dep. Comunicação e Arte, University of Aveiro, Portugal e-Review of Tourism Research (eRTR), Aveiro, Portugal, 2013.)

tourism is developing new mobile applications or even mobile websites. m-Tourism is a very promising branch of tourism, after e-tourism.

A quick scan of recent work done in this field suggests that, within the m-tourism context, until now the core issues researched are related with interaction design and its intrinsic concerns such as usability, accessibility, and the search for the flawless ubiquitous system (Kimber et al., 2005; Bortenschlager et al., 2010; Canadi et al., 2010). What is suggested in this paper is that, with the aid of what is known from the Web 2.0 experience, participatory culture should also become a core research issue thus granting the possibility of exploring what is beyond what is suggested by the tourism industry.

The concept of m-Tourism 2.0, illustrated in Figure 14.1, establishes as one of its cornerstones the need for solutions that allow the tourists to communicate their perspective of what they are feeling and experiencing in any given moment and being able to share that information with someone who, may or may not, use that information for their own benefit. In Figure 14.2 the tourists, either with the use of their laptops or mobile phones, when trying to access multimedia content (text, photos, videos, audio) about a points of interest (POI), will be presented with K, which is the content provided by the tourism authority responsible for that POI, plus C, which is the content provided by the tourists linked to the community and tailored according to the tourists profile preferences. The (i) n is basically the information provided by the community of users onto the system, which will be moderated, filtered, and correlated before being added to K and shared with other users. With these contributions it is expected that the tourists will be able to have access to additional information that may help enrich their experience; have access to new information in case they decide to repeat the experience in the future; provide their own opinions on matters that may at times seem gray, dull, or too partial and insufficient from an information point of view; allow them

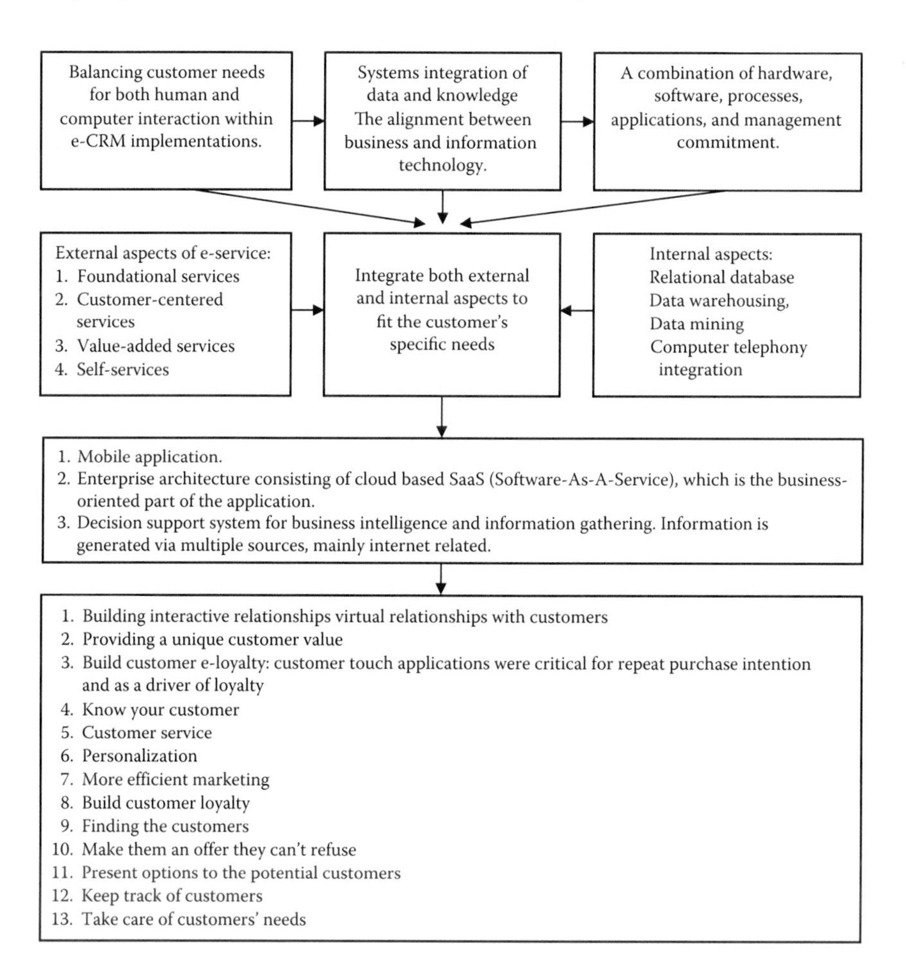

Figure 14.2 An integrated e-CRM and m-CRM model in the tourism industry.

to live a before, during, and after experience and share it with others. It may also allow tourism professionals to learn from the interactions and content provided and, in some cases, rethink current strategies, projects, products, and services (Beça and Raposo, 2013).

14.3 *Integrated e-CRM and m-CRM model in tourism industry*

According to the comprehensive literature review mentioned earlier, the study constructs an integrated e-CRM and m-CRM model in the Tourism industry as Figure 14.2. e-CRM and m-CRM integrates both external and internal aspects to fit the customer's specific needs, the external aspects of e-service include (1) foundational services, (2) customer-centered services, (3) value-added services, and (4) self-services. e-service is "an interactive, content-centered and Internet-based customer service, driven by the customer and integrated with related organizational customer support processes and technologies with the goal of strengthening the customer-service provider relationship (Ruyter et al., 2001)." The internal aspects include modern information technology—from relational database, to data warehousing, to data mining, to computer telephony integration, to Internet delivery channels—to unlock customer profitability.

e-CRM integrates sales, marketing, and service strategies that will create more value propositions for customers. It enables firms to recreate an *old-fashioned* one-to-one relationship with customers along with mass-market efficiencies from selling to millions of customers—the power of mass-customization. e-CRM integrate both external and internal aspects to building interactive relationships and virtual relationships with customers, providing a unique customer value, building customer E-loyalty: Customer touch applications were critical for repeat purchase intention and as a driver of loyalty. A practical e-CRM is being able to know the customer, sending personalized messages, being there at every touch-point involving planning, purchasing, service consumption and poststay, and providing a unique value proposition that leads to increased customer loyalty.

The main components of an e-CRM and m-CRM strategy in travel and tourism:

1. Know your customer
2. Customer service
3. Personalization
4. More efficient marketing
5. Build customer loyalty
6. Finding the customers
7. Make them an offer they cannot refuse
8. Present options to the potential customers
9. Keep track of customers
10. Take care of customers' needs

14.4 Case studies of e-tourism and m-tourism

14.4.1 Travel agencies

It is increasingly geared today to talk about e-tourism as tourism as new technologies are inexorably bound to converge to a completely digital tourism. Therefore, e-tourism includes activities in the tourism sector, on the Internet. For users, e-tourism offers ways to prepare, organize, and book their travel via the Internet: identification of the destination, purchase transportation, development of an itinerary, accommodation booking, and information exchange with other surfers. Since its appearance in 1998, e-tourism is a way of promoting and booking essentials in the tourism and travel industry. Users and consumers of e-tourism are designated by the world of marketing as touristonautes tourists or online. The m-tourism is the declination of mobile phones, e-tourism. There are numerous exhibitions and conferences now dedicated to e-tourism, such as Encounters eTourism institutional Pau, Travel in multimedia (MEC) in St Raphael, Tourism Professionals meeting in Sierre (Valais)—Switzerland. (SEOzoic.com, 2012).

The area of e-tourism brings together many tourism stakeholders as well as new entrants pure play (SEOzoic.com, 2012):

1. Tourism businesses: travel agency online service hotel, airline, railway, or navigation.
2. Editing and publishing: travel guide, magazine, travel blogs (the "infomediaries" websites specializing in travel information).
3. Technologies: Compare prices, aggregator, search engine, and Meta search specializes in travel.
4. Institutional actors: tourist office, CRT, CDT, minister, observatories.

14.4.2 Using mobile technology for e-tourism

1. The mobile tourist guide system has functions to enter and to organize the tourism objects in the database and to make them available for tourists and administrators using the e-tourism management information systems (TMIS) portals in which the tourism process and its activities have been structured.
2. The m-tourism object management systems are now compatible for all kind of user devices, the conventional desktop PC, as well as the mobile devices laptop, PDA, and mobile telephone (Samir and Hosam, 2010).
   - Samir and Hosam (2010) indicated m-tourism means the provision of tourism services on wireless devices: portable computers, PDAs, mobile telephones, and tablet PC. In the m-tourism field, we notice that the tourism services must answer specific conditions. We will have to get around the technical restrictions so we can create good tourism services. The usage of video, audio, clear interfaces, and divided tourism services must contribute to this. Furthermore, we will have to adapt the content to the needs of the mobile user. Because the user has a fragmentized time schedule, we will have to be sure that the tourism objects are not too long. Dividing the knowledge in smaller modules offers a solution.

The advantages, difficulties, and limitations of mobile tourism are as follows (Samir and Hosam, 2010):

1. The advantages of mobile tourism refer mostly to the mobile user: a great flexibility, an improved tourist schedule is possible, increased productivity during dead moments, and just in time tourism.
2. The difficulties and limitations: Nevertheless, they have very small screens, limited memory capacity, and the large diversity of mobile devices obstruct a good tourism experience.

14.4.3 *Use of mobile platform technologies in tourism (m-tourism)*

observatory.gr (2007) indicated one of the most important technological developments in tourism and a fundamental paradigm in the provision of personalized services is in the use of mobile platform technologies in tourism, both from the user's point of view and from the supply side:

1. In the promotion and management of alternative tourism and in city tours, in the form of portable guides. The presence of multimedia and location-based services in mobile platform applications is noteworthy.
2. Also notable is the particular appeal of portable guides with facilities for information storage, organizing, location finding, and combining and scheduling visits.
3. Such applications can organize, schedule, and optimize sightseeing in cities or museums and other places of interest, by suggesting the best combinations to the user. The probability of a destination being selected very much depends on the provision of such facilities.
4. Among the most widespread mobile platform applications are those in the area of alternative tourism, and in particular in ecological tourism, with sightseeing in natural parks, bicycle tours, mountaineering, etc.
5. The countries pioneering in m-tourism applications are the Netherlands, France, Italy, and Finland. The Netherlands is the worldwide leader in mobile platform technologies. An indicative example is the technological solution developed in the framework of the ABEL research project that constitutes a mobile guide for vacations in Twente. The guide operates in a PDA device equipped with GPS, fully customized to the user's individual profile. Also important was the participation of the Netherlands in the WebPark project that was implemented in the Wadden Sea area and the Swiss National Park. The relevant solution exploits GPS services in smartphones and PDAs and, among others, responds to user's queries and provides information for his or her current location.

6. The one delivered by the EU-funded REGEO project in natural parks in Germany, Poland, Czech Republic, and Austria. This project involved the development of an information system combining an application for the design of web maps, a virtual geo-multimedia database, mobile devices, and advanced techniques for the creation of virtual 3D worlds.
7. A case in Italy, in particular the Mountain biker application. This is a mobile tourist guide for mountain biking in the southern Tyrol area. The visitor is given a GPS device to put on his or her bike. The device assists the mountain bikers during their tours and also provides facilities for room e-booking, in cooperating hotels.

14.4.4 *m-Tourism: The case of the Paris transportation applications*

The applications of Metro Paris Subway are shown as follows*:

Augmented reality: Your New Eye is an augmented reality functionality that enables the traveler to locate the nearest subway station with iPhone's camera live view. The application displays elements located at maximum 1 km away from where the person stands. The iPhone user can display his phone in the horizontal position to get a list of all nearest stations. When a station is selected, a red arrow (compass) is displayed to show the direction and the distance to that point. He can also position his phone vertically to obtain a 360-degree view of all stations nearby your location, updated in real time as you walk in the streets.

Flashcodes: The RATP (the Paris transportation company) has been displaying more than 20,000 flash codes over the past few months that are 2D bar codes, on the overall tram and bus network. When scanning this bar code with a smartphone, it is possible to get all information about the network in real time. There is however one inconvenience: The application is compatible with few mobiles.

POI (point of interest database): This application can be added to the Metro Paris Subway (receiving Subway alerts in real time). It allows the traveler to activate Food and Drink POI, Restaurants POI, Coffee, Pastry and Ice Cream POI, Leisure POI, and Accommodation POI in all over France. It enables to locate all nearby POI on Google Map.†

* www.metroparisiphone.com/index_en.html.

† https://itunes.apple.com/us/app/metro-paris-subway/id297404959?mt=8.

14.4.5 m-Tourism: The case of Lion's travel

Ten years of *networked period* for Lion Travel rapidly expand the scope in 2000–2009, not only open a channel with consumers, using the strategy which both virtual and physical channel, also offers 24 h service to consumers, as the Call Center service centers of pioneered tourism, and in order to nurture talent, Lion's senior managers are also encouraged to enter the EMBA program. Furthermore, since 2010, the *mobility period*, in the face of the use of the rise of Mobile, Lion also began to import Mobile Commerce, and the product furthered audience fragmentation, with different themes for different target customers, the main push being customization of goods, and even a cultural and creative inject into the tourism industry, and contributed to the development of sophisticated travel. Lion's corporate culture is the most important dialogue with consumers, from leading the industry to explore new travel trends, and that *consumer satisfaction* as a concept, then innovation and investment, both in the products or marketing, to take diversified development, and in order to ensure a leading position in the market, Lion also actively involved in developing new talent to face the future knowledge economy era, the organization wants to expand, the importance of talent will also be increased. In order to enter the direct sale market, Lion also import Internet, open e-business models, so that consumers can be 24 h and zero-day to obtain information and purchase products, and with the rise of mobile devices, mobile technology in recent years, the Lion also import the development of mobile positioning and real-time services. In 1990, e-devotion, in 2000, networked, in 2010, put into action the clouds, after 10 years of their vigorously rectifying, the establishment of ICT application of technology as the core competitiveness of Lion.

On one hand, Lion Travel obtains an advantage in the virtual channel, on the other hand, it did not ignore the physical channel to the actual situation in hand strategy and in response to corporate globalization, and gradually in the United States, Canada, Hong Kong, New Zealand, Australia, Thailand, Japan, and mainland China and other places to set up local subsidiaries. And in 2006, it opened the first flagship store in order to provide 24 h of travel services for consumers and SMEs. Currently, it has 71 entities worldwide service locations that become Taiwan's largest travel agency outlets channel, providing consumers with the most complete services. Travel agencies offer cultural space, cafe? Business social media, content development focus? Lion Travel into the past few years or even life, cultural, and creative industries of operation, combined with the tourism outlets "Lion Hin auditorium," "Lion Café culture space"; *shaping* Life Style of complex cultural and creative space, auditorium content not limited to travel, and further extends to architecture, and culture, music, food, etc., intended to establish a sense of cultural and creative experience and aesthetics living area. Lion Group Chairman Wang Wenjie said, "Tourism into the school of modern tourism has become a Lifestyle!" So the tourism industry is no longer simply an increase in value, but to show the force of Taiwanese culture, lion upheld the "Travel is life" philosophy, and turned the *tourism industry* into a *life estate*.

Lion Travel enters into the period of mobilization since 2010, faces the rise of the use of MOBILE application, Lions do not fall in the trend, the first to import mobile commerce, the product further massed in full congregation for the market, different for different ethnic groups launched Focus commodities, while pushing the *customized products, theme travel, professional people to lead the way*, inject vitality of the cultural and creative tourism products, the development of sophisticated tourism-oriented, in order to improve the quality of Taiwan's tourism effort. In the community's economy, how to unite the divided, exchange and sharing, the tourism share expanded to life, is the Lion Travel direction of current efforts, Wang Wenjie stressed lion growing healthy and strong, insisted from the consumer point of view, with a depth of integration and cultural heritage of the content (Content), Focus formula for community management (Community), all aimed at niche markets looking to create unlimited business opportunities (Commerce). Lion Travel shares current updates through the entity "Lion Hin Lecture" virtual "Joy Media," "Hin Tourism," and communities Facebook etc., the actual situation in hand, will present significant learning community concept, through the process of gathering and focus to become a platform for sharing and exchange among travelers. From a global point of view, the development of the tourism industry, travel agencies, and life has become a trend in fashion industry combined, for example, Japan's largest travel agency JTB (Japan Travel Bureau), would be "Your Global Lifestyle Partner" with the corporate philosophy and the message of "Lifestyle" combined with travel ideas.

14.5 Conclusions and further research

In travel today, where firms' only communication with its customers often occur over the Internet, it is more important than ever to have a robust e-CRM and m-CRM strategy. Firms do not have to spend huge amounts of money in order to start an e-CRM and m-CRM program for their travel and hotel needs. Addressing these main aspects of firms' e-CRM and m-CRM program: Gaining better knowledge of past and current customers, enhancing all aspects of customer service (both online and offline), and personalizing firms' marketing messages with information that is relevant to firms' customers will result in more premium-driven loyalty for firms' travel and hotel products. The study has proved the factors that affect the tourism and travel industry adopting e-CRM and m-CRM strategy, but the industries still need to account their situations in different context to ensure their best performance.

The tourism and travel industry is one of most high Labor-intensive industry, if e-CRM and m-CRM strategy adoption, and the study believes that no matter on technique or interaction with customers, it would get better than the before. The study also accredit other industries will follow this trend of e-CRM and m-CRM and join the parade. Is e-CRM and m-CRM in Travel a Database Management or a Business Strategy to Engage the Customer? The most common misconception in the industry is that CRM in travel and hotel is synonymous with database management. Many hotel/travel companies, authors, and conference speakers alike treat CRM as some kind of a technology application. The truth is that CRM in travel is much more than technology or database management. CRM and its online applications, e-CRM and m-CRM, are business strategies aiming to engage the customer in a mutually beneficial relationship.

Within this context, here is the best and most universal description of e-CRM and m-CRM for any travel company, travel supplier, or online travel agency: e-CRM and m-CRM, in the context of Internet distribution and marketing in the travel vertical, is a business strategy supported by web technologies. It allows travel companies (travel suppliers and online travel agencies—OTAs) to engage customers in strong, personalized, and mutually beneficial interactive relationships, increase conversions and sell more efficiently.

The CSFs of an e-CRM and m-CRM strategy in travel and hotel are as follows:

1. Developing customer-centric strategies
2. Redesigning workflow management systems
3. Reengineering work processes
4. Supporting with the right technologies
5. Know your customer
6. Customer service
7. Personalization
8. More efficient marketing
9. Build customer loyalty
10. Finding the customers
11. Make them an offer they cannot refuse
12. Present options to the potential customers
13. Keep track of customers
14. Take care of customers' needs

Establishing mutually beneficial interactive relationships with customers is the ultimate goal of any e-CRM and m-CRM initiative. Outcomes from this research will indicate that companies implement e-CRM and m-CRM for different reasons, and that e-CRM and m-CRM implementation brings both tangible and intangible benefits to the companies.

Mobile services and web application world is emerging very fast, and in a couple of years they will have a significant value in the market share, may be even surpassing the use of desktop applications. This will have a huge impact in every area of the economy, technology, and service industry. Companies who adapt to fast-growing mobile market will have a significant advantage over those who fail to act fast and adapt quickly. A number of companies that have realized the power of mobility and web interconnection have already created advanced mobile services and applications and have made significant profits, increasing their market shares. All this can be applied to the travel and tourism sector too, making it more modern and more profitable.

The real challenge is not the implementation of technology-based e-CRM and m-CRM solutions itself but rather the adoption of the necessary technological, organization, and cultural changes within the companies. e-CRM and m-CRM adoption must be coupled with customer intelligence systems. Concerning the organization itself, they must be data integration interdepartmental communication and links with distributors and resellers. With regard to the corporate culture, the organization must be customer-centric in order to provide the best service. All the ingredients would allow e-CRM and m-CRM to contribute to the productivity and profitability of the tourism organizations.

References

Abowd, G.D. (1997). Cyberguide: A mobile context-aware tour guide, *ACM Wireless Networks*, 3, 421–433.

Alhaiou, T., Irani, Z., and Ali, M. (2009). The relationship between e-CRM implementation and eloyalty at different adoption stages of transaction cycle: A conceptual framework and hypothesis. *Proceedings of the European and Mediterranean Conference on Information Systems*. Izmir, Turkey: EMCIS.

Android. (2011). Android developers, http://developer.android.com (accessed on March 2011).

Anton, J. and Postmus, R. (1999). The CRM performance index for Web based business. Available: www.benchmarkportal.com (accessed on January 23, 2004).

Antonio, P., Krunoslav, Z., and Mario, M. (2011). Conceptual model of mobile services in the travel and tourism industry. *International Journal of Computers*, 5(3), 314–321.

Apple App Store. (2011). Apple Apple's app store downloads top 10 billion. http://www.apple.com/pr/library/2011/01/22Apples-App-Store-Downloads-Top-10-Billion.html (accessed on December 2011).

Bachman, J. (March 21, 2011). Russia eyes import duty for phones without GLONASS, Reuters, [Online]. Available: http://www.reuters.com/article/2010/10/27/telecoms-russia-dutiesidUSLDE69Q1KX20101027 (accessed on October 27, 2010).

Beça, P. and Raposo, R. (2013), m-Tourism 2.0: A concept where mobile tourism meets participatory culture. Dep. Comunicação e Arte, University of Aveiro, Portugal e-Review of Tourism Research (eRTR), Aveiro, Portugal, 7/02/2013.

Bigus, J.P. (1996). *Data Mining With Neural Networks: Solving Business Problems from Application Development to Decision Support*. McGraw-Hill, New York.

BlackBerry App World. (2011). BlackBerry App World v2.1.1.2 now available for download, http://crackberry.com/blackberry-app-world-v2-1-1-2-now-available-download (accessed on December 2011).

Bortenschlager, M., Häusler, E., Schwaiger, W., Egger, R., and Jooss, M. (2010). Evaluation of the concept of early acceptance tests for touristic mobile applications. *Information and Communication Technologies in Tourism*, 2010, 149–158.

Broadcom. (2011). 2011 Annual report, https://materials.proxyvote.com/Approved/111320/20120319/AR_121321/HTML2/default.htm (accessed on February 2012).

Bruner, G. and Kumar, A. (2003). Explaining consumer acceptance of handheld Internet devices. *Journal of Business Research*, 58, 115–120.

Buhalis, D. (2003). *eTourism: Information Technology for Strategic Tourism Management*. Essex, U.K.: Person Education Limited.

Buhalis, D. and Law, R. (2008). Progress in information technology and tourism management: 20 years on and 10 years after the Internet—The state of e-Tourism research. *Tourism Management*, 29, 609–623.

Caccomo, J.-L. and Solonandrasana, B. (2001). L'innovation dans l'industrie touristique: Enjeuxet stratégies (Paris, L'harmattan).

Camponovo, G., Pigneur, Y., Rangone, A., and Renga, F. (2005). Mobile customer relationship management: An explorative investigation of the Italian consumer market. *Proceedings of 4th International Conference on Mobile Business*, Sydney, New South Wales, Australia, July 11–13, 2005.

Canadi, M., Höpken, W., and Fuchs, M. (2010). Application of QR codes in online travel distribution. *Information and Communication Technologies in Tourism* 2010, 137–148.

Carlsson, C. and Walden, P. (2002). Mobile commerce: A summary of quests for value-added products and services. *15th Bled International Conference on E-Commerce*. Bled, Slovenia.

Cech, P. and Bures, V. (2009). Advanced technologies in e-tourism. *Proceedings of the 9th WSEAS International Conference on Applied Computer Science*, Moscow, Russia, August 20–22, 2009, pp. 85–92.

Chiu, D.K. et al. (2009). Towards ubiquitous tourist service coordination and process integration: A collaborative travel agent system architecture with semantic web services. *Journal of Information Systems Frontiers*, 11(3), 241–256.

Conci, M., Pianesi, F., and Zancanaro, M. (2009). Useful, social and enjoyable: Mobile phone adoption by older people. *Lecture Notes in Computer Science*, 5726, 63–76.

Constantelou, A. (2002). Emerging trends in customer relation management using ICT: The travel industry. STAR issue report, No. 22.

Cooper, J. (2003). Fractal assessment of street level skylines: A possible means of assessing and comparing character. *Urban Morphology*, 7(2), 73–82.

Cyr, D., Head, M., and Ivanov, A. (2006). Design aesthetics leading to m-loyalty in mobile commerce. *Information and Management*, 43(8), 950–963.

de Ruyter, K., Wetzels, M., and Kleijnen, M. (2001). Customer adoption of e-service: An experimental study. *International Journal of Service Industry Management*, 12(2), 184–207.

Dìez-Dìaz, F., Gonz´alez-Rodr´ıguez, M., and Vidau, A. (2007). An accessible and collaborative tourist guide based on augmented reality and mobile devices. *Lecture Notes in Computer Science*. Springer-Verlag, Berlin, Heidelberg, Germany, pp. 353–362.

Dobre, C. and Constantin, A. (2010). The experiences of the online buyers. *Proceedings of the 5th WSEAS International Conference on Economy and Management Transformation* (*Volume II*), Timisoara, Romania, October 24–26, 2010, pp. 824–830.

Facchetti, A., Rangone, A., Renga, F., and Savoldelli, A. (2005). Mobile marketing: An analysis of key success factors and the European value chain. *International Journal of Management and Decision Making*, 6(1), 65–80.

Facebook. (2010). A new help center for 2010, https://www.facebook.com/notes/225000527130 (accessed on December 2010).

Feinberg, R. and Kadam, R. (2002). e-CRM web service attributes as determinants of customer satisfaction with Websites. *International Journal of Service Industry Management*, 13, 432–451.

Feiner, S., Macintyre, B., and Seligmann, D. (1993). Knowledge-based augmented reality. *Communication of ACM*, 36(7), 53–62.

Finn, M., Elliott-White, M., and Walton, M. (2000). *Tourism & Leisure Research Methods: Data Collection, Analysis and Interpretation*. Pearson Education, Longman, Harlow, U.K.

Fjermestad, J. and Romano, N. (2003). Electronic customer relationship management: Revisiting the general principles of usability and resistance–An integrative implementation framework. *Business Process Management Journal*, 9, 572–591.

Gensollen, M. (2003). Biens informationnels et communautés médiatisées. *Revue d'Economie Politique*, 113, hors série, 9–40.

Gibbons, G. (March 7, 2011). China GNSS 101 compass in the rearview mirror. Inside GNSS, [Online]. Available: http://www.insidegnss.com/auto/jan-feb08-china.pdf (accessed on January/February 2008), pp. 62–63.

Google. (December 22, 2010). [Online]. Available: http://www.google.com/

Google Android Market. (2011). A new android market for phones, with books and movies, http://googlemobile.blogspot.tw/2011/07/new-android-market-for-phones-with.htmlTuesday (accessed on July 2011).

Grelier, T. (March 7, 2011). Initial observation and analysis of compass MEO satellite signals. Inside GNSS, [Online]. May/June 2007, pp. 39–43. Available: http://www.insidegnss.com/auto/IG0607CompassFinal.pdf.

Guerley, W. (2000). Making sense of the wireless web. Fortune. Available: www.fortune.com (accessed on August 15, 2000).

Hampe, J.F., Swatman, P.M.C., and Swatman, P.A. (2000). Mobile electronic commerce: Reintermediation in the payment system. *Proceedings of the 13th Bled International Electronic Commerce Conference*, Bled, Slovenia, June 19–21, 2000, pp. 693–706.

Han, J. and Han, D. (2001). A framework for analyzing customer value of internet business. *Journal of Information Technology Theory and Application*, 3(5), 25–38.

Helm, S. (2000). Viral marketing—Establishing customer relationships by word-of-mouse. *Electronic Markets*, 10(3), 158–161.

Herrera-Viedma, E., Martınez, L., Mata, F., and Chiclana, F. (2005). A consensus support system model for group decision-making problems with multigranular linguistic preference relations. *IEEE Transaction on Fuzzy Systems*, 13(5), 644–658.

Herrera-Viedma, E., Peis, E., and Morales-del-Castillo, J.M. (2005). Gathering information on the Web using fuzzy linguistic agents and semantic web technologies. *Proceedings of the 7th Conference of the European Society for Fuzzy Logic and Technology* (*EUSFLAT—LFA*). pp. 1243–1247.

Hsu, C.-L. and Lin, J.C.-C. (2008). Acceptance of blog usage: The roles of technology acceptance, social influence and knowledge sharing motivation. *Information & Management*, 45(1), 65–74.

Iansiti, M. and Kim, B.C. (1994). Integration and dynamic capability: Evidence from product development in automobiles and mainframe computers. *Industrial and Corporate Change*, 3(3), 557–605.

Internet World Stats. (2011). Internet usage statistics, the internet big picture world internet users and population stats, http://www.internetworldstats.com/stats.htm (accessed on October 2011).

Ip, C., Leung, R., and Law, R. (2010). Progress and development of information and communication technologies in hospitality. *International Journal of Contemporary Hospitality Management*, 23(4), 533–551.

Issler, J. et al. (March 11, 2011). Galileo frequency and signal design. GPS World, [Online]. Available: http://www.gpsworld.com/gnss-system/galileo/galileofrequency-signal-design-810 (accessed on June 14, 2003), pp. 30–37.

ITU. (January 23, 2011). The world in 2010: ICT facts and figures. ITU ICT Statistics, [Online]. Available: http://www.itu.int/ITU-D/ict/material/FactsFigures2010.pdf.

Jahn, A. et al. (2003). Evolution of aeronautical communications for personal and multimedia services. *IEEE Communications Magazine*, 41(7), 36–43.

Jarvinen, J., DeSalas, J., and LaMance, J. (March 20, 2011). Assisted GPS: A low-infrastructure approach. GPS World, [Online]. Available: http://www.gpsworld.com/gps/assisted-gps-a-lowinfrastructure-approach-734 (accessed on March 1, 2002).

Julier, S., Lanzagorta, M., Baillot, Y., and Brown, D. (2002). Information filtering for mobile augmented reality. *Proceedings of the 3rd International Symposium on Mixed and Augmented Reality: Let's Go Out: Workshop on Outdoor Mixed and Augmented Reality*. Orlando, FL.

Keen, P. and Mackintosh, R. (2001). *The Freedom Economy: Gaining the M-commerce Edge in the Era of the Wireless Internet*. Osborne/McGraw-Hill, Berkeley, CA.

Kimber, J., Georgievski, M., and Sharda, N. (2005). Developing usability testing systems and procedures for mobile tourism services. In: *Annual Conference on Information Technology in the Hospitality Industry*, HITA 2005, June 19th and 20th, 2005, Hospitality Information Technology Association, Los Angeles, CA, pp. 12–23.

Kotler, P. and Armstrong, G. (2010). *Principles of Marketing*, 13th ed. Pearson Education, Harlow, U.K.

Kotorov, R. (2002). Ubiquitous organizational design for e-CRM. *Business Process Management Journal*, 8, 218–232.

KPMG. (2001). Customer relationship management (CRM) and the global travel industry. http://www.kpmg.cz/dbfetch/52616e646f6d49560d910130babfbd69896d05863247cdcb/crm.and.the.global.travel.industry, 2008 (accessed on December 2012).

Kumar, S. and Zahn, C. (2003). Mobile communications: Evolution and impact on business operations. *Technovation*, 23(6), 515–520.

Larsen, J., Urry, J. and Axhausen, K.W. (2007). Networks and tourism: Mobile social life. *Annals of Tourism Research*, 34(1), 244–262.

Law, R. and Cheung, C. (2005). A study of the perceived importance of the overall website quality of different classes of hotels. *Hospitality Management*, 25, 515–531.

Lee, D., Park, J., and Ahn, J. (2001). On the explanation of factors affecting e-commerce adoption. *Proceedings of the 22nd International Conference on Information Systems*, New Orleans, LA, December 16–19, 2001, pp. 109–120.

Leiper, N. (2000a). Are destinations 'the heart of tourism'? The advantages of an alternative description. *Current Issues in Tourism*, 3(4), 364–368.

Lidija, P. (2008). E-CRM in the tourism sector. *Journal of Tourism*, 5, 14–19.

Liu, J. et al. (2010). Tourism emergency data mining and intelligent prediction based on networking autonomic system. Presented at the *International Conference on Networking, Sensing and Control (ICNSC)*, Chicago, IL, April 10–12, 2010.

Luck, D. and Lancaster, G. (2003). E-CRM: Customer relationship marketing in the hotel industry. *Managerial Auditing Journal*, 18(3), 213–231.

Maervoet, J. et al. (2008). Tourist decision support for mobile navigation systems: A demonstration. *Applied Artificial Intelligence*, 22(10), 964–985.

Maurer, C. (2007). The role of e-CRM in an integrated eMarketing strategy for tourism organisations. *In Tourism-Review On-Line Magazine*, 2(2007), 42–43.

Microsoft Store. (2011). Sales and support, http://www.microsoftstore.com/store/msusa/en_US/DisplayHelpContactUsPage/ (accessed on December 2011).

Milovic, B. (2011). *Razlike CRM Ie-CRMposlovnestrategije. Xmeđunarodninaučno-stručniSimpozijum INFOTEH-JAHORINA.* ElektrotehničkifakultetIstočno Sarajevo, Sarajevo, Bosnia and Herzegovina, pp. 720–724.

Mobile Marketing Association. (2004). *What is Mobile Marketing*. Mobile Marketing Association Ltd, Denver, CO.

Newell, F. and Lemon, K. N. (2001). *Wireless Rules*. New York: McGraw-Hill Companies, Inc.

Nysveen, H., Pedersen, P.E., Thorbjornsen, H., and Berthon, P. (2005). Mobilizing the brand: The effects of mobile services on brand relationships and main channel use. *Journal of Service Research*, 7(3), 257–276.

observatory.gr (2007), Evaluation, Synthesis and Formulation of Proposals for the Use of ICT in the Tourism Sector, ICAP S.A. (Subcontractors: ITMC S.A., TAYLOR NELSON SOFRES ICAP S.A.., Panteion University—Academic Research Regional Development Institute).

O'Reilly, T. (February 7, 2011). What is web 2.0—Design patterns and business models for the next generation of software. [Online]. Available: http://oreilly.com/pub/a/web2/archive/what-is-web-20.html (accessed on September 30, 2005).

Otto, J. and Chung, B. (2000). A framework for cyber-enabled retailing: Integrating e-commerce retailing with brick and mortar retailing. *Electronic Markets*, 10, 185–191.

Paelke, V. and Sester, M. (2010). Augmented paper maps: Exploring the design space of a mixed reality system. *ISPRS Journal of Photogrammetry and Remote Sensing*, 65(3), 256–265.

Palen, L., Salzman, M., and Youngs, E. (2001). Discovery and integration of mobile communications in everyday life. *Personal Ubiquitous Computing*, 5(2), 109–122.

Pedersen, P.E., Methlie, L.B., and Thorbjørnsen, H. (2002). Understanding mobile commerce end-user adoption: A triangulation perspective and suggestions for an exploratory service evaluation framework. *Proceedings of the 35th Hawaii International Conference on System Sciences*, Big Island, HI; Los Alamitos, CA: IEEE Computer Society Press, January 7–10, 2001.

Pettey, C. and Goasduff, L. (March 1, 2011). Gartner says worldwide mobile device sales to end users reached 1.6 billion units in 2010; Smartphone sales grew 72 percent in 2010. Press release, [Online]. Available: http://www.gartner.com/it/page.jsp?id=1543014 (accessed on February 9, 2011).

Phelps, E. A., O'Connor, K. J., Cunningham, W. A., Funayama, E. S., Gatenby, J. C., Gore, J. C., and Banaji, M. R. (2000). Performance on indirect measures of race evaluation predicts amygdala activation. *Journal of Cognitive Neuroscience*, 12, 729–738.

Pocket Gear Mobile Application Store. (2011). Mobile app store pocketgear rebrands as Appia, Goes White Label, http://readwrite.com/2011/02/02/mobile-app-store-pocketgear-rebrands-as-appia-goes-white-label (accessed on December 2011).

Porter, L. and Golan, G.J. (January 12, 2011). From subservient chickens to brawny men: A comparison of viral advertising to television advertising. *Journal of Interactive Advertising*, [Online], 6(2), 2006. Available: http://jiad.org/article78.

Qualcomm. (2011). Qualcomm Inc/De Form 10-K (annual report) filed 11/02/11 for the period ending 09/25/11, http://files.shareholder.com/downloads/QCOM/3303429723x0x523322/6EBEB924-8E0D-4829-B233-2ED3D7FD3625/2011_Annual_Report_on_Form_10-K.pdf (accessed on September 2011).

Rajamaky, J. et al. (2007). Laurea POP indoor navigation service for the visually impaired in a WLAN environment. *Proceedings of the 6th WSEAS International Conference on Electronics, Hardware, Wireless and Optical Communications*, Corfu Island, Greece, February 16–19, 2007, pp. 96–101.

Riegner, C. (2007). Word of mouth on the web: The impact of web 2.0 on consumer purchase decisions. *Journal of Advertising Research*, 47(4), 436–447.

Riel, A.C., Liljander, V., and Jurriens, P. (2001). Exploring customer evaluations of e-services: A portal site. *International Journal of Service Industry Management*, 12, 359–377.

Rowley, J. (2002). Reflections on customer knowledge management in e-business. *Quantitative Market Research: An International Journal*, 5, 268–280.

Russian Institute of Space Device Engineering. (March 7, 2011). Global navigation satellite system GLONASS, Navigational radiosignal in bands L1, L2. Interface control document (Edition 5.1), [Online]. Moscow, Russia, 2008. Available: http://rniikp.ru/en/pages/about/publ/ICDGLONASS eng.pdf. (accessed on October 2013).

Samir A. El-Seoud and Hosam F. El-Sofany (2010), Mobile Tourist Guide—An Intelligent Wireless System to Improve Tourism, using Semantic Web, Conference ICL2010 September 15–17, 2010, Hasselt, Belgium.

Schierholz, R., Kolbe, L.M., and Brenner, W. (2006). Mobilizing customer relationship management—A journey from strategy to system design. *Proceedings of the 39th Annual Hawaii International Conference on System Sciences* (HICSS'06). Kauai, HI.

Schmalstieg, D. and Wagner, D. (2007). ARToolKitPlus for pose tracking on mobile devices. In: *Proceedings of the 12th Computer Vision Winter Workshop*, St. Lambrecht, Austria, 2007, pp. 139–146.

Schmandt, C. and Marmasse, N. (2004). User-centered location awareness. *Computer*, 37(10), 110–111.

Senn, J.A. (December 2000). The emergence of m-commerce. *Computer*, 33(12), 148–150.

Sideco, F. (January 21, 2011). Market for mobile communications gear nears quarter-trillion-dollar mark. Market research press release, December 22, 2010. [Online]. Available: http://www.isuppli.com/Mobile-and-Wireless-Communications/News/Pages/Marketfor-Mobile-Communications-Gear-Nears-Quarter-Trillion-Dollar-Mark.aspx.

Singh, M. (2002). e-services and their role in B2C. *Managing Service Quality*, 12, 434–446.

Sigala, M. (2004). Designing experiential websites in tourism and travel: A customer-centric value approach.

Sigala, M. (2004). The ASP-Qual model: Measuring ASP service quality in Greece. *Managing Service Quality*, 14(1), 103–114.

Sigala, M. (2005a). Integrating customer relationship management in hotel operations: Managerial and operational implications. *International Journal of Travel Management*, 24(3), 391–413.

Spizzichino, F. (1991). Sequential burn-in procedures. *Journal of Statistical Planning and Inference*, 29, 187–197.

Srivastava, J. and Cooley, R. (Spring 2003). Web business intelligence: Mining the web for actionable knowledge. *INFORMS Journal on Computing*, 15(2), 191–207.

Stevens, H. and Pettey, C. (March 1, 2011). Gartner says worldwide smartphone sales reached its lowest growth rate with 3.7 per cent increase in fourth quarter of 2008. Press release, [Online]. Available: http://www.gartner.com/it/page.jsp?id=910112 (accessed on March 11, 2009).

Suh, Y., Shin, C., Woo, W., Dow, S., and Macintyre, B. (2011). Enhancing and evaluating users' social experience with a mobile phone guide applied to cultural heritage. *Personal and Ubiquitous Computing*, 15(6), 649–665.

Takada, D., Ogawa, T., Kiyokawa, K., and Takemura, H. (2009). A context-aware AR navigation system using wearable sensors. In: *Proceedings from the 13th International Conference Human-Computer Interaction*, pp. 1–10. Berlin, Germany: Springer-Verlag.

Tarasewich, P. (2003). Designing mobile commerce applications. *Communications of the ACM*, 46(12), 57–60.

The Forrester Report. (2001). *Custom Chemicals Materialize*. Cambridge, MA: Forrester Research, Inc.

The Nielsen Company. (2011, January 21). Mobile youth around the world. The Nielsen Company Market Report, [Online]. Available:http://www.nielsen.com/content/dam/corporate/us/en/reportsdownloads/2010%20 Reports/Nielsen-Mobile-Youth-Around-The-World-Dec-2010.pdf (accessed on December 2010).

Thomas, J.S. and Sullivan, U.Y. (2005). Managing marketing communications with multichannel customers. *Journal of Marketing*, 69(4), 239–251.

Tremblay, R. E. (1999). The development of aggressive behaviour during childhood: What have we learned in the past century? *International Journal of Behavioural Development*, 24(2), 129–141.

Tribe, J. (1997). The indiscipline of tourism, *Annals of Tourism Research*, 24, 638–657.

Twitter. (2010). Twitter: 2010, the year in review, https://2010.twitter.com/ (accessed on December 2010).

Virrantaus, K. et al. (2001). Developing GIS-supported location-based services. *Proceedings of WGIS2001 First International Workshop on Web Geographical Information Systems*. Kyoto, Japan, pp. 423–432.

Virtanen, A. (2001). Navigation and guidance system for the blind. *Proceedings of Interactive Future and Man'01*, Tampere, Finland, November 2001.

Virtanen, A. and Koskinen, S. (2004). NOPPA navigation and guidance system for the blind. *Proceedings of the 11th World Congress and Exhibition on ITS*, Nagoya, Japan, October 2004.

Voss, C. (2000). Developing e-service strategy. *Business Strategy Review*, 11, 21–33.

Wilhelm, A. (2011, January 17). How big is the internet. [Online]. Available: http://thenextweb.com/shareables/2011/01/11/infographichow-big-is-the-internet.

Wong, K. and Choi, Y. (2007). A simple location-based service on urban area. *International Journal of Computers*, 4(1), 291–295.

chapter fifteen

Security intelligence for healthcare mobile electronic commerce

Joseph M. Woodside and Mariana Florea

Contents

15.1 Mobile healthcare environment

In the healthcare environment of today, individuals are increasingly connecting with one another using mobile devices such as tablets and smartphones (Camlek, 2011). Clinicians and patients demand current information at their fingertips during all phases of healthcare delivery to save time, reduce errors, and improve outcomes. As a result of the growth of mobile technology in healthcare, there arises a security challenge to the administrators of the environment to ensure high levels of security and control.

The healthcare environment is a prime target for data and identity theft due to the available content and detection capabilities. Mobile devices can provide unwanted access to a variety of data, including contacts, texts, calls, e-mail, calendars, internal systems, credit card information, and clinical data. The increases in healthcare security breaches can be tied to regulation requirements, automation increases, social media development, and human errors. The economic burden created by these data breaches in healthcare is estimated at $7 billion annually, with $1 million per organization annually in case of a breach. Risks of breaches are expected to continue to grow along with mobile technology usage (IDExperts, 2012).

Security is a major priority to healthcare organizations, given that patients entrust their detailed information to the organizations. When security monitoring systems are in place, this information is not always up to date, generate false positives, and the results vary between vendors. This chapter provides an overview of security intelligence, with three modules of mobile security, security methods, and application of a security defense.

15.2 Mobile security

With an increased usage of healthcare information technology such as e-prescribing, electronic health records (EHRs), personal health records (PHRs), social media networks, health information exchanges, and mobile devices, the potential risk of data or information loss has increased (Keckley, 2011). Specifically, with the growth in mobile technology, the results of a 2012 study that reviewed breaches reported to the US Department of Health and Human Services in response to Health Insurance Portability and Accountability Act (HIPAA) mandates showed that over 60% of breaches occurred due to mobile devices that had been lost or stolen (Kruger and Anschutz, 2013).

One of the main issues of the use of information technology and mobile devices in healthcare is the high risk of data loss, a security and privacy aspect that concerns healthcare staff as well as patients. Defined by the US Department of Health and Human Services as "an individual right to control the acquisition, uses, or disclosures of his or her identifiable health data," health information privacy is an important aspect of the patient's experience. In healthcare, there are at least two main categories of medical records: the EHR, the record created and managed by the care provider, and the PHR, the record created and managed directly by the patient. The most popular PHR devices are Microsoft Health Vault and, previously, Google Health. These products allow users to perform multiple operations "such as deleting, editing, and sharing their protected health information (PHI) with multiple entities including family, friends, and health care professionals," but besides these useful features, mobile PHR devices have their share of risk (Avancha et al., 2012).

While these methods exist and can be used to reduce the risk of information theft, a report made in December 2012 in Ponemon's Third Annual Benchmark Study on Patient Privacy and Data Security revealed that "81 percent of healthcare organizations permit employees and medical staff to use their own devices to connect to their organization's networks or enterprise systems" and 54% of these people said they were not confident that the devices they were using were secure (Kruger and Anschutz, 2013). At the same time, 66% of the nurses declared they used their smartphones for clinical communication while 95% of them said that the hospital IT departments did not support their devices "fearing security risks" (Kruger and Anschutz, 2013).

Despite initial security risks and concerns over the use of personal devices, the growth in "bring your own device" and "bring your own technology" has been increasing and driven primarily by cost reductions, productivity improvements, and employee satisfaction through allowing the use of personal smartphone devices. In addition, mobile devices are able to access cloud-based platforms for storing information and utilized for daily operational items including notes, documents, messages, other information that often contains PHI in a healthcare environment, where clinical staff may take notes on a patient or postclinical documents for clinical review and monitoring (Trend Micro, 2012).

15.3 Security methods

In order to ensure that adequate security is in place for the user of mobile devices in healthcare, security methods must be employed. In security methods, or the tools and techniques used to prevent security issues, there are three main categories: (1) authentication and authorization, (2) prevention and resistance, and (3) detection and response (Baltzan and Phillips, 2012).

15.3.1 Authentication and authorization

Authentication and authorization deals primary with people, which is often the greatest source of security breaches. This includes people both inside the organization who may misuse or distribute their access and outside the organization who may include social engineering to access information (Baltzan and Phillips, 2012). The danger can be diverse and can come from outsiders who are unauthorized to get the information, from insiders who are authorized users, or even from the patient himself with the misuse of technology. A report made in 2011 showed that "71 percent of healthcare organizations suffered one or more ePHI breaches in the course of one year- most of which originated from insiders in one form or another" (Kruger and Anschutz, 2013).

The possible threat includes stealing the patient's identity, an unauthorized access into the system, or the disclosure of information with negative impact. Identity fraud can be reduced by mechanisms that are used to authenticate the patient, the provider, or the devices that request access to the network. In health-related cases, correct authentication is particularly important because "in other clinical scenarios, the patient may also be motivated to cheat by attaching the sensor to a healthier friend or relative" (Avancha et al., 2012).

Access threat concerns include inappropriate access but also involve modification of health records that may be done by mistake or done for *illegal or malicious purpose*. The access topic has two main perspectives: one regarding the role of the patient and the other one concerning the devices that are allowed to access records, since both these aspects can have important implications. Disclosure threats, including *data at rest and data in transit*, can result in the release of information due to the allowance of "data disclosure beyond what was intended by the act of sharing" or motivated by "financial gain or to embarrass the patient" (Kruger and Anschutz, 2013).

In summary, "anyone with potential access to ePHI could pose a potential threat to the security of the information" (Kruger and Anschutz, 2013). Strong information security policies and security plans such as password and log-on requirements can help prevent these types of issues. Authentication confirms the user's identity, whereas authorization provides the user with an access to the environment. Devices operated using password such as smart cards, tokens, and biometrics are used for the authentication of the user (Baltzan and Phillips, 2012).

15.3.2 Prevention and resistance

All the advantages that information technology is bringing to the healthcare system come with multiple challenges regarding the data security and privacy. The traditional approach for data protection, the perimeter approach that describes the internal network of an organization as a perimeter defined network, is nowadays obsolete due to the extended usage of mobile devices. Not only that the perimeter can no longer be defined, but also that the "perimeter-only security ignores the inside threat that exists when a hospital's own staff or others with access to the organization's ePHI maliciously or nonmaliciously access or leak protected health information" (Kruger and Anschutz, 2013).

Prevention and resistance deals primarily with data and technologies, including encryption, content filtering, and firewalls. A firewall is a hardware or software device that analyzes information to detect unauthorized use. Content filtering prevents e-mail and spam from being received or transmitted. Encryption requires a special key to decode the information and make it readable; this is used to keep financial-related information secure (Baltzan and Phillips, 2012).

Encryption is one efficient method of maintaining the safety and privacy of the data used by many healthcare providers. This is a relatively low-cost data protection method averaging at about $55 per computer. According to a survey conducted in 2012 by the Healthcare Information and Management Systems Society, only 64% of healthcare organizations were using encryption when transmitting data. The rest of 36% were not using encryption even though since 2005 all data handlers have been required to follow the HIPPA security standards (Kruger and Anschutz, 2013; Conn, 2013).

The risk in transmitting or even storing unencrypted data is high. One example is the case of the stealing of four computers, in July 2012, from Advocate Medical Group located in Park Ridge, Illinois, which was followed by another similar case that took place in November 2009 affecting the medical records of 812 patients of Advocate. Although the data contained in the stolen computers was varied and not highly private, "we are certainly not trying to state that this information couldn't be used inappropriately," said Kelly Jo Golson, senior vice president and chief marketing officer for Advocate Health Care, based in Downers Grove, Illinois. Multiple existing methods can be used to make it possible for data to be "decrypted only when a legitimate user on a known device using an approved application opens them," which can significantly reduce the risk of data theft (Kruger and Anschutz, 2013).

15.3.3 Detection and response

Detection and response deals primarily with attacks by analyzing suspicious activities such as password attempts or file access. Intrusion detection software will monitor and alert if patterns are detected and can even shut down part of the network as warranted (Baltzan and Phillips, 2012). Organizational users need advanced tools similar to those of malicious users being used to compromise the systems. Security Intelligence and event monitoring systems analyze network, user, application, and datasets to identify trends, behaviors, and incidents (LogRhythm, 2012).

For dataset inputs, these include firewall, network, system, application, rules, and other event logs. These logs are then normalized to a standard format for review. Once standardized, the data are analyzed for patterns, and alerts are generated for user review. Examples of analysis methods include aggregation and categorization of logs and events, time of events and directions, statistical log, source and host information, and top items within various categories for further detailed drill-down and analysis. Examples of analysis output include the ability to detect unusual application behaviors, unusual network connections, user behavior, network baseline deviations, and compromised credentials. Other methods analyze historical data to recreate scenarios for auditing and also generate detailed and summary reporting output for security professionals, compliance officials, or other end users to review. Organizations are beginning to establish a security center in which monitoring and investigations occur (LogRhythm, 2012).

15.4 Security defense

Healthcare security breaches have become a common catastrophe. In a recent study, 94% of healthcare organizations suffered a data breach in the last 2 years, and nearly half experienced more than five data breaches in the last 2 years. In 2010, a flash drive with PHI of 280,000 members was stolen from a health plan. In 2006, a laptop and a disk with PHI of 26.5 million veterans were stolen from an employee's home. Since the enforcement of the HIPAA, over 11,000 HIPAA violations have been reviewed, and 7 million patients have been impacted (Keckley, 2011).

Despite security risks, there are many benefits to healthcare mobile usage. At ASAN Medical Center in Seoul, the staff is using smartphones and laptops to improve productivity, diagnostics, and problem solving up to three times faster by keying and accessing information directly on their mobile devices (HP, 2012). In 2009, the American Recovery and Reinvestment Act (ARRA) was enforced, which also included components from the Health Information Technology for Economic and Clinical Health (HITECH) Act. These improved HIPAA enforcement for privacy and security safeguards, along with increased penalties and accountability (Keckley, 2011).

Healthcare organizations must implement safeguards to address new security concerns as a result of increased electronification as well as comply with enforced laws (CMS, 2007). The primary components of a security defense include physical, technical, and organizational safeguards.

Technical safeguards include unique user identification, automatic log-off, encryption, having a responsible person to authorize and verify passwords, strong passwords, locking accounts after invalid log-ins, and deactivating employee accounts after termination. From a mobile perspective, endpoint access should also be verified and permitted or prevented from accessing the network, including monitoring and notification of an unauthorized device. Wireless threats should be detected and prevented through security policies and location tracking. Security compliance should be kept by administrators to verify personally owned and operated devices to ensure compliance (Iron Mountain, 2010; HP, 2012; Thrasher and Revels, 2012).

Physical safeguards include protecting facility access against unauthorized entry, as well as security workstations, transportation, and storage of media and information. However, physical safeguards also apply well behind the walls of an organization. Stanford's Lucille Packard Children's Hospital announced a breach of 57,000 patients due to a stolen laptop from a physician's vehicle. Gibson Hospital in Indiana also reported a breach of 29,000 patients due to a laptop being stolen from an employee's home.

Organizational safeguards include business associate agreements (BAAs), customer requirements, and policies and procedures. BAAs should be updated to eliminate breaches and enforce liability that may not be covered under the law. A business associate is anyone who works on behalf of a healthcare entity and uses or discloses PHI (Keckley, 2011).

References

Avancha, S., Baxi, A., and Kotz, D. (November, 2012) Privacy in mobile technology for personal healthcare. *ACM Computing Surveys*, 45(1): 1–54. Retrieved from Business Source Complete, EBSCO. Retrieved from http://ehis.ebscohost.com.proxy.stetson.edu.

Baltzan, P. and Phillips, A. (2012) *Business Driven Technology*, 5th edn. McGraw-Hill Irwin, New York, 2012.

Camlek, V. (2011) Healthcare mobile information flow. *Information Services and Use*, 31(1–2): 23–30.

CMS. (2007) Security Standards.

Conn, J. (September 9, 2013) Unprotected data. *Modern Healthcare*, 43(36): 16–16. Retrieved from Business Source Complete, EBSCO. Retrieved from http://ehis.ebscohost.com.proxy.stetson.edu.

Hewlett-Packard Development Company. (2012) Bring your own device in healthcare. Palo Alto, CA.

IDExperts. (December, 2012) Ponemon study reveals ninety-four percent of hospitals surveyed suffered data breaches.

Iron Mountain. (2010) HIPAA best practices checklist: Best practices that go beyond compliance to mitigate risks. White Paper, Boston, MA.

Keckley, P.H. (2011) *Privacy and Security in Health Care: A Fresh Look*. Deloitte, Washington, DC.

Kruger, D. and Anschutz, T. (February, 2013) A new approach in IT security. *Healthcare Financial Management*, 67(2): 104–106, 108. Retrieved from Business Source Complete, EBSCO.

LogRhythm. (December, 2012) Security intelligence: Can "Big Data" analytics overcome our blind spots? White Paper, Boulder, CO.

Thrasher, E. and Revels, M. (2012) The role of information technology as a complementary resource in healthcare integrated delivery systems. *Hospital Topics*, 90(2): 23–32.

Trend Micro Incorporated. (2012) Embracing BYOD: Are you exposing critical data?

chapter sixteen

Automated teller machine and mobile phone interface in a developing banking system

Abel E. Ezeoha and Anselm Nkalemu

Contents

16.1 Introduction: Emergence and growth of mobile phone and automated teller machine penetration in Africa

Two major visible revolutions that have very directly changed the ways and manners business was conducted in Africa over the last one and a half decade are the growth in the popularity of electronic payment (via automated teller machines [ATM]) and the rapid explosion in mobile phone penetration. It is interesting to note that before the year 2000, ATM terminals, for instance, were nonexistent in most parts of the sub-Saharan Africa (SSA). Today, this channel of bank service delivery has been fully integrated into the economic and social fabrics of the entire African society. Unlike the era when ATM card ownership was restricted to high-end bank customers, now, in most parts of the continent, having a bank account overwhelmingly qualifies an individual to be issued with an ATM card. Similarly, before the 1990s, telephone services and telephone line ownership was an exclusive preserve of rich individuals and businesses in Africa. Then, only fixed landline telephones were in existence; and the average telephone penetration rate in the SSA region was just 0.86 per 100 persons in 1985, compared with 1.48 and 1.57 per 100 persons in 2000 and 2005, respectively (based on the data from World Development Indicators). The recorded increase was largely due to the emergence and increase in the popularity of the mobile telephone system in most parts of the continent. The erstwhile barrier against telephone ownership and access has consequently been dismantled—thus making mobile phone emergence an effective tool for bridging the digital divide between the rich and the poor (Helton, 2012).

No doubt, the value and importance of the payment and telecommunication revolutions continues to be on the increase. There are for instance, a number of empirical studies linking each to economic growth and development. Abraham (2007) used the Indian case to demonstrate how the mobile phone improved the supply chain and enhanced the living standard of Indian fishermen. Another study by GSM Association and Deloitte revealed that a doubling of mobile data use brought about an "increase of 0.5 percentage points in the GDP per capita growth rate across selected 14 countries, and that countries with higher level of data usage per 3G connection had seen increases in GDP per capita growth exceeding a percentage point" (GSM Association, 2012). Regarding the impact of the ATM, the summary of a 56-country study on the impact of electronic payment on economic growth by Moody's Analyst (2013) showed that (1) the increased usage of electronic payment products added $983 billion in real (US) dollars to GDP in the countries studied; (2) card usage raised consumption by an average of 0.7% across the 56 countries; (3) that consumption contributed to average additional growth in GDP of 0.17% point per year for this group of countries. The study further indicated that real global GDP grew by an average of 1.8% during that time period in those countries; the additional GDP growth was realized solely by increased card usage and penetration was equivalent to creating 1.9 million jobs during the period of study; and that "a future 1% increase in card usage across the countries in the study would produce an annual increase of 0.056% in consumption and a 0.032% increase in GDP".

In addition to this interesting revelation, industry practitioners and policymakers have remained constantly in search of optimal strategies that can guarantee more inclusive financial and economic systems, as well as efficient business operations. The argument in this direction is that an inclusive financial/economic system is capable of lifting a huge percentage of the world's population from poverty.

While in some countries in Africa, telecom adoption is still at the experimental stage, in a good number of others, mobile telephone is fully integrated into the mainstream banking and business operations. Nigeria is one of the countries where ATM and mobile telephone interface is speedily evolving, with both having a very recent history. The first ATM in the country, for instance, was introduced for the high net worth customers of the defunct Societe Generale Bank in 1989. However, the adoption of the InterSwitch ATM system in 2003 brought about an increase in the number and rate of its use in the mainstream banking business in the country. Since then, there has been a massive increase in the number and usage of ATMs, which is largely due to regulatory reforms and market dynamisms. A report from InterSwitch, the largest of the four major global card operators in the country (others are QuickCash, ETranzact, and VPay), indicates that the number of ATM cards grew from 7 million in 2010 to 11 million in 2011 and 26 million in 2012 (Oketola, 2012).

In the same vein, mobile (GSM) phone system was introduced in the country in the year 2000, and has since then been increasingly used as a support tool for business operations. Banks have also taken advantage of the increasing popularity of mobile telephone as an essential business support tool to develop new mobile banking products. Thus, as in the case of most other developing countries, the boom in the mobile banking services effectively coincided with the era of the increasing wave of mobile telephone penetration.

Customers' mobile phone numbers are today effectively linked to their individual bank accounts. This provides opportunities for the customers to access banking services via dedicated telephone lines from the comfort of homes and offices, check account balance, authorize interbranch money transfer, and get transaction alerts. With mobile telephone banking, customers can instantly get notification via short message service (SMS) or e-mail on every transaction that takes place in their respective accounts.

Specifically, this chapter discusses issues on the relationship between ATM and mobile phone penetration; the benefits of an effective ATM–mobile phone interface; and the regulatory and policy framework for such technology interface in an emerging financial system such as Nigeria. To do this, extensive theoretical and empirical literature review on the link between ICT innovation and banking development is carried out. This is followed by a review of mainstream publications on electronic banking and banking regulations and telecommunication development in Nigeria, and a statistical analysis of data on ATM and mobile phone penetration in a sample of SSA countries. Broadly, the chapter draws inference from and builds on the existing literature on inclusive banking system, electronic banking, information and communication technology adoption, and penetration, as well as the emerging challenges of technology-induced banking regulations.

16.2 *Nature and benefits of ATM and mobile phone interface*

There are well documented empirical and theoretical evidence that effective ICT deployment generally promotes and facilitates business growth (Parent and Cruickshank, 2009; Byrne et al., 2011; Vu, 2011; Bilbao-Osorio et al., 2012; Sassi and Goaied, 2013). Arguably, ICT protocols that have more significantly influenced economic and financial system development in recent times are the Internet and mobile phone. Among others, the Internet specifically assists businesses through the reduction in information costs and the expansion of business opportunities and platforms (Comptroller's Handbook, 1999; Calisir and Gumussoy, 2008; Ezeoha, 2010; Alonso et al., 2013). Evidence on the economic impact of mobile phone is equally well documented (see for instance Michalakelis et al., 2008; Singh, 2008; Funk, 2009; Porter et al., 2012; Asongu, 2013; Qureshi, 2013). There are also arguments that favor the existence of a significant interactive impact of mobile banking and mobile phone, especially in the developing economies. Beshouri and Gravråk (2010) have, for instance, posited that the emerging high mobile phone penetration and limited branch network of commercial banks and microlenders is capable of driving the growth of mobile money in the emerging markets of the Middle East, Asia, and Africa, where it is expected to cross one billion users; and specifically in Africa where it is estimated to grow to a US$22 billion industry across Africa by 2015.

Mobile banking has however been shown to offer more feasible benefits in an environment where there is a reasonable degree of mobile phone penetration. Rajnish and Stephan (2007) highlight that the operational scope of a telecom-based mobile banking can cover both the conduct of a bank's account system administration, customer services, and even stock market operations. Specifically, Pousttchi and Schurig (2004) have argued that the expansion in mobile banking operations around the world coincided with the advent of mobile telephone. It is this coincidence that has revolutionalized the platform for banking services delivery and payment systems in different developing economies (Sullivan, 2007). A unique outcome of this has been an increased sophistication in mobile banking and the growth from mere informational usage to more interactive applications. This development has some positive implications on the well-being of the overall banking system, especially with regard to:

- The transaction cost reduction channel—which creates a platform for moderating the deterring high cost features that are associated with the conventional banking technique. Such an interface is capable of reducing the cost of basic banking services and making them affordable to customers (Datta et al., 2001).

- Information dissemination channel—this provides medium for wider and effective dissemination of information, awareness, and outreach.
- The Security channel—which can happen because such interface helps bridge the communication gap.

Specifically on the benefits to end users, Goswami and Raghavendran (2009) identify the advantages of mobile banking to include: (1) secure authentication, transaction and data transmission, and easy deleting of content in the event of handset loss; (2) icon-driven, user-friendly interface; (3) contactless payment that offers quicker checkout at the point of sale and replaces all current payment solutions; (4) dynamic credit facility and innovative point-of-sale offers; (5) dynamic account monitoring and around-the-clock alerts; (6) convenience of micropayments (parking meters, vending machines); (7) real-time access to account information, outstanding debt, and bill payment; (8) ubiquitous access to banking services (personal ATM).

In all, three key arguments can be drawn from the existing literature to demonstrate how the link between ATM and mobile phone could provide the opportunities for enhancing inclusive banking in Africa. These include pluralism, business efficiency, and economic development arguments.

16.2.1 Pluralism argument

Mobile phone use is rapidly becoming an important part of everyday life in most societies, even in poor households (Porter et al., 2012). This implies that a banking service delivery protocol that is built on the mobile phone channel is capable of encouraging the inclusion of greater percentage of the bankable population (Mishra and Bisht, 2013). Karjauoto (2002) illustrate that among the factors that have helped to popularize mobile banking are the low fees, time saving, and freedom from time and place that characterize most handheld devices. Aided by the emergence of mobile phone and ATM, banks are now able to serve customers outside the banking hall. ATM, consequently, performs some of the traditional functions of bank cashiers and other counter staff—among which include withdrawing cash from one's account, making balance inquiries, transferring money from one account to another, payment of bill, printing of statement, etc. An emerging proposition, therefore, is that linking personal GSM telephone lines to ATM cardholding can help encourage inclusive banking by drawing more prospective customers into the banking system and by widening the catchment areas of banks (Alonso et al., 2013).

16.2.2 Business efficiency argument

Mobile phone use brings about significant reduction in business operations and transaction costs (Aker and Mbiti, 2010; Qureshi, 2013). This benefit provides a strong premise for increased acceptability of mobile phone platforms by banks and the sustainability of their acquisition and use by customers (Gupta, 2013). In addition, the adoption of ATM technology is also argued to have increased community efficiency through improvement in quality of service delivery and enhancement of value-added service to customers (Kamel, 2005). The increasing wave of mobile technology innovations thus helps moderate the effects of traditional banking challenges. This in itself holds great opportunities for banks having difficulty providing profitable services to minority customers through the traditional channels (Mishra and Bisht, 2013). However, the capacity of the banking system to take advantage of the potentials of such telecom innovations is dependent on the quality and stability of banking regulation, the strength of the overall financial sector, the state of infrastructural development, and the level of ICT knowledge base in the affected countries (Sarokolaei et al., 2012).

16.2.3 Economic development argument

There is empirical evidence that mobile phone adoption and penetration aid growth in both the upper-low-income and the low-income countries in Africa (Chavula, 2013). Mobile phone can also serve as a vehicle for facilitating economic activities and growth in the medium to long term. Because a huge and vibrant informal cash economy exists in almost all SSA countries and because the economies are dominated largely by unbanked small and microscale businesses, broad-based business platform, such as mobile banking and telecommunication, provides a magnetic framework for an inclusive economic practice. Channeling funds in the informal cash sector through the formal banking system is therefore capable of bolstering economic development (Global Financial Development Report, 2014). In summary, Aker and Mbiti (2010) identify five potential mechanisms through which mobile phone aids economic development, namely:

1. The capacity of mobile phone to improve access to and use of information
2. The capacity to provide opportunity for business efficiency through enhanced supply chain
3. The capacity to provide income-generating opportunities to rural and urban population

4. The possibility of facilitating information sharing and reduced households' risk exposures
5. The possibility of facilitating essential services delivery, especially in the areas of health, education, and agriculture

Essentially, transaction-based and nontransaction-based mobile banking exist. Transaction-based includes funds transfer/remittances and payment via the mobile device. However, by their nature, transaction-based services require more advanced ICT facilities such as smartphones, iPods, and Blackberry. These advanced mobile phones can be used as Internet access points and can result in the provision of two fundamental customer services: instant access to an account, and the ability to make payments and transfers remotely. More generally, the mobile phone device and wireless connectivity bring the Internet terminal into the hands of otherwise unbanked customers. On the other hand, nontransaction mobile banking is more or less informative in nature and is specifically dedicated for balance inquiries and short bank statement issuance, transaction alerts, product communication, loan account follow-up, and other information-sharing activities between the bank and the customers. Figure 16.1 illustrates this trend and reveals the interactive capacity of mobile phone devices at each of the different developmental stages.

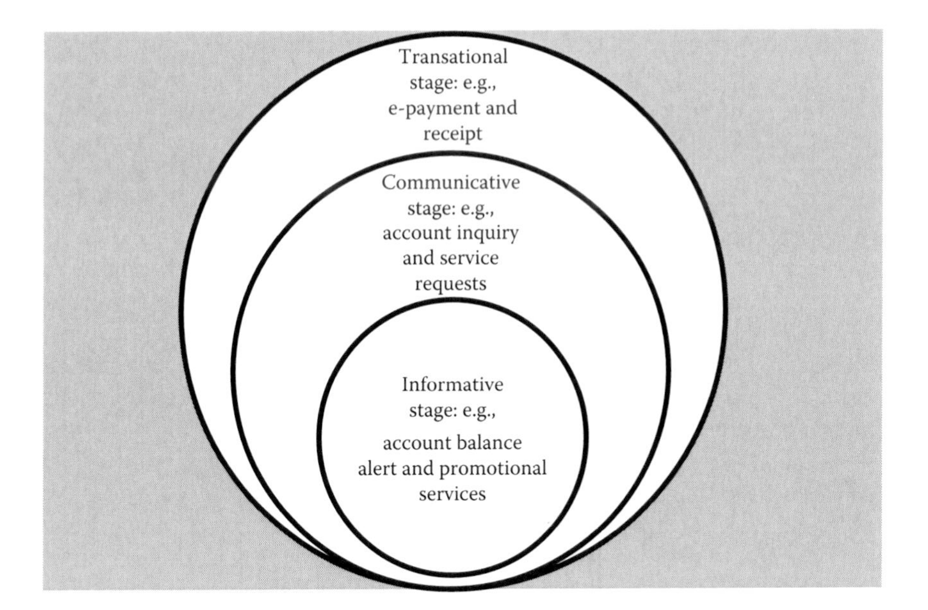

Figure 16.1 Stages of mobile phone banking development.

16.3 Pre-e-banking environment in Nigeria

Historically, banking business started in Nigeria in 1892 as an ordinary commercial venture with no particular government emphasis on regulation or control. On the other hand, modern banking system took off in the country in 1952 following the enactment of the Banking Ordinance of 1952. The Ordinance itself was the first major attempt by the government to regulate the business of banking in the country and was designed to prescribe legal basis for the issuance of banking license, lay down rules and procedures for the conduct of banking business, and minimum capital and reserve structures for banks (Ezeoha, 2007).

Until the mid-1990s, Nigerian banks operated under a system that was paper-based and could be described as purely armchair banking. The use of computer and other electronic platforms in the delivery of bank services and management of banking operations was then visibly absent. Such banking era could, in the context of this paper, be classified as part of preelectronic banking history and was essentially characterized by a number of regulatory and structural issues. First, the Structural Adjustment Programme (SAP) adopted by the Nigerian government in 1986 was to be one of the most radical regulatory events in financial system development in the country. On the positive side, SAP widened the financial system landscape by increasing the level of participation by both domestic and foreign banks, and by providing significant incentives for financial product development and outreach. On the negative side, the program brought with it an extensive deregulation of the Nigerian economy, created opportunities for outright evasion of banking and financial laws in the country, and made it possible for some new banks to survive and prosper by mainly buying and selling foreign exchange (Uche, 2000). One way SAP-motivated reforms injected instability in the system was through the relaxation of the entry conditions of new banks. By loosening the regulatory requirements for the granting of bank licenses, for instance, promoters borrowed funds on short-term basis to float banks, and only for the new banks to be stripped of the capital shortly after the acquisition of their licenses (Central Bank of Nigeria, 2001). The visible outcome was the early collapse of such banks and incessant financial distress in the system, which in itself culminated in the Central Bank of Nigeria placing embargo in 1991 (up until 1998) on the licensing of more banks in the country (Ezeoha, 2010). It was consequent to this development that SAP was discontinued in 1996 as part of the strategies to instill sanity not only in the financial system but also in the economy as a whole. Yet, not much success was recorded considering that in 1998, for instance, as many as 26 (out of 115 licensed banks) had been liquidated by the Central Bank of Nigeria (Ezeoha, 2007).

Specifically, attempts to curb the persistent distress in the financial system and minimize the rising level of intra-industry squabbles among operators also led to the prescription and adoption of universal banking model by the Central Bank in 2001. Among other numerous consequences of this model were the eradication of the usual regulatory and operational differences between merchant and commercial banking in the country; increased fragility and rigidity due to weak regulatory capacity and overlaps on the side of the CBN and the resultant system abuses and vulnerability (Uche, 2001). By implication, the universal banking practice was unable to resolve the predicated issue of persistent distress in the financial system. Instead, as argued by Ezeoha (2010, 2007), it brought with it more chronic challenges such as:

1. Rigidity in the ownership structures of banks—resulting in a good number of the banks being privately and closely held, with thick family and personal links
2. Visible disconnection between the banking industry and the capital market—which resulted in a lack of access to the capital market, very high information asymmetry, overreliance on public sector deposits, illegal and unprofessional competition for funds, and loss of enough public confidence required to run efficiently
3. Uneven distribution in the size and structure of the industry—leading to the emergence of too many small banks that were neither well capitalized nor better managed and the resultant dominance of the industry by just about 5 out of the 89 banks
4. Deficiency in the intermediation process—which reflected hugely in the dominance of short-term loans and deposits, as well as increased monetization of the economy
5. Regulatory inconsistence and increased incidence of frauds, especially in the areas of fraudulent transfer and withdrawals of funds, presentation of forged checks, granting of unauthorized credits, posting of fictitious credits, check and cash defalcation, outright theft, and bank robbery

It was indeed those challenges that led to the structural reforms that took place between 2004 and 2005. The banking consolidation reform, as it was popularly referred to, provided for the increase in minimum capital base of depositing money in banks from 2 billion to 25 billion Nigerian Naira, convergence of the industry from 89 to 25 banks, eradication of government ownership influence in the banking industry, reemergence of foreign banks, increased participation of banks in the capital market, as well as diversification of Nigerian banks into other African countries (Ezeoha, 2007). On the downside, the structural reform was unable to induce the required level of financial system inclusion necessary to ensure effective and adequate funding of small- and medium-scale enterprises and other minority sectors such as agriculture and manufacturing (Ezeoha and Amaeshi, 2010).

The desired level of banking efficiency could also not be attained largely due to the inability of the reform to account for systemic challenges such as poor integration of governance and operational protocols, nationwide infrastructural deficiencies, and poor culture of regulation enforcement.

Consequent to the 2007/2008 global economic and financial crisis, the CBN had in 2009 raised alarm that the banking industry was highly infested with systemic distress and risks. The regulatory authority noted then that 8 out of 25 banks that survived the 2004/2005 industry consolidation were technically distressed. The financial situations of the affected banks were so bad that the central bank spent as much as 620 billion Nigerian Naira (or US$4 billion) to bail them out.

A major regulatory goal that remained relatively unchanged amid the regimes of structural reforms in the country mentioned earlier was the need to make the banking industry more competitive and attractive to investors. More so, the timing of some of the industry reforms coincided with the period of ICT revolution in the country. For instance, massive computerization of banking process and operations in the 1990s was occasioned by the fact that right from the late 1980s, banks had started suffering serious deterioration in their levels of operating efficiency, leading to the consequent rise in operating cost over and above operating income (Ezeoha, 2010). Thus, to survive the stiff banking competition during that period, banks had to take advantage of the rising profiles of electronic banking protocols that were capable of offering cost-reduction opportunities. It was also the search for cheaper operational media that brought about the adoption and popularization of ATMs by Nigerian banks during the same period. However, issues relating to the cost of adoption and requisite expertise posed a major constraint on the capacity of banks to fully computerize their operations in the 1990s. This is reflective in the fact that as of June 2004, only about 7 out of the 89 banks then in operation offered ATM services. Apart from issues relating to cost and technical expertise, fraud was also a key constraining force for electronic banking services in the country. The 419 fraud, for instance, ranks among the top cyber frauds in the world, and remains the most popular of Internet frauds globally linked to Nigeria, with a very serious negative threat to public confidence in electronic business transactions (Ezeoha, 2010). This was so much so that the Nigerian government in 2006 enacted the Advanced Fee Fraud (419) Act, dedicated specifically to deal with such economic crime. The banking public were also skeptical in transacting businesses through electronic media because of the alleged vulnerability of such platforms to fraud and technical failures. the CBN in recognition of the high incidence of card frauds in the country, in a February 2011 circular on "the Need to Combat Card Fraud" directed banks, among others, to apply proper KYC (know your customer) for issuance of cash cards; set limit and ensure second level authentication for card to card transfers,

POS (point of purchase), and web payments; give cardholders options to choose channels for which their cards will be applied; and restrict cash card usage for payment of services specifically to the agreed schemes.

Notwithstanding the threat posed by this peculiar electronic business fraud in Nigeria, there have been other system-wide developments that made the business model indispensable and appealing, especially in the banking industry. One of such development was the emergence of GSM in 2001, which coincided with the later introduction of universal banking in the country. The emergence of GSM itself facilitated information flow and outreach, and equally reduced the geographical barriers against financial intermediation. Taking advantage of the opportunities offered by the telecommunication development was in itself facilitated by the general banking industry reform of 2004/2005. The reform provided surviving banks with huge financial resources needed for comprehensive ICT redeployment. As argued by Ezeoha (2010), the consolidation exercise offered banks the opportunity to increase their capital bases and be able to update their technological infrastructures. With the advancement in technology awareness and improvement in the telecommunication system, banks in Nigeria went on to install a number of electronic banking protocols and service delivery platforms that could more efficiently link them to the customers. With mobile phone introduction, the deployment and use of ATM as a service delivery channel, for instance, shifted from a strategic focus on privileged customers to a policy effort for attaining inclusive financial system.

Certain other features of the Nigerian banking system that facilitated the shift from traditional to modern technology-based banking emanated from the inherent structural rigidities in the system. The industry itself was characterized by exclusionary practices arising from a prevalent low banking density in the country and the resulting exclusion of a greater percentage of economic agents from the system; high rigidity challenges, including delay in essential service delivery due to manual system of operation; poor policy transmission and response to global dynamics, as well as the inability to inclusively attend to the banking needs of the citizens. Before the introduction of electronic payment into the system, customers had to walk into the banking hall to do transactions of all kinds. They had to queue up and spend more hours to talk to a teller and other bank staff to make their transactions. Thus, any policy with promises of resolving those background issues was expected to enjoy favorable public reception.

The inherent features and benefits of ATM channel consequently facilitated acceptability and assimilation by the bank customers. Broadly, ATM functions prominently as a way of easing the human traffic in banking halls and as well respond to the changing nature of modern banking operations. It also offers high level of flexibility and convenience to customers. In addition, some regulatory push from the CBN has equally helped in improving the rate of ATM penetration in the country. There is for instance, at the moment, a policy directive requiring banks to limit the amount of over-the-counter cash payments to bank customers and requesting bank customers withdrawing a certain threshold amount to do so through ATM or pay charges for withdrawing in the banking hall. Despite the advantages and policy incentive that go with the ATM operations, some have continued to avoid the use of such electronic banking modes, largely due to the persistent incidence of fraud and technical failures associated with such operations.

16.4 Emergence of ICT-based banking regime in Nigeria

The use of ICT in the delivery of banking services is generally referred to as electronic banking (e-banking). The application of its concepts, techniques, policies, and implementation strategies to banking services has become a subject of fundamental importance and concerns to all banks. It has also become a prerequisite for local and global banking competitiveness.

In the case of Nigeria, major Information and Communication Technology (ICT) revolution did not occur until the late 1990s. In fact, the use of ICT as a complementary business model became more visibly in place from 2004. This was made more prominent by the licensing of GSM operators earlier in 2000 and the consequent issuance of GSM telephone lines to the general public in 2001. By 2011, a GSM Association report on Africa had revealed Nigeria as having the highest number of mobile phone subscriptions in Africa—standing at more than 93 million, representing 16% of the continent's total mobile subscriptions (Saliu, 2011).

Mobile phone operation in the country was further boosted by the introduction of a National Policy on Information and Technology in January 2002. The aim of the policy, among others, was to lay emphasis on the need to effectively deploy information technology in the promotion and support of private sector industrial growth in Nigeria. Consistent with this critical mandate, section 8.2 of the policy specifically identified the objectives of the country's IT development to include the need

1. To develop a transparent, stable, and effective legal operating environment that promotes private sector business and investment in IT
2. To cultivate a culture of electronic commerce, which makes business transactions easy, quick, and cost-effective, for both national and international transactions

3. To positively raise the local and international visibility of Nigerian businesses
4. To encourage foreign and domestic private sectors investment to build information infrastructure and related assets and develop subsequent downmarket activities
5. To stimulate the proliferation of information technology services led by the private sector and consequently generates meaningful employment opportunities for Nigerians

Another major development within the period was the formation, in 2003, of the Nigerian Communication Commission (NCC), which was established by the Act of Parliament (the Nigerian Communication Commission Act of 2003), with the core mandates to establish a regulatory framework for the Nigerian communications industry and for this purpose to create an effective impartial and independent regulatory authority; and to promote the provision of modern, universal, efficient, reliable, affordable, and easily accessible communications services and the widest range thereof throughout Nigeria (Chapter 1, No. 1, Nigerian Communication Commission Act, 2003). Specifically, Chapter 4, No. 1 of the Act mandates the NCC to, among others, carry out the following functions:

1. The facilitation of investments in and entry into the Nigerian market for provision and supply of communications services, equipment, and facilities
2. The protection and promotion of the interests of consumers against unfair practices including but not limited to matters relating to tariffs and charges for and the availability and quality of communications services, equipment, and facilities
3. The promotion of fair competition in the communications industry and protection of communications services and facilities providers from misuse of market power or anticompetitive and unfair practices by other service or facilities providers or equipment suppliers
4. Granting and renewing communications licenses whether or not the licenses themselves provide for renewal in accordance with the provisions of this Act and monitoring and enforcing compliance with license terms and conditions by licensees
5. Fixing and collecting fees for grant of communications licenses and other regulatory services provided by the Commission
6. The development and monitoring of performance standards and indices relating to the quality of telephone and other communications services and facilities supplied to consumers in Nigeria with regard to the best international performance indicators
7. Making and enforcing such regulations as may be necessary under this Act to give full force and effect to the provisions of this Act

8. Management and administration of frequency spectrum for the communications sector and assisting the National Frequency Management (NFM) Council in developing a national frequency plan
9. Management and administration of frequency spectrum for the communications sector and assisting the National Frequency Management (NFM) Council in developing a national frequency plan
10. Encouraging and promoting infrastructure sharing amongst licensees and providing regulatory guidelines thereon
11. Examining and resolving complaints and objections filed by and disputes among licensed operators, subscribers, or any other person involved in the communications industry, using such dispute-resolution methods as the Commission may determine from time to time including mediation and arbitration

The consequence of the developments mentioned earlier in the ICT sector, according to data from the Nigerian Communication Commission, is the significant rise in the level of teledensity from a level of 0.51 in 2000 to 63.11 in 2010 and 80.85 in 2012—a trend ascribed to the drastic decline in the subscription cost per GSM line from about US$95 in 2001 to just about US$1 in 2009 (Ezeoha, 2010). This is also supported by the growth in the use of other ICT protocols by Nigerians, especially with regard to the Internet that recorded a rise in penetration rate from 0.10% in 2000 to 28.3% as at March 2011.

Figures 16.2 through 16.6 show the huge progress recorded by Nigeria in the telecom sector within the last decade. As shown in Figures 16.2 and 16.4, with the exception of the year 1993/1994, Nigeria was clearly absent

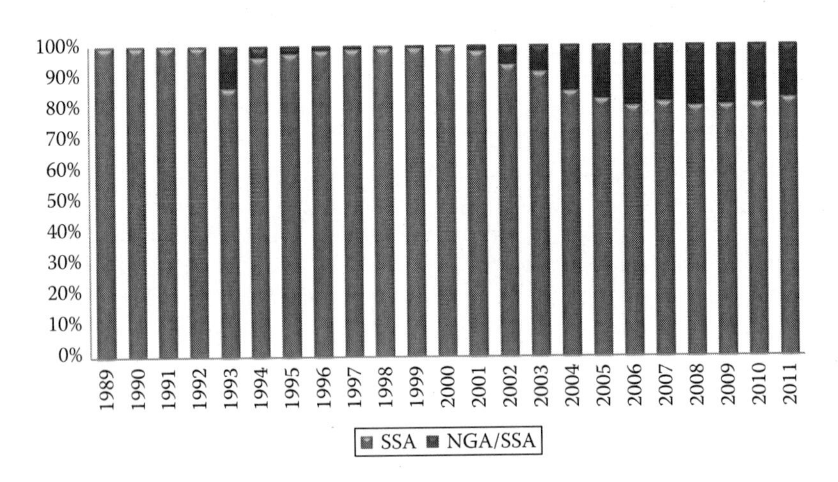

Figure 16.2 Number of mobile cellular telephone subscriptions in Nigeria and in SSA countries compared.

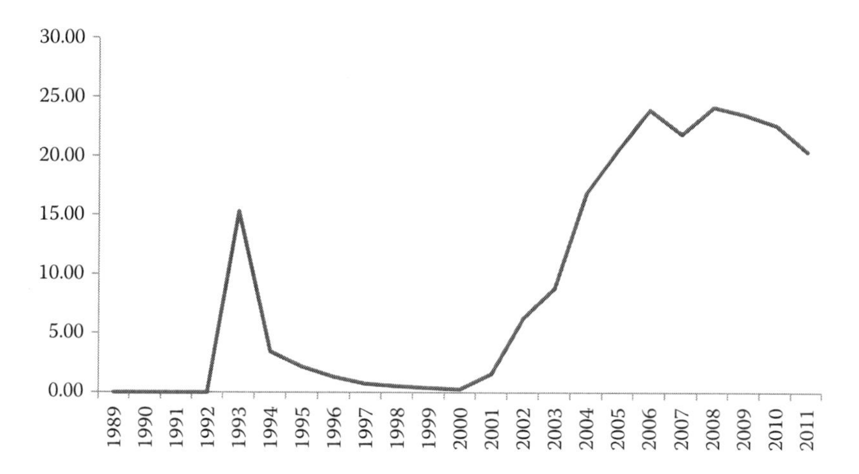

Figure 16.3 Trends on mobile cellular telephone in Nigeria (*Note*: Numbers in the vertical axis are in percentage).

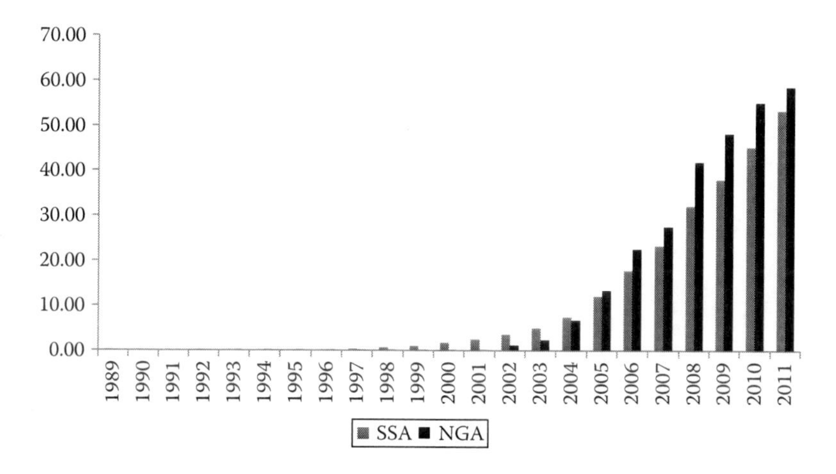

Figure 16.4 Rate of mobile cellular telephone penetration in Nigeria and in SSA countries compared (*Note*: Numbers in the vertical axis are in percentage).

in the African cellular telephone market up until 2001 when the GSM formally took off in full force in the country. Figure 16.3 even reveals that from 2005, the rate of mobile cellular phone penetration in Nigeria continued consistently to surpass the SSA average. In similar vein, increased rate of Internet penetration in the country maintained a higher trend than that of SSA average beginning from 2005. The country also recorded a higher fit, in terms of ATM penetration during the same period. The recorded coincident in the growth patterns in mobile phone, Internet, and ATM penetrations in both Nigeria and other SSA countries creates a

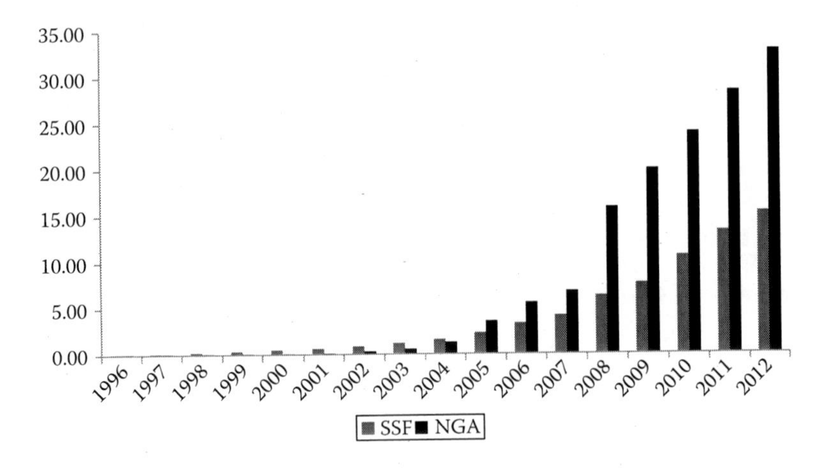

Figure 16.5 Internet penetration rates in Nigeria and in SSA countries compared (*Note*: Numbers in the vertical axis are in percentage).

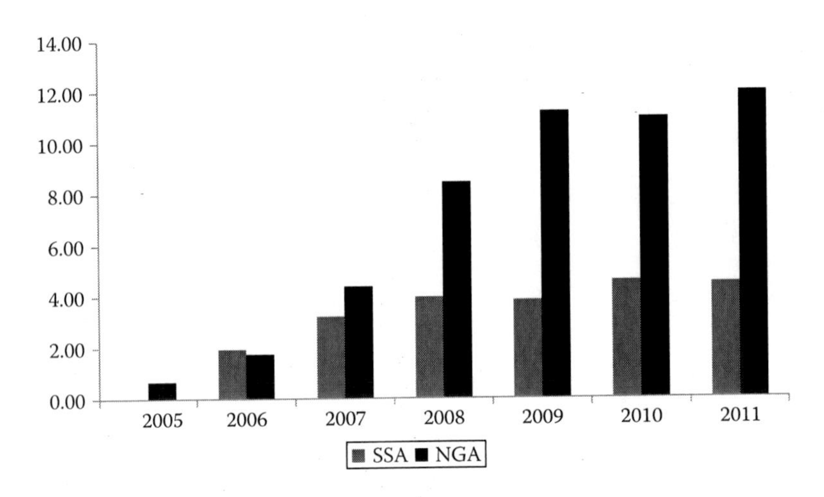

Figure 16.6 ATM penetration rates in Nigeria and in SSA countries compared (*Note*: Numbers in the vertical axis are in percentage).

cautious impression that there is a somewhat causal relationship among them, which can be confirmed through an empirical examination. Due to the lack of long series data, a robust and efficient country-level empirical analysis in this regard is unfeasible. To test this foregoing hypothesis and based on data availability, we however constructed a panel data from 27 SSA countries over the years 2004–2011 and carried out both pooled and panel regression analysis. The procedure for and results of this empirical exercise are reported in section 7.

Through the use of ICT, Nigerian banks now employ different channels such as Internet technology, video banking technology, mobile telephone banking, ATM, and WAP technology to deliver most of their services (Ayo et al., 2010). Existing statistical records, for example, reveal that ATM remains the most widely used e-payment instrument in Nigeria. For instance, data on various e-payment channels according to Central Bank of Nigeria (2012) indicated that ATM with a total number of 10,221 as of end of June 2012 was the most patronized, accounting for 96.4%. This is followed by mobile payments with 1.3% and POS terminals with 1.2%. The web (Internet banking) was the least patronized, accounting for only 1.1% of total e-payment transactions. The result of research by Okechi and Kepeghom (2013), on empirical evaluation of customers' use of electronic banking systems in Nigeria equally shows that among all e-banking systems, ATM has the highest level of usage. Also, Adesina and Ayo (2010) stated that ATM is responsible for about 89% (in volume) of all e-payment instruments between 2006 and 2008.

The growth in ATM usage rate also reflects the practice of multiple accounts holding by bank customers. Thus, a bank customer with accounts in two or more banks would expectedly have a corresponding number of ATM cards. The implication is that in some cases, the number of ATM cards in circulation may not effectively represent higher baking density or inclusion.

16.5 ATM operational environment in Nigeria

ATM applications dominate the spectrum of today's Nigerian banking system. Societe Generale Bank (SGBNS), now Heritage Bank, was the first bank to introduce ATM (NCR machine) in 1989 at their Broad Street and Apapa Branches in Lagos. The trade name for SGBN's ATM was "Cash Point 24." First Bank of Nigeria Plc came on stream with its own ATM in December 1991, a year behind SGBN. It also gave a trade name "FIRST CASH" to its ATM. At the introduction of the ATM in 1989, the machines looked complex and preserved for the top-edge of banking customers. Today, all the banks in Nigeria have adopted the technology and have invested huge sums of money in the deployment of the ATM and toward the issuance of ATM debit cards. A study by Adesina and Ayo (2010) found that all members of the Nigerian banking industry had engaged the use of Information and Communication Technology (ICT) as a platform for efficient conduct of financial transactions. At date, the core ATM providers are InterSwitch, VPay, ETranzact, and QuickCash. InterSwitch today has all banks in the country connected to its network. This actually makes it possible to use their cards in all bank branches nationwide and in almost all machines. The ATM service providers are also interconnected. InterSwitch, for example, supports Verve, Visa, and MasterCard on her

ATM machines and vice versa. According to 2010 InterSwitch statistics, Nigeria has 30 million ATM card holders who conduct over 100 million transactions on the machines every month. As at second quarter of 2013, Nigeria's 24 banks operate over 11,000 ATM machines across the country's 36 states and Federal Capital Territory.

No doubt, the growing popularity of ATMs has pushed further development in mobile banking in the country. There is at the moment a functional mix between mobile banking and mobile telephone penetration. As in most parts of Africa and other developing areas, mobile telephone banking initially started in Nigeria with the SMS (the second most common use of mobile phones, after voice calls) and is expanding through a combination of Internet protocols enabled by the emergence of smartphones and iPads. The dominance of the SMS channel is necessitated by the fact that majority of the mobile phone devices affordable to most Nigerians are those that can only perform basic operations. At the moment, just a little fraction of the population has access to smartphone devices mostly due to issues relating to acquisition costs and maintenance. It is a common knowledge for instance that whereas with less than US$20 one can buy a mobile handset in the country, the price of an average smartphone ranges from US$150 to as high as US$500. With the high poverty rate in the country, it becomes clearer why the basic mobile handsets are more in use and why the SMS–ATM interface offers huge opportunities for inclusive banking practice in the country. Interestingly, the linkage between bank accounts and GSM telephone handsets are more prevalent among the lower-income class. A study by VISA showed that ATM use is more common with the individuals within the lower/middle-income classes, suggesting a negative correlation between income size and ATM usage. According to the results of the study, "the more Nigerians earn, the less they use their cards for online purchase."

The situation mentioned notwithstanding, over time, the flexibility and the policy support for ATM operations by the Central Bank of Nigeria have endeared more people toward its acceptability. This has consequently led to the expansion in its transactionary usefulness from mere cash withdrawal and account balance enquiries to cash and check deposits, mini-statements of accounts, as well as purchase of mobile phone recharge airtime and tickets. However, as in the cases of other e-banking development, the emergence of the mobile phone banking platforms in the country appears to have been constrained by the consequent high level of frauds in the system (Ezeoha, 2010). Specifically, the Nigerian case is a reflection of how weak institutional and regulatory capacity can undermine the intensity of e-banking and ICT innovations support economic growth and development. The rising incidences of mobile banking frauds, for instance, are argued to be responsible for the underoptimization of the opportunities created by global e-business revolution.

Rising rate of ATM frauds in the country has equally undermined the growth and expansion in the usage rate/penetration. Among the key fraudulent acts penetrated through the ATM platform are card theft, card skimming, pin compromise, card trapping, and transaction reversal. Other types or forms of ATM fraud that are permeating the country's ATM practice are shoulder surfing, Lebanese Loop, use of stolen cards, card jamming, use of fake cards, duplicate ATMs, and card swapping (Adeoti, 2011). There are also cases of both operational distortions by banks and system abuses by users of ATM services in the country. The menace of ATM frauds has, for example, been blamed on the indiscriminate issue of ATM cards by banks without regard to the customer's literacy level (Obiano, 2009; Adeoti, 2011). According to Obiano, one of the frequent causes of fraud is carelessness in the handling of ATM cards by customers, as well as customers' response to unsolicited e-mail and text messages. Power outage is another major challenge facing ATM users in Nigeria. There are also issues relating to the lack of cooperation among banks, especially with regard to sharing vital information and data on ATM usage, operational distortions, and frauds, which makes the curbing of fraud very difficult.

The fact that Nigeria has a vast unbanked population with high mobile phone penetration signals a somewhat disconnect in the entire mobile banking arguments. Although the country possesses the potential to be the leading mobile money market in Africa, that is not the case currently. Barton (2013) reported that there are nearly 110 million mobile subscribers in Nigeria but only 56 million people have bank accounts, giving the market massive growth potential for mobile money services. Based on the report of the Enhancing Financial Innovation & Access (EFInA), an organization committed to deepening financial inclusion in the country, only one in five Nigerian adults has a bank account (Beshouri and Gravråk, 2010). A CBN 2012 report has it that a total of 39.2 million adult Nigerians (or 46.3% of the adult population of 84.7 million) were financially excluded in 2012 (Central Bank of Nigeria, 2012). On the average, in the case of the SSA region, less than 25% of adults reportedly have bank accounts (Ondiege, 2012). At the moment, there is a conscious policy effort by the CBN to encourage financial innovations through the mobile telecom channel. This policy effort is expected to foster financial inclusion of the unbanked populace, especially among the rural and urban-poor population.

The deliberate effort mentioned earlier by the CBN offers some insights as to why both the pattern and depth of ATM deployment have very much been induced more by regulatory policies. The CBN, for example, in 2010 issued standards and guidelines for the operations of the ATM services in Nigeria. The policy document was targeted primarily at guaranteeing the efficiency of ATM services, and protecting bank customers in the conduct of ATM operations in Nigeria (Central Bank of Nigeria, 2010). Among the key policy issues addressed in the guidelines are ATM technology and specification standards, ATM deployment (including the ownership, the licensing, the location, safety, and privacy), the operations, maintenance, security, dispute resolution process, and liability shifting, regulatory monitoring and penalties for default (Central Bank of Nigeria, 2010).

In the same vein, consequent to the inability of the banks to address customers' concerns over safety of and access to ATM terminals, the CBN issued a circular in 2011 mandating banks and other financial institutions to adopt appropriate and effective mechanisms to address the spate of complaints, enhance public confidence, and guarantee customer satisfaction. Prior to the circular, banks were widely accused of using the ATM facilities to exploit customers, by engaging in acts relating to excess charges, unauthorized deductions, excess commission on turnover, and other frivolous charges. February 7, 2011 circular titled "Penalty for Non-Compliance with CBN Circulars and Guidelines on ATM Operations in Nigeria" was principally aimed at creating the requisite conducive environment necessary for an efficient ATM system and to make the platform more attractive to the general public. Among its core provisions was the spelling out of a number of penalties targeted at mandating banks to facilitate ATM deployment and security. Some of such penalties, which are expected to serve as stringent measures, are

- Depositors wishing to withdraw less than ₦60,000 over the counter to pay a surcharge of ₦100, irrespective of the amount, effective July 31, 2008
- Increase in the maximum ATM withdrawal limit from ₦4,000 in 2008 to ₦200,000 as of date
- The imposition of a daily fine of ₦50,000 on banks for failure to establish ATM help desk to address customer complaints
- The imposition of a daily fine of ₦50,000 on banks for failure to respond to customer complaint after 72 h of such compliant
- The imposition of a weekly fine of ₦50,000 on banks for noncompliance with Payment Card Industry Data Security Standards, PCIDSS, until compliance is established
- The imposition of a weekly fine of ₦50,000 for bank's failure to provide audit trails and journals for ATM transactions
- A refund of ATM surcharge to customers if such surcharge is not disclosed to the affected customers
- The imposition of a daily fine of ₦50,000 on banks for the lack of online monitoring mechanism and backup power (inverter) for ATM
- An ATM without a camera installed to attract a fine of ₦50,000 and deactivation of the ATM pending when the camera is installed

- A fine of ₦50,000 per day will be applied for the late submission of returns/data on ATM frauds when required
- Failure to resolve any ATM dispute with evidence of resolution within 14 days, the deployer to refund the total amount involved in the fraud

Another positive development that is helping to improve ATM acceptance rate in the country is the scrapping of ₦100 interbank charge by the Bankers' Committee in November 2012. Before then, some banks had gone ahead to introduce a ₦100 monthly ATM card maintenance fee, which did not go down well with the regulatory authority and as such was abolished in March 2013. An earlier circular referenced BOD/DIR/CIR/2009/GEN/10 and dated December 18, 2009 was also issued by the CBN directing all deposit money banks to establish ATM help desk to handle all consumer complaints. Whereas this latter circular did little in resolving much of the complaints raised by customers, the CBN in a follow-up circular referenced FPR/DIR/CIR/GEN/01/020 dated August 16, 2011 observed some rising pattern in the number of complaints it received from customers of financial institutions. The complaints ranged from allegations of excess charges, unauthorized deductions, excess commission on turnover, other frivolous charges, and frauds. Ever since, the CBN has had to embark on sustainable campaign and policy inducements to make ATM operations more conducive and user-friendly. The overall goal, from the standpoint of the CBN, is to reform the payment system in the country through the adoption of e-payment and cashless payment policies.

16.6 Statistical tests of the relationship between ATM and mobile phone penetration rates

To provide some empirical justifications to the ATM–mobile phone interface, we pooled data from 27 African countries over a period from 2004 to 2011 and formulated two alternative estimation equations—an ATM and a Mobile Phone Penetration Equations. The first equation theorizes that ATM adoption is dependent on the level of mobile phone, Internet access, banking development, economic growth, and investments in telecommunication. That is,

$$\begin{aligned}\mathrm{ATM}_{it} = \alpha_0 + f\big(&\mathrm{Mobile\,Phone}_{it}, \mathrm{Internet\,Access}_{it}, \mathrm{Banking}_{it},\\ &\mathrm{Growth}_{it}, \mathrm{Investments}_{it},\big) + \mu_{it}\end{aligned} \tag{16.1}$$

The second equation alternatively theorizes that the level of mobile phone penetration is a function of the level ATM penetration, Internet access, banking development, economic growth, and investments in telecommunication, and can be represented as follows:

$$\text{Mobile Phone}_{it} = \alpha_0 + f\left(\text{ATM}_{it}, \text{Internet Access}_{it}, \text{Banking}_{it}, \text{Growth}_{it}, \text{Investments}_{it},\right) + \mu_{it} \quad (16.2)$$

Equations 16.1 and 16.2 are estimated using pooled, fixed-effect, and random-effect regression techniques, with a Hausman test for model efficiency showing the robustness of the fixed-effect model over the other two. The variables' definitions and sources of data are contained in Table 16.1.

The statistical estimates are reported in Tables 16.2 through 16.4. Specifically, Table 16.2 shows that ATM penetration is positively and

Table 16.1 Variable Definitions and Measurements

Variable	Measurement	Definition	Source
ATM	Automated teller machines (ATMs) (per 100,000 adults)	Automated teller machines are computerized telecommunications devices that provide clients of a financial institution with access to financial transactions in a public place.	World Development Indicators
Mobile phone	Mobile cellular subscriptions (per 100 people)	Mobile cellular telephone subscriptions are subscriptions to a public mobile telephone service using cellular technology, which provide access to the public switched telephone network.	World Development Indicators
Internet subscription	Internet users (per 100 people)	Internet users are people with access to the worldwide network.	World Development Indicators
Banking development	Money and quasi money (M2) as % of GDP	A measure of the size and depth of a country's monetary system.	World Development Indicators
Economic growth	GDP growth (annual%)	A measure of the annual growth in the overall national economy.	World Development Indicators
Telecom investment	Investment in telecoms (current US$) scaled by the GDP	Investment in telecoms with private participation (current US$).	World Development Indicators

Table 16.2 Descriptive Statistics on ATM, Mobile Phone Penetration, and the Control Variables

	ATM	Mobile phone	Internet subscription	Banking development	Economic growth	Telecom investment
Mean	10.176	38.867	7.621	37.821	5.44	0.01
Standard deviation	13.514	32.804	8.649	23.315	4.44	0.013
ATM	1					
Mobile phone	0.7153[a]	1				
Internet subscription	0.6883[a]	0.6758[a]	1			
Banking development	0.8112[a]	0.4631[a]	0.6170[a]	1		
Economic growth	−0.1885[a]	−0.132	−0.1027	−0.1057	1	
Telecom investment	−0.2941[a]	−0.1411[a]	−0.0941	−0.2091[a]	−0.0152	1

Notes: [a] Represents significant at 5 level.

Table 16.3 Regression Results for the ATM Equation

Variable	Pooled regression	Fixed-effect regression	Random-effect regression
Mobile phone	0.179[c] (0.023)	0.063[c] (0.020)	0.077[c] (0.020)
Internet	0.050 (0.081)	0.463[c] (0.074)	0.432[c] (0.076)
Banking deviation	0.310[c] (0.027)	0.042 (0.046)	0.160[c] (0.036)
Economic growth	−0.112 (0.127)	−0.127 (0.089)	−0.085 (0.095)
Telecom investments	−183.69 (49.965)	32.265 (36.295)	−9.589 (38.431)
Constant	−6.293[c] (1.397)	3.539[a] (1.969)	−1.585 (2.019)
R-square	0.808	0.630	0.743
F statistic/Wald test (prob.)	138.96[c]	37.56[c]	234.07[c]
Hausman test (prob.)	N.A.		16.83 (0.0048)
No. of observations.	165	165	165
Years	2004–2011	2004–2011	2004–2011
No. of SSA countries	27	27	27

Notes: Heteroskedasticity consistent asymptotic robust standard errors are given in parentheses. [a] and [c] indicate that coefficients are significant at the 10%, 5%, and 1% levels, respectively.

significantly correlated with mobile phone access, Internet access, and banking development. On the other hand, both economic growth and investments in telecommunication are negatively and significantly correlated with the level of ATM penetration—a result that appears to contradict theoretical expectation. The fact that economic growth in the sampled countries

Table 16.4 Regression Results for the Mobile Phone Equation

Variable	Pooled regression	Fixed-effect regression	Random-effect regression
Mobile phone	1.549[c] (0.198)	1.139[c] (0.352)	1.225[c] (0.254)
Internet	1.369[c] (0.214)	2.002[c] (0.312)	1.784[c] (0.257)
Banking dev.	0.284[c] (0.105)	0.454[b] (0.193)	−0.098 (0.125)
Economic growth	−0.081 (0.373)	0.173 (0.382)	0.004 (0.364)
Telecom investments	−240.954 (151.787)	164.159 (153.776)	243.483[b] (146.849)
Constant	21.407[c] (4.016)	−9.001 (8.416)	13.635[b] (5.881)
R-square	0.671	0.630	0.743
F statistic/Wald test (prob.)	67.80[c]	38.64[c]	227.58[c]
Hausman test (prob.)	N.A.		15.64 (0.0079)
No. of observations	165	165	165
Years	2004–2011	2004–2011	2004–2011
No. of SSA countries	27	27	27

Notes: Heteroskedasticity-consistent asymptotic robust standard errors are given in parentheses. [b] and [c] indicate that coefficients are significant at the 10%, 5%, and 1% levels, respectively.

is mainly driven by natural resources and is characterized by huge dominance of the informal sector can be an arguable justification for this unusual negative relationship between economic growth and ATM penetration. Some competitive trend also exists among investments in other technology-related infrastructure and electronic banking technology. This is especially so in Africa, where the infrastructures were not originally there and where funding difficulties are constraining the development capacity of countries.

Consistent with the outcome of the correlation analysis, the regression results shown in Tables 16.3 and 16.4 reveal strong causal relationship between ATM and mobile phone. In Table 16.3, which represents the outcome of the ATM regression equation, mobile phone penetration has positive and significant impact on ATM penetration in all the estimation models. In the fixed-effect model, which is found to be more consistent and robust, the rate of Internet penetration also positively and significantly affects ATM penetration.

Results contained in Table 16.4 (representing the mobile phone equation) also reveal a corresponding positive and significant impact of ATM on mobile phone penetration rate. Both Internet access and banking development have similar positive and significant effect, especially in the pooled and fixed-effects regression models. In all the equations, there is no empirical evidence confirming any consistent impact of economic growth and investment in telecommunication in the SSA region. In summary, both the correlation analysis and the regression results provide a confirmation of the existence of a strong and systematic interface between the rate of ATM and mobile phone penetrations in the region. Such empirical outcome is consistent with the theoretical arguments and literature reviewed in this paper. Although the estimated results are based on panel data, the conclusion can be country-specific for two reasons—because most of the SSA countries share similar economic and institutional characteristics.

16.7 Conclusion

This chapter reveals the existing literature evidence on the overwhelming importance of increased mobile telephone penetration, as well as the emerging economic relevance and the influence of mobile/electronic banking, especially in developing countries. Mobile phone enhances communication and facilitates business transactions. It provides opportunity for linking customers' phone numbers to their individual bank accounts as a strategy for a more flexibly access to banking services. Existing theoretical evidence has also revealed how the link between ATM and mobile phone can strongly enhance inclusive banking in developing economies. There is not only a consensus, for instance, that such interface is capable of inducing inclusive banking, efficient financial practices, and economic growth, but that optimizing the benefits of such interface is only possible in an environment with minimal incidence of fraud and technical ineptitude.

This chapter synthetically highlights the theoretical, regulatory, and empirical sides of the ATM–mobile phone interface debate. It specifically reviews the developmental and regulatory patterns of both mobile phone and ATM emergence in Nigeria. It makes use of a panel data from 27 SSA countries to test the empirical validity of the hypothesis. The outcome confirms the existence of a strong and systematic interface between the rates of ATM and mobile phone penetration in the SSA region. It also shows that the relationship between the two is causal and that both are positively and significantly influenced by the level of Internet access in each of the countries. This outcome has important theoretical and policy implications, especially with regard to the need to ensure that ICT deployment in developing countries is guarded toward realizing the goal of inclusive banking and efficient financial system. From the issues highlighted in this chapter, it is also important to note that an efficient ATM–mobile phone interface can only be attained if the constraining factors, in the likes of the prevalence of frauds and technical ineptitude, are more systematically addressed. In Nigeria, as shown in this chapter, a lot of policy and regulatory efforts are being made by the Central Bank of Nigeria in this direction.

References

Abraham, R. 2007. Mobile phones and economic development: Evidence from the fishing industry in India. *Information Technologies and International Development*, 4(1): 5–17.

Adeoti, J.O. 2011. Automated teller machine (ATM) frauds in Nigeria: The way out. *Journal of Social Science*, 27(1): 53–58.

Adesina, A.A. and Ayo, C.K. 2010. An empirical investigation of the level of users' acceptance of e-banking in Nigeria. *Journal of Internet Banking and Commerce*, 15(1): 1–13.

Aker, J.C. and Mbiti, I.M. 2010. Mobile phones and economic development in Africa. *Journal of Economic Perspectives*, 24(3): 207–232.

Alonso, J., de Lis, S.F., Hoyo, C., López-Moctezuma, C., and Tuesta, D. 2013. Mobile banking in Mexico as a mechanism for financial inclusion: Recent developments and a closer look into the potential market, BBVA Working Paper No. 13/20, Mexico City, Mexico, June.

Asongu, S.A. 2013. How has mobile phone penetration stimulated financial development in Africa? *Journal of African Business*, 14(1): 7–18.

Ayo, C.K., Adewoye, J.O., and Oni, A.A. 2010. Design of a secured e-payment in Nigeria: A case study. *African Journal of Business Management*, 4(9): 1753–1760.

Barton, J. 2013. Raising mobile money awareness will bear fruit in Nigeria, http://www.developingtelecoms.com/raising-mobile-money-awareness-will-bear-fruit-in-nigeria.html. Accessed September 20, 2013.

Beshouri, C. and Gravråk, J. 2010. Capturing the promise of mobile banking in emerging markets. McKinsey report, February, http://www.csr-weltweit.de/uploads/tx_jpdownloads/MCKinsey_Article_on_Mobile_Phones.pdf. Accessed September 20, 2013.

Bilbao-Osorio, B., Dutta, S., and Lanvin, B. 2012. The global information technology report growth and jobs in a hyperconnected world. World Economic Forum, Geneva.

Byrne, E., Nicholson, B., and Salem, F. 2011. Information communication technologies and the millennium development goals. *Information Technology for Development*, 17(1): 1–3.

Calisir, F. and Gumussoy, C.A. 2008. Internet banking versus other banking channels: Young consumers' view. *Journal of Information Management*, 28(3): 215–221.

Central Bank of Nigeria. 2001. The effects of economic crimes in the financial industry. Central Bank of Nigeria Banking Supervision Annual Report, pp. 55–63.

Central Bank of Nigeria. 2010. Standards and guidelines on automated teller machine (ATM) operations in Nigeria. Central Bank of Nigeria, Abuja, Nigeria.

Central Bank of Nigeria. 2012. National financial inclusion strategy report. Abuja: Roland Berger Strategy Consultant.

Chavula, H. 2013. Telecommunications development and economic growth in Africa. *Information Technology for Development*, 19: 1.

Comptroller's Handbook. 1999. Internet banking. Comptroller of the currency. Administrator of National Banks.

Datta, A., Pasa, M., and Schnitker, T. 2001. Could mobile banking go global? *The McKinsey Quarterly*, 4: 71–80.

Ezeoha, A.E. 2007. Structural effect of banking industry consolidation in Nigeria: A review. *Journal of Banking Regulation*, 8(2): 159–176.

Ezeoha, A.E. 2010. Internet banking in a highly volatile business environment—The Nigerian case, in *E-Banking and Emerging Multidisciplinary Processes: Social, Economical and Organizational Models*, Sarlak, M.A. (ed.), Hershey, PA: IGI Publishing, pp. 64–99.

Ezeoha, A.E. and Amaeshi, K. 2010. Banking development, small businesses and minority lending in Nigeria. *International Journal of Financial Services Management*, 4(4): 281–297.

Funk, J.L. 2009. The emerging value network in the mobile phone industry: The case of Japan and its implications for the rest of the world. *Telecommunication Policy*, 33(1–2): 4–18.

Global Financial Development Report. 2014. Financial Inclusion, Washington, DC: International Bank for Reconstruction and Development.

Goswami, D. and Raghavendran, S. 2009. Mobile-banking: Can elephants and hippos tango? *Journal of Business Strategy*, 30(1): 14–20.

GSM Association. 2012. What is the impact of mobile telephony on economic growth? GSM association report, http://www.gsma.com/publicpolicy/wp-content/uploads/2012/11/gsma-deloitte-impact-mobile-telephony-economic-growth.pdf . Accessed September 25, 2013.

Gupta, S. 2013. The mobile banking and payment revolution. *The European Financial Review*, February 20, http://www.europeanfinancialreview.com/?p=6199. Accessed September 20, 2013.

Helton, D.A. 2012. Bridging the digital divide in developing nations through mobile phone transaction systems. *Business Quest*, 17: 1–19.

Kamel, S. 2005. The use of information technology to transform the banking sector in developing nations. *Information Technology for Development*, 11(4): 305–312.

Karjaluoto, H. 2002. Selection criteria for a mode of bill payment: Empirical investigation among Finnish Bank Customers. *International Journal of Retail & Distribution Management*, 30(6): 331–339.

Michalakelis, C., Varoutas, D., and Sphicopoulos, T. 2008. Diffusion models of mobile telephony in Greece. *Telecommunication Policy*, 32(3–4): 234–245.

Mishra, V. and Bisht, S.S. 2013. Mobile banking in a developing economy: A customer-centric model for policy formulation. *Telecommunications Policy*, 37(6–7): 503–514.

Moody's Analyst. 2013. The impact of electronic payments on economic growth, Moodys Economy White Paper, February, http://corporate.visa.com/_media/moodys-economy-white-paper.pdf.

Obiano, W. 2009. How to fight ATM fraud. *Online Nigeria Daily News*, June 21, p. 18.

Okechi, O. and Kepeghom, O.M. 2013. Empirical evaluation of customers' use of electronic banking systems in Nigeria. *African Journal of Computing & ICT*, 6(1): 7–20.

Oketola, D. 2012. Nigeria now has 26 million e-Payment cards—Interswitch, Punch Newspaper, November 19.

Ondiege, P. 2012. Mobile financial services in Africa: Reaching all sections of the population. *The African Development Bank Group NY CTED Conference Paper*, Abu Dhabi, UAE, February 12–13.

Parent, I. and Cruickshank, N. 2009. The growth of the internet and knowledge networks, and their impact in the developing world. *Information Development*, 25(2): 91–98.

Porter, G., Hampshire, K., Abane, A., Munthali, A., Robson, E., Mashiri, M., and Tanle, A. 2012. Youth, mobility and mobile phones in Africa: Findings from a three-country study. *Information Technology for Development*, 18(2): 145–162.

Pousttchi, K. and Schurig, M. 2004. Assessment of today? Mobile banking applications from the view of customer requirements. MPRA paper no. 2913.

Qureshi, S. 2013. What is the role of mobile phones in bringing about growth? *Information Technology for Development*, 19(1): 1–4.

Rajnish, T. and Stephan, B. 2007. *The Mobile Commerce Prospects: A Strategic Analysis of Opportunities in the Banking Sector*. Hamburg, Germany: Hamburg University Press.

Saliu, Y. 2011. Mobile phone: A lucrative business, http://observer.gm/africa/gambia/article/mobile-phone-a-lucrative-business. Accessed August 6, 2013.

Sarokolaei, M.A., Rahimipoor, A., Nadimi, S., and Taheri, M. 2012. The investigating of barriers of development of e-banking in Iran. *Procedia—Social and Behavioral Sciences*, 62: 1100–1106.

Sassi, S. and Goaied, M. 2013. Financial development, ICT diffusion and economic growth: Lessons from MENA region. *Telecommunications Policy*, 37(4–5): 252–261.

Singh, S.K. 2008. The diffusion of mobile phones in India. *Telecommunication Policy*, 32 (9–10): 642–651.

Sullivan, N.P. 2007. *You Can Hear Me Now: How Microloans and Cell Phones Are Connecting The World*. San Francisco, CA: John Wiley & Sons.

Uche, C.U. 2000. Banking regulation in an era of structural adjustment: The case of Nigeria. *Journal of Financial Regulation and Compliance*, 1(2): 157–169.

Uche, C.U. 2001. The adoption of universal banking in Nigeria. *Butterworths Journal of International Banking and Financial Law*, 16(9): 421–428.

Vu, K.M. 2011. ICT as a source of economic growth in the information age: Empirical Evidence from the 1996–2005 period. *Telecommunications Policy*, 35(4): 357–372.

section six

Mobile electronic commerce and social, economic, and environmental aspects

chapter seventeen

Mobile content and applications value networks

Evidence from the Italian mobile telecommunications market

Antonio Ghezzi, Raffaello Balocco, and Andrea Rangone

Contents

17.1 Introduction

The mobile telecommunications industry has a relatively short but undoubtedly intense history, made of significant advancements at both the infrastructure and service levels. On a network standard level, the industry evolved through a well-orchestrated multistep transition from global system for mobile communications (GSM) to general packet radio service (GPRS) and universal mobile telecommunications system (UMTS), which is currently leading to the development of 4G—the so-called long-term evolution (Barnes, 2002; Li and Whalley, 2002; Ballon, 2004; Ibrus, 2013).

At the same time, on a service level, the innovation initiative of mobile network operators (MNOs), device manufacturers (DMs), and a growing set of third parties catalyzed the development of a wide and appealing offer of value-added, nonvoice mobile digital media services—which include handset browsing, mobile social networking, mobile applications, mobile games, mobile music, mobile video, mobile TV, ringtones, wallpapers, and infotainment alerts—pertaining to the so-called mobile content and applications market segment (Peppard and Rylander, 2006; Kuo and Yu, 2006; Jung et al., 2013).

The mobile content and applications market relevance in the overall mobile telecommunications landscape is rising dramatically, as analysts expect its global value, only partially hindered by the ongoing recession, will exceed $140 billion by 2014 (Idate, 2012).

Within the global Mobile Telecommunications, Italy holds a key position, thanks to its world's highest service penetration and diffusion rates; its positioning at the forefront in industry innovation at the global level; and the strong international presence of its key players (Idate, 2012; Informa, 2012).

As the market grows and its structure evolves, involving a larger number of heterogeneous firms—characterized by firm-specific activities performed—and giving rise to a complex set of relationships between them, the need of rigorously identifying and consequently analyzing its constitutive value-creating or value-destroying activities has emerged, deserving attention from both researchers and practitioners. In particular, given the complexity of the aforementioned interfirm links and of the resulting thorough value system structure, the present study claims that the activity analysis process should benefit from the adoption of a value network perspective, which extends the traditional value chain model through a refocus on interorganizational, nonsequential, and multilayered relationships.

Therefore, the study aims at identifying which are the core activities that constitute the mobile content and applications market, how they are interrelated, and how they can be internalized and combined by different market players to define and shape their value domain.

Conceptually, this is achieved through applying the value network and the strategic network theoretical frameworks to the Italian market for mobile content and applications. Since different activity combinations within a given key actor's perimeter may arise, five alternative configurations of mobile content value networks are provided—characterized by different roles of the involved actors, in terms of activities covered—representing the most significant cases that emerged and could be inferred from the information collected through the literature review and the empirical analysis, based on multiple case studies. As a final step of the research, coherently to the perspective taken, the proposed configurations are evaluated with reference to a set of key variables or drivers derived from value networks and strategic networks theories, in order to delineate and compare their different characteristics.

17.2 Literature review on value networks and strategic networks

Since the renown value chain model was introduced by Michael Porter (1985), the research stream focusing on the analysis of internal activities within firms and external relations between them has been to a great extent *chained to the value chain* (Normann and Ramirez, 1994): in the attempt of individuating and interpreting the performance differentials of firms, the latter were typically studied as stand-alone, atomistic entities.

In the mobile content and applications market, the view of value creation as a well-defined linear sequence of value-adding activities pushed the adoption of market strategies which aimed at obtaining a vertical control of the chain: the *walled garden* solution for mobile portals can be seen as an example of the application of such approach by MNOs (Peppard and Rylander, 2006).

However, the applicability of the traditional value chain model has been questioned by several authors (Hakansson and Snehota, 1989; Normann and Ramirez, 1994; Anderson, 1995; Gulati, 1995; Campbell and Wilson, 1996; Tapascott et al., 2000; Stabell and Fjeldstad, 2002; Alle, 2003; Fjeldstad et al., 2004; Schieffer, 2005; Huemer, 2006; Peppard and Rylander, 2006; Pil and Holweg, 2006), as it emphasizes the concept of competition and does not take into fair consideration the more and more complex networks of both horizontal and vertical relations existing among firms.

Now that in many industries—including the mobile content and applications market—products and services are virtual and the chain of activities is not any more characterized by a physical dimension, the key element of a model meant to capture the drivers of value creation cannot be the mere position held by a firm within the value system but shall be replaced by the concept of *interdependencies* on which the whole network of relationships is built (Huemer, 2006). Moreover, those interdependencies are often far from being linear, and can be structured on several levels or layers: the sequence of activities constituting the value creation process is essentially multidirectional—horizontal, vertical, diagonal, retroactive, parallel, and simultaneous (Pil and Holweg, 2006).

All the previous types of durable and strategically significant interdependencies are embraced by the so-called value network model, which extends the value chain model by stressing the concept of network of relationships a firm builds within its boundaries—in terms of transversal processes—and outside its perimeter, and therefore claiming that value is created through interorganizational streams of activities.

Value networks and strategic networks literature has focused on the identification of variables or drivers capable of supporting a thorough description of a network, both from a *static* point of view—that is, in terms of its structural characteristics—and from a *dynamic* one that considers it as an evolving system subject to both endogenous and exogenous forces that determine some changes in time (Eggert et al., 2005).

Taking from a wide literature review (Jarrillo, 1988; Hakansson and Snehota, 1989; Burt, 1992; Hinteruber and Levin, 1994; Hobday, 1994; Normann and Ramirez, 1994; Anderson, 1995; Anderson et al., 1995; Gulati, 1995; Campbell and Wilson, 1996; Keil et al., 1997; Parolini, 1999; Gulati et al., 2000; Tapascott et al., 2000; Li and Whalley, 2002; Stabell and Fjeldstad, 2002; Alle, 2003; Antoniou and Ansoff, 2004; Chiesa and Toletti, 2004; Fjeldstad et al., 2004; Eggert et al., 2005; Schieffer, 2005; Hagedoorn et al., 2006; Huemer, 2006; Okamura and Vonortas, 2006; Peppard and Rylander, 2006; Pil and Holweg, 2006; Schoenmakers and Duysters, 2006; Gilsing et al., 2007; Marjolein et al., 2008; Ahuja et al., 2009; Desarbo et al., 2009; Funk, 2009; Gulati et al., 2009; Lin et al., 2009), a set of seven variables to describe a value network were identified. The five key *static* or structural variables are the following:

- *Network focal*—It refers to the firm positioned in the center of the network, controlling the original source of value, and linking the *peripheral* firms.
- *Critical network influences*—It refers to the most significant value-creating relations between firms.
- *Structural equivalences*—It refers to the condition where two or more members hold a similar position within the network.
- *Structural holes*—It refers to the situation where two or more firms within a network are connected only through the focal firm.
- *Revenue streams*—It refers to the direct or indirect exchanges of revenues between network members.

Concerning the dynamic variables, the model focuses on entangling the endogenous rather than the exogenous forces—the latter being mainly related to environmental changes, like the convergence of the IT, telecommunications, and media industries mentioned earlier (Gulati et al., 2000; Wirtz, 2001; Li and Whalley, 2002)—and considers two main phenomena:

- *Lock-in and lock-out effect*—It refers to the condition where the establishment of a relation with a given firm sets constraints to the creation of further relations with other firms.
- *Learning races*—It refers to the case where firms involved in a relation find themselves competing in a race for internalizing the partner's assets and resources, before leaving the alliance. This is most likely to happen when private benefits acquirable by any of the partners after they have learnt from the other exceed the common benefits of the alliance (Hamel et al., 1989; Wirtz, 2001).

These key variables are employed to describe the value network configurations proposed later on.

17.3 Research methodology

The present research is based on a wide literature analysis on strategic networks and value network theories, integrated with the adoption of multiple qualitative interviews as a research method (Rubin and Rubin, 1995; Yin, 2003).

Qualitative research methodology was chosen as particularly suitable for reaching the research objectives, which aim at understanding the complex phenomenon of value network definition within a given industry—that is, mobile content and applications, and at thus building new theory—or extending existing theories—on it.

At first, the literature review allowed identifying the variables to be employed for building the value network models and for assessing their characteristics in both static and dynamic terms. In addition to this, from January 2012 to September 2012, 94 qualitative interviews on firms operating in the Italian mobile content and applications market were conducted. The interview process involved three senior researchers; interviews lasted 1 h 39 min on average.

Consistently with the research methodology employed (Yin, 2003), the firm sample was not randomly selected, but firms were selected as they conformed to the main requirement of the study, while representing both similarities and differences considered relevant for the data analysis. Leveraging on the research carried out by the 2012 Mobile Internet, Content and Applications (MICA) Observatory, which has been focusing on the Italian market for mobile content since its rise in the early 2000s, the theoretical sample quasi-exhaustively covered the key actor categories belonging to the Italian Mobile ecosystem. The main categories of the interviewed firms were the following: MNOs (4 players); mobile service providers (MSPs) (24 players); mobile content and applications providers—media companies, web companies; web editors, developers—(31 players); mobile technology providers (MTPs) (20 players); DMs (6 players); and advertisers (9 players). The sample accounted for 90% of the universe of actors identified as involved in the market under scrutiny, and the analysis of secondary sources on the excluded firms allows inferring that their positioning and perspective was to a great extent similar or totally overlapped to that of other firms included in the sample—this grants the minimal loss of information due to the sampling process. Whenever possible, for each company, the following group of informants was interviewed:

- Chief Executive Officer
- Chief Information Officer
- Mobile content business area manager (this category assumed different denominations in different companies: the most common were mobile content business area manager; value-added services manager; mobile portal manager; and mobile third-parties manager)
- Product Managers

The semistructured nature of the interviews made it possible to start from the key issues identified through the literature review (such as the network theory's assessment variables) and also to let any innovative issue emerge from the open discussion.

The interview's general scheme of analysis was built around four investigation building blocks.

At first, each informant was asked to identify the mobile content and applications market's most significant activities to be carried out, from service conception to service delivery. Thanks to this first round of questions, in the subsequent analysis and elaboration phase it was possible to identify the most significant activities performed to generate value within the market under scrutiny, thus shaping a generalized *mobile content and applications value network* (see Section 1.4).

Second, informants were asked to describe the role their company currently took on in the mobile content and applications market, in terms of key activities covered (with questions such as "How would you describe your company's core business?" "What are the activities your company performs?" "Among the activities carried out, where does your company outperforms competitors, thus grounding its competitive advantage?"). In the analysis and elaboration phase, this collection of information allowed identifying the span of activities typically covered by each and every actor; also, it allowed spotting any strategic misalignment or uncommon positioning within the same actor category considered in the sample, to support a cross-case analysis and feed the value–strategic network scenario generation step of the study.

Third, informants were asked to provide information on the existing relationships established between their company and other players involved in the market (with questions such as "What activities directly impacting on your core business do you outsource or are depending on other actors?" "What resources and assets, either tangible or intangible, do you share/exchange with your business partners?" "What kind of contractual agreements and revenue generation models do you set up with them?"). Data collected with this round of questions helped in understanding the relative positioning of different actors in the whole value system, highlighting the most significant business interrelations and value–strategic network interconnections; it also helped identifying any coverage overlapping that could determine competitive attritions between different actor categories.

The last investigation area dealt with the companies' future strategies and business plans, in terms of objectives, roles, expected market trends, etc. (with questions such as "What are your company's short–medium term, and long-term objectives?" "Where are your strategies expected to lead your company?" "Are you eager to either strengthen or modify your current strategic positioning and role, and why?" "Do you perceive any attempt by other actors—either partners or competitors—to modify their positioning as well, and how do you interpret their moves?"

"According to your perspective, what are the market's evolutionary trends you expect to consolidate in the near future?"). This last set of questions allowed integrating the analysis with a dynamic view, while making insightful inferences about the most relevant alternative value–strategic networks that are disrupting the traditional value system configuration.

A multiple case study approach reinforced the generalization of results (Meredith, 1998) and allowed performing a cross-analysis on the relevant variables drawn from the literature, to highlight any differential in terms of their combination—to see which variables changed and which remained constant going from one value network configuration model to another—due to the presence of extreme cases, polar types or niche situations within the theoretical sample (Meredith, 1998). The unit of analysis for each case study was the firm operating in the market.

As the validity and reliability of case studies rest heavily on the correctness of the information provided by the interviewees and can be assured by using multiple sources or "looking at data in multiple ways" (Eisenhardt, 1989; Yin, 2003), multiple sources of evidence or research methods were employed: direct interviews, analysis of internal documents, study of secondary sources—research reports, websites, newsletters, white papers, databases, and international conferences proceedings. This combination of sources allowed obtaining *data triangulation*, essential for assuring rigorous results in qualitative research (Bonoma, 1985).

Throughout the research, theory—represented by the literature review and the original model proposed—was used as *part of an iterative process of data collection and analysis* (Eisenhardt, 1989), meaning that it was employed as an initial guide to design the study and the process of data gathering, though it was never intended to constrain emergent issues coming from the qualitative analysis, so as to preserve the suggested considerable degree of openness to the field data (Walsham, 1995; Yin, 2003).

17.4 Mobile content and applications value-adding activities

The market for mobile digital content is undergoing a process of value system reconfiguration (Gulati et al., 2000; Li and Whalley, 2002; Stabell and Fjeldstad, 2002; Fjeldstad et al., 2004; Huemer, 2006; Peppard and Rylander, 2006; Funk, 2009; Ibrus, 2013).

The overall value network is resulting from the juxtaposition of different major value chains (Wirtz, 2001):

- Mobile telecommunications
- Information technology
- Media
- Electronic commerce

Such process is made of a first phase of *unbundling*, where each value chain is divided into its elementary activities, followed by a *rebundling* phase, where the activities are recombined to create a new structure. The *value configuration* (Stabell and Fjeldstad, 2002) coming from the combination of different activities shall be analyzed in terms of not only length but also depth: value is created through the vertical interaction of parallel and coexisting activities located on distinct levels, in what Huemer (2006) defines a *layered architecture* made of overcurrent and undercurrent activities.

In the light of the previously mentioned concepts, the mobile content and applications value system here proposed, related to the process of creation, management and delivery of mobile digital content, is composed of four parallel but interconnected layers:

1. *Content and Service Layer,* covering the activities related to the life cycle management of mobile digital content and services—content creation, content packaging, content publishing, content management, portal provisioning, advertising bundling, content delivery and market making, content charging, content billing and accounting, and customer relationship management
2. *Device Layer,* covering the activities related to the design, manufacturing, and delivery of devices for content and service fruition—device design, device manufacturing, and device provisioning
3. *Platform Layer,* undercurrent to the previous layer, which comprises the activities of designing, producing, and operating the middleware platforms for mobile content management and delivery—platform design, platform manufacturing, platform provisioning, platform operations, platform management (Firtman, 2013)
4. *Network Layer,* encompassing the cross activities related to the installation and operations of the mobile network infrastructure

The interconnection between the first two layers becomes evident with the activity of Content Publishing on the middleware platform. The content and service layer can be further divided into an *upstream chain*, encompassing the activities from content creation to its preparation for delivery, and a *downstream chain* considering the stages which follow the content commercialization (Figure 17.1).

The identified core value-adding activities can be performed by several actor typologies belonging to the network: since a different attribution of activities covered can give rise to alternative network configurations with specific strategic implications, this study proposes five feasible configurations of value networks for the mobile content and applications market. Such models are hence analyzed and assessed on the basis of key variables drawn from the existing theories on value networks and strategic networks.

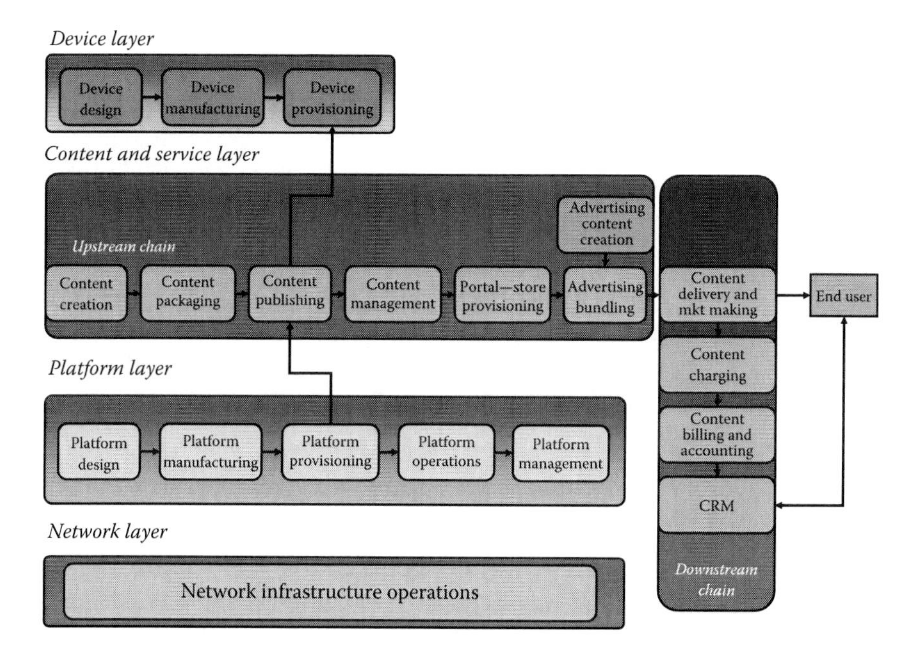

Figure 17.1 The mobile content and applications value-adding activities.

17.5 *Mobile content and application value network alternative configuration*

To fulfill the main objective of the study, that is, creating a model capable of portraying the mobile content and applications value network, the core value-adding activities in the value system of mobile digital content are to be assigned to the key actors belonging to the system itself.

The main actor typologies considered as the market's key player derive from the integration of the literature review and the empirical data collection analysis, and include the following seven broad categories:

1. MNO—owns the network and is responsible for the provisioning of its functionalities.
2. MSP—is mainly active on the overall management of content and services, taking on an intermediary role between MNOs and mobile content providers (MCPs).
3. MCP or Developer—concentrates on the creation of digital content, service, or application. In a broad sense, media–web companies and single developers also belong to such categories.
4. MTP—focuses on the provision of the content and service delivery platforms (CSDPs) to manage and deliver digital content.

5. DM—manufactures and provides the mobile devices (e.g., cellular phones, smartphones, and personal digital assistants (PDAs)) through which the content, service, or application is accessed by the end user.
6. Advertiser—is any kind of firm interested in leveraging on the mobile channel to promote their products/services.
7. End user—final customer of mobile digital content.

As the different allocation and combination of activities within an actor's domain can give rise to a different overall value network, the present research proposes five feasible noteworthy configurations, whose characteristics and strategic implications are further investigated through the static and dynamic variables derived from the literature review.

17.5.1 Full-walled garden configuration

In the *full-walled garden* configuration, the MNO represents the *network focal*, the actor benefiting from a central positioning and a deeper *embedding* within the network structure. In this situation, the MNO takes on a pervasive role, integrating most of the activities pertaining to the content and service layer—ranging from content aggregation to content delivery and market making; it also takes care of the *charging–billing–accounting* (CBA) process as well as of CRM—in addition to its natural presidium of the network layer (Whalley and Curwen, 2013).

The *network influences* or value ties for the MNO are those built with MCPs to feed the value added services (VAS) portfolio and with one or more MTPs providing the CSDP through which delivering a range of value-adding functionalities to the end customer—for example, personalization, interactivity, context awareness, localization, and so on.

In terms of revenue streams—identified in Figure 17.2 by the arrows connecting the actors' domains—the MNO pays out to the CPs either a content-based or a transport-based fee for original content provisioning, and receives from them a fee for the exploitation of its proprietary platform's functionalities; it also receives a fee from the advertiser for the services offered. Being the only responsible for content delivery and market making, the MNO retains both the content cost and the transport cost billed to the end customers, and pays a share of it to the MTP, who is in charge of the platform's operations.

The central position of the MNO creates some *structural holes* among actors not directly tied neither between themselves—for example, MTPs and MCPs, the DMs and the CPs—nor to the end customer: this allows the operators to benefit from higher profits (Burt, 1992).

Besides, some structural equivalence can be found, as a single operator can relate with multiple MCPs and advertisers that hold a similar role

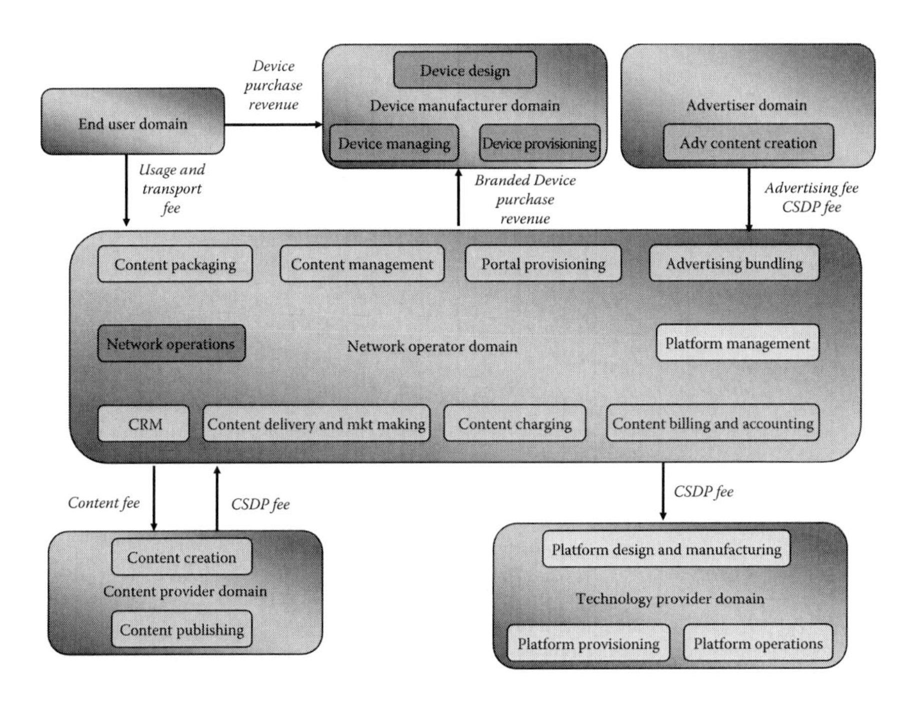

Figure 17.2 Full-walled garden configuration.

and somehow find themselves competing for the possibility of establishing a strong relationship with the focal firm.

From a dynamic point of view, lock-in and lock-out effects can emerge, since the contracts stipulated between the MNO and the MCPs for content provisioning, or between the MNO and the MTPs for platform provisioning, can be exclusive: these *monogamous alliances* shape the network structure, and can include or exclude some players from a given network graph for medium–long periods of time.

As a whole, this full-walled garden configuration is only feasible with regard to elementary services, and does not seem to meet the current needs of the operators themselves, interested in developing the mobile content and applications segment through leveraging on third parties' propositions, as shown in Figure 17.3.

This consideration gives rise to a second possible configuration, where the role of the MSP as an intermediary emerges.

17.5.2 *Intermediated content delivery configuration*

The *intermediated content delivery* configuration sees the rise of a further actor typology, the MSP, controlling some downstream activities previously performed by the MNO, like content delivery and market making.

Static or structural assessment

- Network focal
  - MNO
- Critical network influences
  - MNO–MTP
  - MNO–MCP
  - MNO–DM
- Structural equivalences
  - MCP; DM; Advertisers
- Structural holes
  - MTP–MCP; MTP–Advertiser
  - MCP–End User
  - DM–MCP; DM–TP
  - DM–End User (for Content Offering)
- Revenue streams
  - Direct for MNO, DM (for Device Offering)
  - Indirect for MCP, MTP

Dynamic assessment

- Lock-in and lock-out effects
  Relations between:
  - MTP–MNO
  - MCP–MNO
  - DM–MNO
- Learning races
  - Absent, due to the MNO's overcontrolling positioning

Figure 17.3 Full-walled garden configuration: static and dynamic assessment.

Acting as intermediaries, the MSPs establish a direct contact with the end user, thus enhancing their role and bargaining power within the network.

The MNO maintains coverage of the main upstream activities, for example, content aggregation and management, portal provisioning, and so on. It also keeps managing the relations with multiple MCPs and one or more MTPs, adding to this portfolio of alliances the MSPs, from which it receives a consistent share of revenues coming from the value-added services commercialization.

The MTPs concentrate within their boundaries the activities belonging to the platform layer, and can direct their CSDP offer to both the MNO and the MSPs. The latter relation can even result in a *learning race*, where the Service Provider tries to steal technological resources and competencies to internally design and operate the platform.

The DMs keep on being related to the MNO and the end users only, though the direct revenue streams they receive from the latter are related to the selling of devices rather than the commercialization of content and service.

The focal actor remains the MNO, but as the network structure becomes less concentrated because of the presence of MSPs, some structural holes characterizing the first configuration are filled—like the one between MSPs and advertisers, MSPs and MTPs, and MSPs and end users.

The main structural equivalences noted are related to the roles of MSPs, CPs, and advertisers—such factor increases the level of competition on the corresponding activities covered—while the most significant network influences for the operator in terms of value creation are those with the MCPs for widening the content and services portfolio, with the MSPs

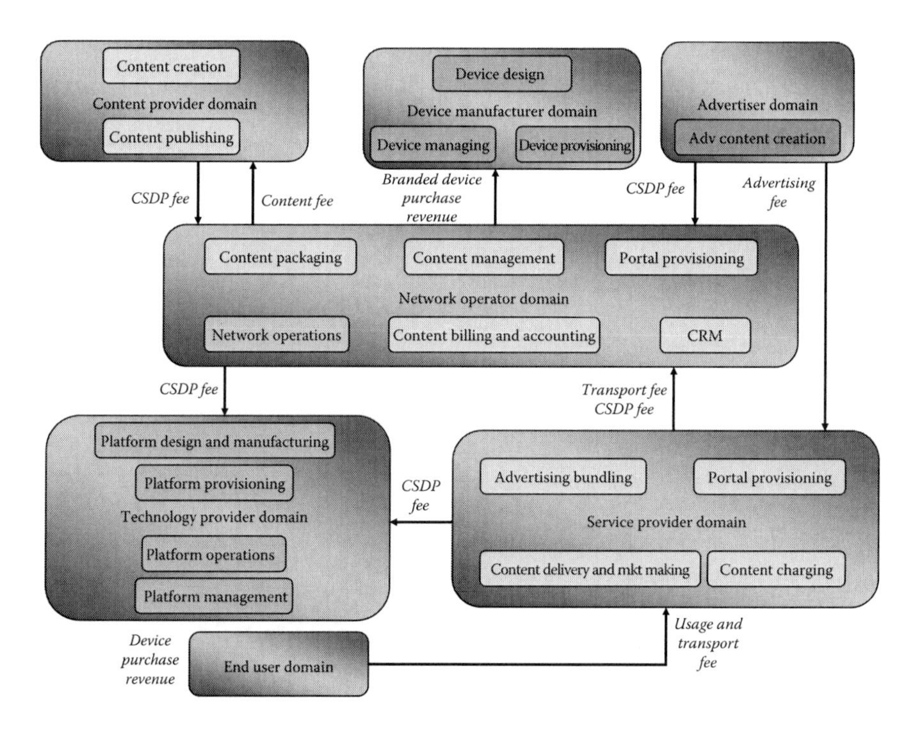

Figure 17.4 Intermediated content delivery configuration.

to leverage on their assets and resources for better serving a larger pool of customers and with MTPs providing best-in-class platform solutions that could influence the content management and delivery performances.

Moreover, even in this configuration the peripheral actors could find themselves locked in or out of an MNO's strategic group, with heavy impacts on their profitability and chances of success or survival (see Figure 17.4).

In sum, this configuration is related to the large majority of content and services published on-portal (e.g., infotainment): the MNO externalizes the downstream activities of content market making to the MSP, through a model implying higher openness toward peripheral actors, which paves the way to a third parties–driven development of mobile content and applications (see Figure 17.5).

17.5.3 Full open garden configuration

The *full open garden* configuration represents a possible development and degeneration of the second alternative, where the Service Provider replaces the operator as the network focal, substantially relegating the latter to the role of carrier and network manager.

Static or structural assessment

- Network focal
  - MNO
- Critical network influences
  - MNO–MTP
  - MNO–MCP
  - MNO–MSP
  - MNO–MSP
- Structural equivalences
  - MCP; MSP; Advertisers
- Structural holes
  - MTP–MCP; MTP–MSP; MTP–Advertisers
  - MCP–End User
  - DM–MCP; DM–MTP; DM–End User (for Content Offering)
- Revenue streams
  - Direct for MNO, MSP, DM (for Device Offering)
  - Indirect for MCP, MTP

Dynamic assessment

- Lock-in and lock-out effects
  Relations between:
  - MTP–MNO
  - MCP–MNO
  - MCP–MNO
  - DM–MNO
- Learning races
  - Between MSP–MTP for technological competencies

Figure 17.5 Intermediated content delivery configuration: static and dynamic assessment.

The MSP extends its control to key upstream activities, directly relating to MCPs, MTPs and advertisers as well as to the MNO; in the downstream section of the content and service layer, the MSP strengthens the ties bonding it to the end customers.

Concerning revenue streams, the MSP pays a fee to the MCPs for the exploitation of their content and pays a CSDP fee to the MTPs; it receives direct revenues from the advertisers—for the enablement of their campaigns on the mobile channel—and from the end users. A consistent part of these revenues from content commercialization is redirected to the MNO, for the provisioning of the network infrastructure, of the CBA system, of the mobile portal, and of the customer-related information. In fact, in a situation where the MNO holds a secondary role within the network, it should leverage on its core assets like the infrastructure and the 3G licenses, the billing and accounting systems, and the control over customers' information and usage data (Kuo and Yu, 2006).

The structurally equivalent firms are now the competing MSPs, MCPs, DMs, and MTPs. The main structural holes become those separating the MNO from MCPs and advertisers, while the key network influence for the MSP is the relation with the operator, which ensures the possibility of reaching the end customers.

In terms of network dynamics, peripheral firms could be locked in or locked out of the strategic groups originated by the network's core dyads MNO–MSP. Learning races can also arise between MSPs and MCPs, where the former try to acquire core competencies related to content creation to become capable of internally developing a white label offer, or between

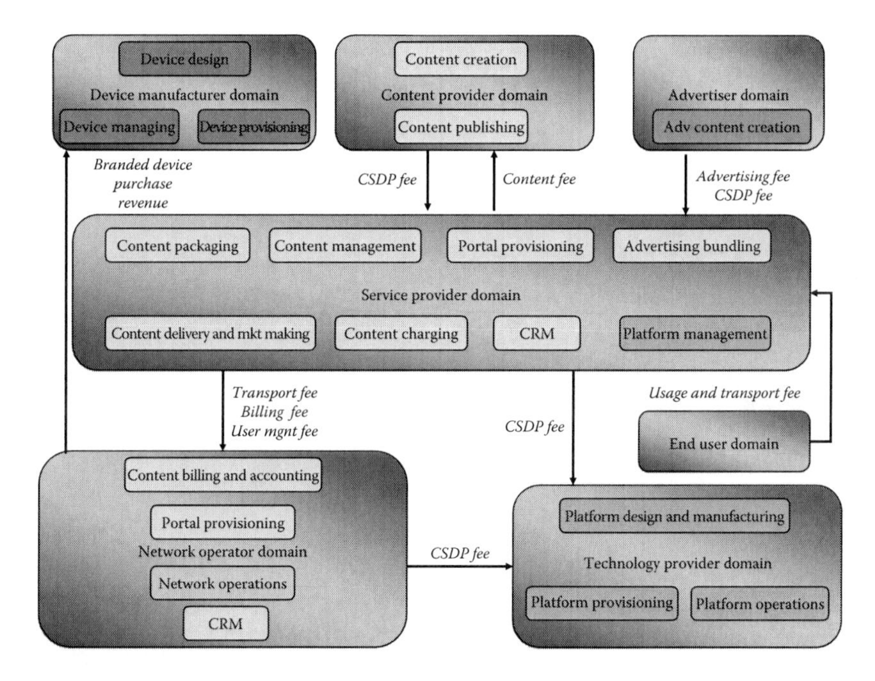

Figure 17.6 Full open garden configuration.

MSPs and MTPs, where Service Providers could be interested in stealing the know-how necessary to design and operate a middleware platform to exploit it for their exclusive benefit, and afterwards breaking the alliance, shown in Figure 17.6.

The configuration modeled here is surely characterized by a disruptive nature, as it takes to the extremes the trend of openness toward third parties, reshaping the network structure to a point where the substitution of the MNO as focal firm occurs, shown in Figure 17.7.

17.5.4 Technology and service provider configuration

The *technology and service provider* configuration is alternative to the previous one, since it sees the rise of the MTP as the leading actor of the network; the MSP is no longer present, as its activities are taken on by the technology vendor. The configuration can be labeled as disruptive because it reduces the MNO's control over the value system and extends the MTP's coverage to activities not entangled by its traditional core business—that is, the activities belonging to the platform layer.

Being the network focal, the MTP directly relates with MCPs, advertisers, and end user, as well as to the MNO for exploiting its network infrastructure functionalities. The MTP benefits from direct revenue

Static or structural assessment

- Network focal
  - MSP
- Critical network influences
  - MSP–MNO
  - MSP–MTP
  - MSP–MCP
- Structural equivalences
  - MCP; MTP; Advertisers
- Structural holes
  - MNO–CP; MNO–Advertiser
  - MTP–CP; MTP–Advertiser
  - DM–MCP; DM–MTP; DM–End User (for Content Offering)
- Revenue streams
  - Direct for MSP, DM (for Device Offering)
  - Indirect for MNO, MCP, MTP

Dynamic assessment

- Lock-in and lock-out effects
  Relations between:
  - MSP–MNO
  - MSP–MTP
  - DM–MNO
- Learning races
  - MSP–MTP for technological competencies
  - MSP–CP for content creation

Figure 17.7 Full open garden configuration: static and dynamic assessment.

streams coming from the end customer, and partly redirects those streams to the MNO as a payoff for the transport, CBA, and user management services. For all these reasons, it is evident that the actor incorporates the activities performed by the MSP, becoming a *technology and service provider.*

Within the network originating from the rise of this actor, competitive attrition manifests itself among the different graphs centered on the dyads MTP–MNO, to which the peripheral firms are connected: the *relational space* thus created influences the firms' performances and behavior.

As it happened in Configuration 3, the network's structural gaps are those separating the MNO from the peripheral firms—that is, advertisers and MCPs. The critical network influence is the relationship between the MTP and the MNO, essential for the actual deployment of the rich media content and services to the end user, shown in Figure 17.8.

The shared ownership of end customers between the MTP and the MNO could even result in a learning race for the internalization of data and information regarding the user profiles and the usage reports.

As a conclusion, in the final configuration the reach of the MTP is extended to an overall management of both the mobile digital content and the underlying platforms, thus creating competitive dynamics between the MTP and the MSP that could result in the disappearance of the latter actor from the value ecosystem. However, the indications coming from the case studies allow judging such configuration as the least likely to manifest on a large scale, since the existing network sections where the MTP is the *deepest embedded firm* are mostly resulting from market anomalies—for

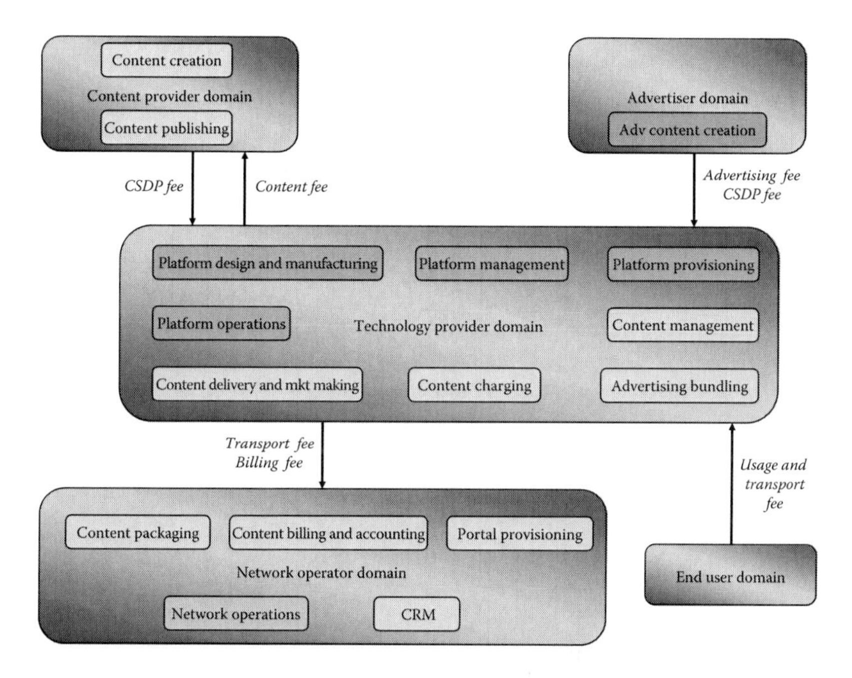

Figure 17.8 Technology and service provider configuration.

Static or structural assessment

- Network focal
  - MTP
- Critical network influences
  - MTP–MNO
  - MTP–MCP
- Structural equivalences
  - MCP; Advertisers
- Structural holes
  - MNO–MCP; MNO–Advertiser
  - DM–MCP; DM–MTP; DM–End User (for Content Offering)
- Revenue streams
  - Direct for MTP, DM (for device offering)
  - Indirect for MNO, MCP

Dynamic assessment

- Lock-in and lock-out effects
  Relations between:
  - MTP–MNO
  - DM–MNO
- Learning races
  - MTP–MNO for user profiling

Figure 17.9 Technology and service provider configuration: static and dynamic assessment.

example, contingent opportunities or long-lasting partnerships the technology vendors have exploited to profit in the short–medium term, rather than from a clear strategic long-term planning.

Figure 17.9 illustrated and describes the static and dynamic characteristics of the Technology and service provider configuration.

17.5.5 *Device manufacturer–driven configuration*

The "*DM-driven* configuration is quite a recent alternative related to the recent and emerging *application store* phenomenon (e.g., Apple Store, Vodafone 360°), where the DM leverages on the mobile device to challenge the incumbent's leadership, bypassing its portal and CBA system (Bergvall-Kåreborn and Howcroft, 2013).

Within such disruptive alternative, the DM takes on a network focal role of content and service management, while the MNO withdraws from direct content management to some customer-related activities, whose value is, however, competed away by the presence of DM-owned alternatives (e.g., CBA provisioning vs. direct billing; Portal Provisioning vs. Store Provisioning) (see Figure 17.10).

From a structural perspective, the configuration's critical network influences are those among the DM and MNOs, and MCPs and MTPs (as the MSP is no longer present in the value network, being replaced by the DM itself); the emerging structural equivalences are related to the roles of MCP, MSP, and advertisers; the structural holes emerge among MNOs, MCPs, and advertisers; and the revenue streams are of a direct nature for DMs (Device Offering and Content Offering), while of an indirect one for MNOs, MTPs, and MCPs.

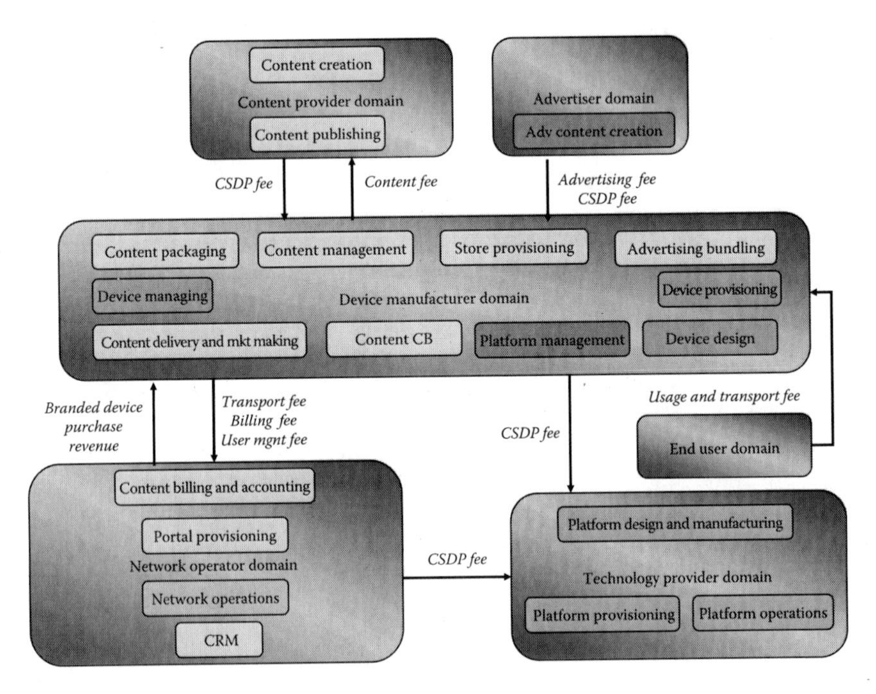

Figure 17.10 DM-driven configuration.

Static or structural assessment

- Network focal
  - DM
- Critical network influences
  - DM–MNO
  - DM–MCP
  - DM–MTP
- Structural equivalences
  - MCP; MTP; Advertisers
- Structural holes
  - MNO–MCP; MNO–Advertiser
- Revenue streams
  - Direct for DM (for device offering and content offering)
  - Indirect for MNO, MTP

Dynamic assessment

- Lock-in and lock-out effects
  Relations between:
  - DM–MNO
  - DM–MCP
  - DM–MTP
- Learning races
  - DM–MNO for customer relationship management
  - DM–MTP for technological competencies

Figure 17.11 DM-driven configuration: static and dynamic assessment.

From a dynamic view, lock-in and lock-out effects could be generated in the relationships between DMs and MNOs and MCPs and MTPs; also, learning races could rise as DMs and MNOs compete for customer relationship management, or DMs and MTPs compete for technological competencies (Figure 17.11).

17.6 Conclusions and future works

The fast-evolving Mobile Content and Applications market results from the juxtaposition of several value chains; therefore, its value analysis requires the development of original models extending the existing literature.

The present research focused on the identification of the market's key value-adding activities and on the further attribution of such activities to the incumbent players' domains: five alternative value network configurations were hence proposed. The configurations were derived from a combination of a wide literature review and the case studies performed: though hybrid, in-between solutions are still plausible, the presented alternatives were selected as the most significant and most likely to arise in network subparts or graphs.

The findings show how varying the set of value-adding activities covered by the different actors shapes the network structure and inner dynamics, with significant impacts on the firms' behavior and strategic options at hand. Specifically, while the first two configurations are characterized by a more *conservative* approach, being dominated by the MNO, the third, fourth, and fifth alternatives represent a disruptive evolution of the current trend of growing openness toward third parties, even resulting in the substitution of the operator as the network focal.

The paper's value for researchers can be brought back to the application of the value networks and strategic networks theories to a further context, where such theories' core variables are found extremely useful to describe and assess the market's feasible configurations, at both a static and a dynamic level.

Moreover, the study emphasizes the relevance of an Italian national market in a highly ICT and innovation-intensive industry.

Value for practitioners lies in the creation of a reference model capable of providing a unified definition of the value-adding activities the market is structured in, while depicting a set of noteworthy network configurations and underpinning the market's main structural and dynamic characteristics as well as their strategic implications.

The study represents a significant contribution to the development of value network theory, with specific reference to the mobile content and applications market. Nevertheless, it mainly focuses on the limited, though relevant Italian national market. Future research will have to address the issue of the generalization of results, through applying the same constructs of analysis to different markets and company samples.

References

Ahuja, G., Polidoro, F., and Mitchell, W. 2009. Structural homophily or social asymmetry? The formation of alliances by poorly embedded firms. *Strategic Management Journal* 30(9), 941–958.

Alle, V. 2003. *The Future of Knowledge: Increasing Prosperity through Value Networks.* Butterworth-Heinemann, Amsterdam, the Netherlands.

Anderson, J.C. 1995. Relationships in business markets: Exchange episodes, value creation, and their empirical assessment. *Journal of Academy of Marketing Science* 23(4), 346–350.

Anderson, J.C., Hakansson, H., and Johanson, J. 1995. Dyadic business relationships within a business network context. *Journal of Marketing* 58, 1–15.

Antoniou, P.H. and Ansoff, I.H. 2004. Strategic management of technology. *Technology Analysis & Strategic Management* 16(2), 275–291.

Ballon, P. 2004. Scenarios and business models for 4G in Europe. *Info* 6(6), 363–382.

Barnes, S.J. 2002. The mobile commerce value chain: Analysis and future developments. *International Journal of Information Management* 22, 91–108.

Bergvall-Kareborn, B. and Howcroft, D. 2013. The Apple business model: Crowdsourcing mobile applications. *Accounting Forum* 37(4), 280–289.

Bonoma, T.V. 1985. Case research in marketing: Opportunities, problems, and a process. *Journal of Marketing Research* 22, 199–208.

Burt, R.S. 1992. *Structural Holes: The Social Structure of Competition.* Harvard University Press, Cambridge, MA.

Campbell, A.J. and Wilson, D.T. 1996. Managed networks: Creating strategic advantage. In *Networks in Marketing* (Iacobucci, D., ed.). Sage Publishing, London, U.K., pp. 125–143.

Chiesa, V. and Toletti, G. 2004. Network of collaborations for innovation: The case of biotechnology. *Technology Analysis & Strategic Management* 16(1), 73–96.

Desarbo, W.S., Grwal, R., and Wang, R. 2009. Dynamic strategic groups: Deriving spatial evolutionary paths. *Strategic Management Journal* 30(3), 1420–1439.

Eggert, A., Ulaga, W., and Schultz, F. 2005. Value creation in the relationship life cycle: A quasi-longitudinal analysis. *Industrial Marketing Management* 35, 20–27.

Eisenhardt, K.M. 1989. Building theories from case study research. *Academy of Management Review* 14(4), 532–550.

Firtman, M. 2013. *Programming the Mobile Web*. O'Reilly Media, Sebastopol, CA.

Fjeldstad, Ø.D., Becerra, M., and Narayanan, S. 2004. Strategic action in network industries: An empirical analysis of the European mobile phone industry. *Scandinavian Journal of Management* 20,173–196.

Funk, J.L. 2009. The emerging value network in the mobile phone industry: The case of Japan and its implications for the rest of the world. *Telecommunications Policy* 33, 4–18.

Gilsing, V.A., Lemmens, C.E.A.V., and Duysters, G. 2007. Strategic alliance networks and innovation: A deterministic and voluntaristic view combined. *Technology Analysis & Strategic Management* 19(2), 227–249.

Gulati, R. 1995. Social structure and alliance formation pattern: A longitudinal analysis. *American Journal of Sociology* 91, 481–510.

Gulati, R., Lavie, D., and Singh, H. 2009. The nature of partnering experience and the gains from alliances. *Strategic Management Journal* 30(11), 1213–1233.

Gulati, R., Nohria, N., and Zaheer, A. 2000. Strategic Networks. *Strategic Management Journal* 21, 203–215.

Hagedoorn, J., Roijakkers, N., and Van Kranenburg, H. 2006. Inter-firm R&D networks: The importance of strategic network capabilities for high-tech partnership formation. *British Journal of Management* 17(1), 39–53.

Hakansson, H. and Snehota, I. 1989. No business is an island: The network concept of business strategy. *Scandinavian Journal of Management* 5(3), 187–200.

Hamel, G., Doz, Y.L., and Prahalad, C.K. 1989. Collaborate with your competitors and win. *Harvard Business Review* 67(1), 133–139.

Hinteruber, H.H. and Levin, B.M. 1994. Strategic network—The organization of the future. *Long Range Planning* 27(3), 43–53.

Hobday, M. 1994. The limits of silicon valley: A critique of network theory. *Technology Analysis & Strategic Management* 6(2), 232–245.

Huemer, L. 2006. Supply Management. Value creation, coordination and positioning in supply relationships. *Long Range Planning* 39, 133–153.

Ibrus, I. 2013. Evolutionary dynamics of media convergence: Early mobile web and its standardisation at W3C. *Telematics and Informatics* 30(2), 66–73.

Idate. 2012. DigiWorld yearbook. The challenges of the digital world. Market Research Report, Idate, Montpellier, France. Available at: http://www.idate.org/en/Digiworld-store/Collection/DigiWorld-Yearbook_9/DigiWorld-Yearbook-2012_730.html. Accessed on January 20, 2014.

Informa. 2012. Mobile content and services (10th edition). Market outlook, revenue opportunities & business models. Market Research Report.

Jarrillo, J.C. 1988. On strategic networks. *Strategic Management Journal* 9(1), 31–41.

Jung, H., Kim, E., Kim, S., and Hwang, J. 2013. How diversification affects innovation by a Korean mobile content firm. *International Journal of Mobile Communications* 10(5), 521–535.

Keil, T., Autio, E., and Robertson, P. 1997. Embeddedness, power, control and innovation in the telecommunications sector. *Technology Analysis & Strategic Management* 9(3), 299–316.

Kuo, Y. and Yu, C. 2006. 3G Telecommunication operators' challenges and roles: A perspective of mobile commerce value chain. *Technovation* 26(12), 1347–1356.

Li, F. and Whalley, J. 2002. Deconstruction of the telecommunications industry: From value chain to value network. *Telecommunications Policy* 26, 451–472.

Lin, Z., Yang, H., and Arya, B. 2009. Alliance partners and firm performance: Resource complementarity and status association. *Strategic Management Journal* 30(9), 921–940.

Marjolein, C.J., Caniels, A., Henny, A., and Romijn, B. 2008. Strategic niche management: Towards a policy tool for sustainable development. *Technology Analysis & Strategic Management* 20(2), 245–266.

Meredith, J. 1998. Building operations management theory through case and field research. *Journal of Operations Management* 16, 441–454.

Normann, R. and Ramirez, R. 1994. *Designing Interactive Strategy: From the Value Chain to the Value Constellation*. John Wiley & Sons, Chichester, U.K.

Okamura, K. and Vonortas, Ns. 2006. European alliance and knowledge networks. *Technology Analysis & Strategic Management* 18(5), 535–560.

Parolini, C. 1999. *The Value Net: A Tool for Competitive Strategy*. John Wiley & Sons, Chichester, U.K.

Peppard, J. and Rylander, A. 2006. From value chain to value network: An insight for mobile operators. *European Management Journal* 24(2), 128–141.

Pil, Fk. and Holweg, M. 2006. Evolving from value chain to value grid. *MIT Sloan Management Review* 47(4), 72–80.

Porter, M. 1985. *Competitive Advantage: Creating and Sustaining Superior Performance*. Free Press, New York.

Rubin, H. and Rubin, I. 1995. *Qualitative Interviewing: The Art of Hearing Data*. Sage Publishing, Thousand Oaks, CA.

Schieffer, A. 2005. Value networks: How organizations really work. *Knowledge Management Research & Practice* 2, 194–199.

Schoenmakers, W. and Duysters, G. 2006. Learning in strategic technology alliances. *Technology Analysis & Strategic Management* 18(2), 245–264.

Stabell, C. and Fjeldstad, Ø. 2002. Configuring value for competitive advantage: On chains, shops, and networks. *Strategic Management Journal* 19, 413–437.

Tapascott, D., Ticoll, D., and Lowy, A. 2000. *Digital Capital: Harnessing the Power of Business Webs*. Harvard Business School Press, Boston, MA.

Walsham, G. 1995. Interpretive case-studies in IS research–Nature and methods. *European Journal of Information Systems* 4(2), 74–81.

Whalley, J. and Curwen, P. 2013. Unravelling complex organisational structures among mobile operators. *Info* 15(4), 3–22.

Wirtz, B.W. 2001. Reconfiguration of value chains in converging media and communications markets. *Long Range Planning* 34, 489–506.

Yin, R. 2003. *Case Study Research: Design and Methods*. Sage Publishing, Thousand Oaks, CA.

chapter eighteen

Segmenting, targeting, and positioning of mobile payment services

Raluca-Andreea Wurster and Cezar Scarlat

Contents

18.1 Introduction

Successful companies in the technological field maintain a leading role only if their innovations are adopted by a majority of consumers. Due to an increasing mobility of today's society and progress in technological infrastructure, the mobile phone technology has been quickly adopted worldwide. The technological advances in the field of near-field contactless communications (NFC) and the development of sophisticated mobile applications have enabled mobile phones to become a potential means of payment.

From the user's point of view, the NFC transactions become popular as they limit the time needed for a user to do a transaction. Conversely, Alimi et al. (2013) studied the use of NFC transactions for the merchant side, proposing "different architectures in order to use a mobile phone as a point of sale." For security reasons, the users are biometrically authenticated through the use of the touch screen on their smart phones.

As NFC-enabled mobile phones, in certain circumstances, could be used as mobile token cloning and skimming platforms (and an attacker could use an NFC mobile phone as such an attack platform by exploiting the existing security control systems), researchers (Francis et al., 2010) paid attention to these aspects as well and developed appropriate security countermeasures for NFC-enabled phones. On the other hand, legitimate transaction information from contactless systems could be used for marketing research purposes.

Overall, the security issues of the mobile payment services (MPS) are addressed by many researchers (Lu et al., 2009; Francis et al., 2010; Vesal and Fathian, 2012); however, the biometric approach seems to be dominant.

In 2009, more than two-thirds of the population worldwide was equipped with a mobile phone. In developed countries, where the market for mobile telephony is mature, mobile payments are seen as a potential source of revenue by mobile operators, who are trying to diversify their services. In developing countries, where mobile telephony is also widely adopted, mobile payments are seen by the governments as an opportunity to improve access to banking services for the unbanked population. The notion of MPS refers to making payments for goods, services, and bills authorized that are initiated and realized by using mobile devices independent from a bank website connection (Schierz et al., 2009). The mobile device can either be a mobile phone, a computer, a personal digital assistant (PDA), or even a wireless sticker that can be attached to any object, such as a ring or a key. The transaction can either be remote (SMS-based, for instance) or be processed locally via contactless technologies such as near field communication or radio-frequency identification (RFID). Mobile money transfers can be remote, in-store, prepaid, or postpaid through reverse billing. Overall, according to a study from Little (2012), so-called mobile payments represented a transaction volume of $250 billion in 2012.

Based on new technologies, MPS are currently diversifying, even mixing them. For example, the mobile traveler's check (MTC) is a new payment instrument that combines the advantages of e-check (security) and e-cash (free use). Ahamad et al. (2012) have developed a mobile payment protocol for MTC, which uses sophisticated algorithms to generate and verify digital signatures (as elliptic curve digital signature algorithm), and for encrypting and decrypting the messages suitable for mobile phones.

However, though mobile payments have attracted a lot of attention, they have so far developed slowly, except in a few countries. While mobile services such as "Apps" quickly became part of consumers' daily lives, growth rates of MPS remained below general expectations, which resulted in withdrawals by telecom companies such as O2 Telefonica from their mobile payment activities. The success of NTT DoCoMo, which launched a contactless payment solution in Japan, is often cited as prime example of the potential upheaval that mobile phones can create in the payment landscape of developed countries. The launch of the M-Pesa service by Vodafone in Kenya, which reached a total of 7.5 million subscribers in 2009, has also generated hope to reduce financial exclusion in developing countries (Bourreau and Verdier, 2014).

As Venkataraman (2008) shows, not all mobile payment implementations are successful: one of the reasons behind the slower growth of m-business is lack of suitable guidance or planning for a successful mobile payment adoption. Venkataraman addressed this issue—from both technical and business perspectives—by proposing a reference framework (roadmap) for mobile payment implementation.

From a legal perspective, the issue of liability (such as whether the payment service user or the payment service provider must bear the financial consequences—in case of fraudulent payment transactions) arises (Steennot, 2011). Edlund (2011) argues that MPS providers "will probably take on strict liability for executing payment instructions, and in some cases, they will bear the risk of payers' insolvencies."

However, a close analysis of these cases reveals that success stories cannot be easily generalized and demonstrates that several criteria are needed for mobile payments to succeed (Bourreau and Verdier, 2014). In this chapter, the authors will analyze what the main criteria are for consumers to adopt or reject MPS and how the players involved in the MPS market could position themselves to gain revenues and market share.

18.2 Background and problem description

> It is important to get away from generic dimensions and identify the content of a firm's specific theme in order to focus on those elements and alignments that are especially important for its situation.
>
> **Fuchs et al. (2000, p. 135)**

This citation is a sound reflection of the challenges the top management is facing today, more than ever. On the one hand, managers have to develop and protect the core business of the company, and on the other hand, they have to explore new products and markets for their company (Kuratko et al., 1990).

This balancing act poses an enormous challenge for the managers since protecting the core business is getting increasingly difficult in the light of radical technological changes associated with new competitors. Entering new markets with either established or new products is equally difficult because it means that the managers have to cope with the industry dynamics that they do not know properly, which is significantly increasing the uncertainty level (Langlois and Steinmueller, 2000).

Nevertheless, most of the companies are obliged to innovate in order to secure their future (Nelson and Winter, 1977). Especially in the telecommunications industry, the companies are constantly searching for new income streams, meaning that they have to either launch new products or develop new markets as mentioned earlier. The rationale behind this strategy is that their established business models are under enormous pressure (Ilinitich et al., 1996, 1998). D'Aveni has coined the word "hypercompetition" to mean a setting of rapidly escalating competition based on price–quality positioning and competition to create know-how and establish first-mover advantages (D'Aveni, 1994, 1996).

Especially, establishing a first-mover advantage is very often considered as the key for new growth and higher margins (Dos Santos and Pfeffers, 1995). This statement neglects, of course, that a first-mover advantage is a result of a successful product launch and not its cause. Only if managers are able to develop and launch an attractive product aligned with defined segments, a company will generate a first-mover advantage (Dosi, 1982). Providing a brief historical perspective on the development of payment card systems in developed countries can be useful to better understand the factors of success and to adopt electronic payment systems. Payment cards were first introduced in the 1950s by closed platforms, such as Diners Club or individual banks, which managed to develop an important acceptation network for their customers. Afterward, some banks decided to create interbank joint ventures, which enabled them to agree on common standards and increase their acceptation networks. The most important joint ventures, which later became Visa (1975) and MasterCard (1979), have a global reach and are now used and accepted in many different parts of the world. In 2006, these companies transformed their organizational structures to become publicly traded. The example of payment card systems shows that at least two criteria are essential to the development of electronic payment systems for mass market. First, the players must cooperate for the development of common standards or specify the conditions for interoperability. Building a joint venture can considerably reduce the costs of incompatibility between different standards. However, the cooperation for the development of MPS seems to be much more complicated than that for the development of payment cards. Second, the players must develop an important acceptation network. The more the consumers adopt a payment instrument, the more the merchants will be willing to accept it, and vice versa. However, one should note that the problems that the players face in the adoption of MPS are very different from the problems in the diffusion of payment cards in developed countries. The reasons why mobile payment has not been fully adopted in European countries such as Germany lie in the payment market itself (TNS Infratest, 2012):

- Paying cash is still the most preferred payment option with over 50% of consumers.
- The German card market is highly competitive and decentralized with more than 30 issuers, 16 acquirers, and 8 card processors.
- Card penetration is high with 156%, although card payments account only for 15% of all cashless payments (Dahlberg, 2007).
- Debit cards are the preferred card payment method (80%); credit card payments are rare (3%).
- Three-fourths of all cards are issued by public and cooperative banks with strong customer ties.

Surveying data from 347 residential cell phone users in Germany, Gerpott and Kornmeier (2009) found that MPS risk assessments and MPS evaluations by social reference groups were important (indirect) determinants of behavioral intentions concerning future MPS adoption.

In summary, the payment market in Germany is competitive and decentralized. Consumers and merchants in developed countries are already well equipped with payment cards, a widely used and accepted electronic payment solution. Hence, if they were to replace the use of payment cards, MPS would have to provide sufficient additional value to be adopted by consumers and merchants. On the consumer side, mobile phones can potentially have a competitive edge over payment cards, as they can be used as an interactive device, in which the consumer can store information. However, consumers are often used to being delivered payment instruments for free, either in a bundle with their bank account or by merchants who try to increase user stickiness. Hence, if the service is to be provided by a nonbank, the provider will have to find a way of recouping its costs of investment in infrastructure and security, which is not necessarily eased by the presence of low-cost services. Another option for mobile payment service providers would be to target niche markets, in which payment cards are absent. Such markets could include person-to-person money transfers and payment services at a low cost or a low risk for the unbanked or *underbanked*. The reasons why some people do not use existing electronic payment instruments are varied; for instance, if a consumer values privacy, he will not necessarily adopt mobile payments more easily than payment cards (Bourreau and Verdier, 2014).

The only window of opportunity telecommunications companies have to successfully enter payment markets with highly developed infrastructures and a consumer preference for conservative payment methods is to merge their mobile phone know-how with new payment products in combination with a new technology. As first retailers and online payment providers such as PayPal have launched mobile payment, it will become another established payment alternative for consumers in a few years.

At the heart of the management problem lies the question "how could the companies involved in the payment market make sure that the newly launched payment products will be accepted and used by its customers?" The answer to this question leads ideally to a first-mover advantage for the provider and subsequently to new and steady income streams.

18.3 Research objective and relevance

The objective of this chapter is to suggest strategic management recommendations for segmenting, targeting, and positioning MPS based on drivers and barriers of MPS adoption that were identified in empirical research related to MPS.

The authors will first give a short review of leading concepts in strategic management theory, that is, competitor-based view (CBV), market-based view (MBV), and resource-based view (RBV), and then provide an overview and classification of the most important variables that influence MPS consumer acceptance and adoption as identified in selected empirical research on MPS. The authors will then derive management recommendations that follow the principles of the segmenting, targeting, and positioning (STP) paradigm.

On a theoretical level, the analysis of leading research streams with regard to their implications of the STP paradigm will lead to new insights and to a much more different picture on a theoretical dimension.

On an empirical level, the influential variables on MPS adoption as identified in empirical research on MPS will be validated, leading to a clear picture of what drives or hinders consumer acceptance of MPS.

On a management level, the authors will help managers to structure and focus their action, especially in the launch of new products in new markets, leading to better management decisions in the long run.

18.4 Review of strategic management streams

In this section, three leading research streams (views) will be reviewed and critically analyzed, namely, the RBV, the CBV, and the MBV. The rationale behind this approach is twofold. On one side, the general theoretical baseline has to be settled, and on the other side, the STP Paradigm has to be theoretically rooted. The common denominator of all views is the search and/or explanation of a sustainable competitive advantage a company should strive for in order to be successful in its market. The core question underlying this chapter is, what are the implications of each view with regards to the launch of a new payment product and has each of the reviewed views potential to theoretically enrich the STP paradigm?

Deriving from the structure–conduct–performance hypothesis (Bain, 1968), which implies that the industry setting (structure) is the main driver for the strategic actions (conduct) of the company and thus for the profitability of the company (performance), Porter developed the CBV, featuring his famous five forces model (Porter, 1980, 1981). The aim of this model is to firstly identify the competitive forces within the industry, which are responsible for its attractiveness. Second, normative actions can be derived from the model as to how the top management can influence the industry structure in its favor (Porter, 1985, 1987). According to Porter's logic, an industry has high and stable profits (and is thus very attractive) when the companies within the industry are able to repel the threat of new entrants, avoid the threat of new substitute products, neutralize the bargaining power of its suppliers, and keep the rivalry among the competitors low. Consequently, the incumbents of the industry have to work on these dimensions in order to keep profits in the industry high (Porter, 1980, 1981, 1985, 1987).

The following comment is a critical extract concerning the logic of the model:

> The central tenet of this paradigm, as summarized by Porter is that a firm's performance is primarily a function of the industry environment in which it competes, and because structure determines conduct (or conduct is simply a reflection of the industry environment), which in turn determines performance, conduct can be ignored and performance can, therefore, be explained by structure.
>
> **Hoskisson et al. (1999, p. 425)**

Despite criticism, the five forces model has become one of the most popular and applied strategic management tools due its simplicity and its broad application spectrum (Cockburn et al., 2000).

A refinement within the industrial organization body of thought is the strategic group concept (Brush et al., 1999). Within the strategic group concept, above-average profits are traced back to mobility barriers like economies of scale and scope or product differentiations within the equivalent strategic group or cluster. The strategic implication for the top management is that it should rather focus on strengthening mobility barriers, for instance, by clever product differentiations, within their strategic groups rather than being focused on overall industry structure as such since this level of analysis is not detailed enough to derive successful strategic action (Hatten and Schendel, 1977). The implication that companies should rather take action on a strategic group level than on an industry level has been amplified by studies such as the one completed by Cool and Schendel (1987). Researchers come to the conclusion that the profitability of companies can be primarily traced down to the competition within the strategic group, meaning that a very high level within the strategic group spoils company profits, rather than intense rivalry on the industry level (Cool and Schendel, 1987). So far, the industry level and the strategic group level have been taken account within the CBV. In markets characterized by hypercompetition (D'Aveni, 1994 and 1996) such as the mobile telecommunications, former rivals may be forced to team up in order to increase their user basis. Coopetition is a business strategy based on a combination of cooperation and competition, derived from an understanding that business competitors can benefit when they work together (Brandenburger and Nalebuff, 1996). The coopetition business model is based on *games theory*, a scientific approach (developed during the Second World War) to understanding various strategies and outcomes through specifically designed games. Traditional business philosophy translates to games theory's *zero-sum game* in which the winner takes all and the loser is left empty-handed; proponents of coopetition claim that it can lead to a *plus-sum game*, in which the sum of what is gained by all players is greater than the combined sum of what the players entered the game with. It starts out with a diagramming process called the value net, which is represented as a diamond shape, with four defined player designations at the corners: customers, suppliers, competitors, and *complementors*. Complementors are defined as players whose product adds value to yours, for example, the way software products gain value because hardware products coexist with them, and vice versa. In comparison, a competitor is defined as someone whose product makes your product less valued, for example, the way a new toothpaste brand entering the market would make the brand that had previously been the only brand on the market become less valued. The game of business is then broken down into its *parts* (players, added values, rules, tactics, and scope) as a means of viewing practices and strategies. *Added values* focus on ways to improve products and services to find ways of making more money from an existing customer base. *Rules* specify ways of attracting customers with strategies such as price matching. *Tactics* are the practices sometimes used to take away a competitor's likely market share, for example, announcing an upcoming (and possibly nonexistent) new and improved product when a competitor's product is released. *Scope*, which is the final part, is used to take a broader prospective and create links between competitor's games and interests and see how coopetition can benefit the players (Brandenburger and Nelebuff, 1996; Rouse, 2005).

> (…) we know much less about how barriers to entry are built: about why this firm and not that one developed the competencies that underlie advantage, and about the dynamic process of which competitive advantage first arises and then erodes over time.
>
> **Cockburn et al. (2000, p. 1123)**

In its early phase, the RBV was primarily a reflex of the dominant thinking of the industrial organisation (I/O) research stream, which manifested itself in the CBV analyzed in the last chapter (Wernerfelt, 1984; Barney, 1991). Leading researchers (Dierickx and Cool, 1989; Collis, 1991a,b; Conner, 1991; Castanias and Helfat, 1991) criticized the one-sided focus on the industry/strategic group level and the ignorance of the resources (managerial, financial, etc.) and processes on the firm level. Today, it is widely accepted that the RBV and the CBV do complement each other since the CBV is focused on external factors such as the industry and strategic groups whereas the RBV focal points are the internal factors of the company (Rasche, 2000). Over the last years, three criteria, namely, nonimitability, lock-in, and nonsubstitutability, to describe successful resources have been established in the research community. Before describing shortly these criteria it has to be remarked that it is an idealized concept, as Montgomery (1995, p. 257) rightly points out:

> Like a Platonic Ideal, this characterization of a perfect form of resources is very useful from both a theoretical and a practical standpoint. At the same time, it is important to know something about the size of the gap between the idealized version of a form and what is seen on reality. This is where the resource-based literature falls dangerously short.

Nonimitability refers to the notion that the resources the competitive advantage of a company rests upon are nonimitable. This means the less a resource can be imitated/copied by the competitors, the more valuable the resource is for the company (Rouse and Daellenbach, 1999). On a theoretical level, this argument is supported by the theory of contestability of markets, which means that companies can protect themselves from their competitors by large irreversible investments in order to discourage them (Schmalensee and Willig, 1989).

Lock-in refers to the fact that there is a strong relation between how deeply ingrained (embedded) the resources are in the company and its success. In other words, when the resources are deeply ingrained in the corporate culture, they become in return enormously nontransparent for the competitors, which in turn makes it very difficult for them to reap the key resources to weaken the competitive advantage significantly (Peteraf, 1993). Since the top management know what resources are key or not, they have to find ways to embed them as deeply as possible in the corporate culture (Peteraf, 1990; Peteraf and Shanley, 1997).

Nonsubstitutability is, in comparison to the first two criteria, the most questionable since "the threat of substitution is omnipresent..." (Collis, 1991a, p. 7). Collis rightly refers, in his quotation, to the certainty that substitution in today's highly technical world is ever present and very hard to assess since normally substitution comes along with a new technology. Especially, the incumbents regularly underestimate the dynamics. For instance, it is, interesting to note that in the emerging mobile payment industry, the banks are not playing a key role but rather companies like Deutsche Telekom and Vodafone, which come from outside the industry but do have the technological access.

"(...) Causal ambiguity exists when the link between the resources controlled by a firm and a firm`s sustained competitive advantage is not understood or understood only very imperfectly" (Barney, 1991). The causal ambiguity concept plays a pivotal role within the RBV. It refers to the notion that very often the relationship between the resources and the competitive position the company has is blurred. It means that a certain part of the success of the company cannot be explained. A good example is the success of Apple. If a researcher is asked to explain the success of Apple, the answers vary from the personality of its founder, Steve Jobs, to the design of its products until its distribution strategy through iTunes and the Apple Stores. The core argument behind the causal ambiguity concept is that the less known the relationship between the resources of the company and its competitive advantage, the higher the imitation hurdles of its competitors:

> When advantage is based on competencies that have causally ambiguous characteristics then it will be difficult for companies to overcome the advantages by imitation. In itself causal ambiguity does not guarantee that a firm will be able to maintain a competitive advantage, but does create a very effective barrier to imitation.
>
> **Reed and DeFillippi (1990, p. 94)**

Critics like Collis rightly point out that if a company does not know exactly what it does right (meaning what their competitive advantage rest upon), any action undertaken is as likely to strengthen its competitive advantage as it is to destroy it (Collis, 1991a, p. 11).

Of all the views analyzed so far, the MBV plays a pivotal role in the launch of a new payment product, as it will be seen. As the name of this

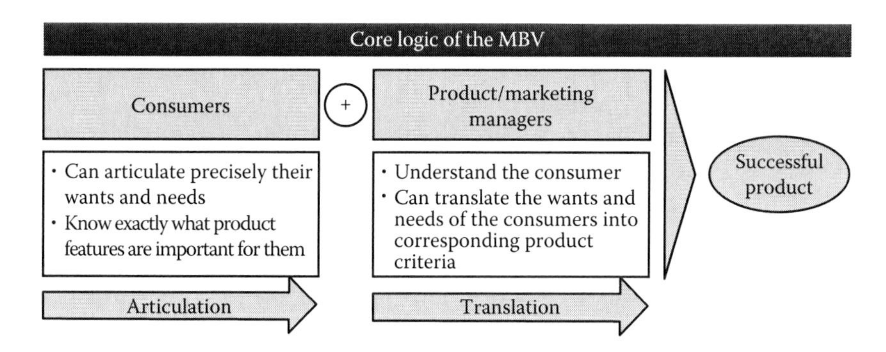

Figure 18.1 Logic of the MBV.

view already implies, the market-based considerations are at its core, and it thus corresponds highly with the marketing concept. The rationale behind the MBV is that—based on an intelligent definition of the relevant market—attractive market segment will be identified, which in turn will be secured and developed by an adequate positioning strategy (segmenting → targeting → positioning) (Kotler, 1989, 1997). The overall aim is to position products and services as idiosyncratic as possible from the consumer perspective to ideally realize a strong position in the market (Abell, 1980, 1999). The underlying hypothesis of this research stream is that the consumers know the preferred product specifications very well and are on top of that in the position to articulate them accordingly (Day, 1990) so that the product and marketing managers are able to develop the corresponding product. Figure 18.1 describes this core logic.

18.5 Preconditions, drivers, and barriers of mobile payment services' consumer acceptance

MPS is a dynamic and growing market segment, regarded as a strategic business field (Bauer, 2012), and thus, potential providers and retailers aim for a competitive advantage if they start offering MPS to their consumers. The main providers are PayPal, Google, Square, and Vodafone. The overall aim is to position the services from the consumer perspective to ideally realize a strong position in the market (Abell, 1980, 1999) as reflected in the MBV. With this logic, the value proposition of a product is a result of how an innovation is accepted by consumers in comparison to products from competitors (Day and Montgomery, 1999).

Pousttchi and Hufenbach (2013) propose a mobile payment reference model that provides not only theory background but is also a tool for marketing research—to be used for data collection and customer preferences analysis. "Discovering preferences transparently means that the marketer learns the customer needs without actually involving them.

When marketers learn customer preferences collaboratively, they engage in a dialogue to help customers to articulate their needs and identify how to meet those needs" (Kahn, 2000, p. 14).

The most established procedure of MPS is the premium rate SMS, whose costs are higher than the regular SMS and is used not only for the purchase of certain ringtones and logos but also for parking fees and purchase of fast-moving consumer goods in selected supermarket chains limited to €25. Because these MPS can be used with a simple mobile phone, it is a common payment method in Africa and India. A payment option that requires a smart phone is to pay mobile via Apps, which is, for instance, a benefit of Starbucks coffee shops. Another option to pay for the owners of smart phones with integrated NFC chip is to just hold the mobile device against an NFC terminal, installed at point of sale (POS), and the transaction occurs automatically for registered customers. This technology has been established in Asia since 2004, while the merchant adoption rate in Western Europe is still low. However, first attempts in this direction have been made by public transport in the underground and bus stations or ticket offices of car parking. What still lacks is the trust of consumers and merchants, leading to hesitation in their adopting this different way of payment (Kreimer and Rodenkirchen, 2010).

With the aim to identify the drivers and barriers of MPS consumer acceptance and adoption, the author selected empirical explorations, each with a different focus, but all contributing unique insights. Tornatzky and Klein's meta study on innovation adoption (1982) provides a fundamental understanding of influential variables of the intention to use innovations in general.

The study by Schierz et al. (2009) explored acceptance determinants of MPS. Mallat (2006) identified MPS drivers and barriers of consumer adoption and, together with Tuunainen (Mallat and Tuunainen, 2008), barriers of merchant adoption. Yang et al.'s (2012) exploration from China is not limited to the *intention to use* but includes results from the postadoption stage of MPS usage. Research results of consumer acceptance in e-commerce (Pavlou, 2003) and m-commerce (San-Martin and Lopez-Catalan, 2012; also see Bauer, 2012) have been incorporated to examine the role of trust for adoption. The samples are heterogeneous concerning age, gender, and educational and professional background.

Bamasak (2011) shows that Saudi Arabians expressed willingness to accept MPS; however, the security of mobile payment transactions and the unauthorized use of mobile phones to make a payment are great concerns. In this respect, some authors (Daskapan et al., 2010) favor short-range MPS as more trustworthy.

In their survey on potential MPS users in Asian rapidly expanding market, Liu et al. (2011) have identified five mobile payment interface usability factors (content, ease of use, promotion, made-for-the-medium, and emotion) that have a positive influence on interaction behavior and intention to use mobile payment.

The authors restructured the variables that are driving MPS acceptance and adoption, according to empirical research, into the following: (1) preconditional drivers of the intention to use MPS, (2) influence factors of the intention to use MPS, and (3) drivers of actual MPS usage. The variables that hinder MPS adoption are summed up under *barriers* in Number 4 (Wurster, 2013).

1. Preconditional drivers of acceptance and of the intention to use MPS (Wurster, 2013) are as follows:
   a. Trust in the system and the provider(s) (Pavlou, 2002; Mallat, 2006; Mallat and Tuunainen, 2009; Bauer 2012; San-Martin and Lopez-Catalan, 2012; Fingerhut, 2014)
   b. Compatibility of the system (Tornatzky and Klein, 1982; Mallat, 2006)
   c. Accessibility to the system (Wurster, 2013; Fingerhut, 2014)
2. Following factors can influence the intention to use MPS, but are depending on the consumer's character and social background or lifestyle and thus cannot be categorized as general drivers according to Wurster (2013):
   a. Social influence of the consumer's peer group or social network community (Tornatzky and Klein, 1982; Schierz et al., 2009; Yang et al., 2012)
   b. Positive attitude toward innovations and individual mobility (Schierz et al., 2009)
   c. Subjective norms as personal risk or control perception (Yang et al., 2012)
3. The providers can drive the MPS adoption rate and the actual usage by offering *more convenience than* alternative payment options (Wurster, 2013) or *"value-added services"* (Fingerhut, 2014), meaning that providers should extend their offers in order to be distinctive vs. card payments (Fingerhut, 2014)
4. Barriers of MPS consumer adoption, in the author's opinion are as follows:
   a. The absence of trust in the security dimensions of a payment system. Most authors and researchers agree upon the dominant role of security issues in the acceptance of online and MPS (Pavlou 2003; Bauer, 2012; Wurster, 2013; Fingerhut, 2014).
   b. Complexity as the opposite of convenience (or simplicity), but in the sense of too many different and long lasting registration procedures. If all providers would agree on the same and most simple technology and would follow the *touch-and-go* logic that

enables transparent procedures and control the money transfer, consumers' fear and risk perceptions would decrease (Ondrus and Pigneur, 2009; GfK, 2011; Wurster, 2013).

c. All factors that hinder merchants to adopt mobile payment systems and thus, prevent from making the services accessible for the consumer (see Mallat and Tuunainen, 2008 for the barriers of merchant adoption).

18.6 Strategic management options for segmenting, targeting, and positioning MPS

For strategically positioning a product as MPS, the authors chose the STP paradigm as it plays a central role in the MBV and the strategic management stream, which is focusing on consumer needs and acceptance within a certain competitive surrounding. The flow displayed in Figure 18.2 summarizes the overall framework.

The ground-laying logic of market segmentation is based on the notion that customers demonstrate heterogeneity in product preferences and buying behavior (Green, 1977; Wind, 1978). This should also be the starting point of the strategic recommendations to launch a new MPS product.

	Segmenting	Targeting	Positioning
Core contents	• Definition of the relevant market • Estimation of the market-volume/potential • Determination of the key segmenting criteria • Identification of attractive segments • Others	• Skimming of the existing target group • Identification of new target groups • Evaluation/selection of the key target groups • Penetration of the key target group • Envisioning new target groups • Others	• Definition of the overall marketing strategy • Finale evaluation of the chosen segments and target groups • Alignment of the selected segments and target groups with the strategy • Start marketing mix activities
Key questions	• In which market are we in? • How attractive is the market? • What are the key segments in this market? • Are the segments independent of each other? • Others	• What is our existing target group and how does it tick? • Are there other target groups we have overseen? • Are there any subtle changes within our target groups? • Others	• What is our overall marketing strategy? • Are we convinced by our chosen segments and target groups? • Are we properly aligned with our marketing strategy? • Others

Figure 18.2 Framework of the MBV.

18.6.1 Segmenting the market properly

For this paper, the definition of market segmentation is adopted from the American Association of Marketing, which states that "market segmentation is concerned with classification of customers and consumption and when enacted, market segmentation usually turns to or is based upon the relationships, which follows" (American Marketing Association, 2014). This definition can be complemented with the following: "it's a creative and iterative process, the process of which is to satisfy customer needs more closely and, in doing so deliver sustainable competitive advantage for the company" (MacDonald and Dunbar, 2012).

In this case, the proven influence of social networks can support marketers. If marketers incorporate *innovators* and early adopters such as *cocreators* (Kotler et al., 2010) of their services constantly interact with their consumers on platforms such as Facebook and use *word-of-mouth* communication, they can transport the *right* benefits, meaning the benefits that are relevant for consumers.

The idea of strategic window such as the outcome of the market segmentation efforts, which takes the discussion to a strategic and tactical level, gives companies the opportunity to increase their profits and allow them as new competitors to challenge and attack traditional market leaders (Abell, 1978). However, this strategic window is reduced to limited periods when key requirements of a market and particular competencies of a company competing in the market are at an optimum (Proctor, 2000). On the tactical side, this means until competitors copy a firm's segmentation routine, the firm has a competitive edge; if the product or service is specific to the segment, then this competitive advantage is multiplied (McBurnie and Clutterbuck, 1988).

Although segmenting is perceived on a theoretical level as an *objective* and straightforward activity, one has to understand that there is an array of issues and challenges when it comes to successfully implementing the identified and selected segments. Indeed many concepts and approaches do exist to segment a market, but when it comes to identifying the key criteria like measurability, accessibility, the strength of the segment as such, and last but not the least the noninterference within segments, especially in terms of buying behavior, the scope of differentiating variables are considerably reduced. The identification and selection of differentiating variables is further complicated as consumers' preferences keep on changing at a much faster rate and the their behavior is increasingly difficult to predict since it is getting more and more hybrid, meaning that consumers cannot be associated with one segment but rather with many. As a result of fast-changing markets and consumer behavior, which is enormously difficult to predict, noninterference within different segments can be hardly granted, which in turn questions the overall segmentation of the company. Due to the increased turbulence of the environment, companies are literally forced to question themselves in terms of their key target segments: Are our segments properly aligned with our strategy or are they contrary? Since the consumer behavior does not take into account *artificial* segments that were made up by the company in order to follow a predefined strategy.

There are varieties of research on dividing the mobile communication market into different target groups. Yang et al. (2012) suggest segmenting potential users according to their personal traits, especially to the degree of their predisposition to adopting innovations and to the degree of their individual mobility (Yang et al., 2012).

Schierz et al. (2009) found that it is crucial to identify the early adopters of MPS as they are likely to adopt innovations first. The positive influence of the early adopters has already been discussed in the Innovation Diffusion Theory (IDT) construct (Rogers, 1983).

Nearly every adult owns a mobile phone and carries it with him; two-thirds of the 1000 test persons in the KPMG study (Kreimer and Rodenkirchen, 2010) have access to the Internet on their mobile phone. However, this would imply to be owner of a smart phone, which accounts only for 25% of the worldwide population and approximately 40% of the German population. Those who possess a smart phone and go online with their mobile phone on a daily basis are much more likely to use MPS in the future than those whose use the Internet less frequently (Kreimer and Rodenkirchen, 2010).

Therefore, a valid segmentation option would be to take the mobile Internet usage behavior of consumers as a basis, as, for instance, done in the mobile communication report 2013 (MCR, 2013) valid for Austria. Although Austria is a small country, its location between east and west Europe is ideal to integrate insights from both western and eastern European consumers. The MCR research revealed three target groups: the digital immigrants (52%), digital natives (36%), and digital outsiders (12%).

The *digital immigrant* include a wide age range and are regular users of mobile web services (76% usage rate), but their usage is selective and goal oriented. Their hesitation is grounded in *security and data protection* issues they have with web services (Klaus Oberdecker, CEO Mind Take Research in: MCR, 2013). Their education level and income are the highest of all three target groups.

Consumers who cannot resist their addiction to the mobile phone belong to the *digital natives*; they are brought up with mobile phones and the Internet and account for 36% of the sample. Their usage rate of mobile web services is 96%. For these people, the mobile phone is essential for their communication. The *digital outsiders* are the ones with most skepticism. Their mobile Internet usage rate is only at 8%; 34% of them are older than 50 years, and the majority of them (81%) have only discovered the web during the last 2 years. The majority of the *digital outsiders* reject using the mobile phone for any communication activities other than to make calls because the display seems too small for them, which reduces convenience, and they are technically too poorly skilled to go online with their phone, even if it is a smart phone (MCR, 2013).

Gender segmentation or segmentation according to the individual risk disposition of consumers would be an additional feasible segmentation strategy for MPS, which would imply to divide potential users into risk-averse and risk-friendly users and targeting them individually, following Mallat's argumentation (2006). Especially in payments, MPS providers need to make sure to increase transparency in information for those users who are risk-averse with special programs individually targeted to them with the aim to make them overcome their fears.

18.6.2 Targeting the right consumers

Generally, the literature embraces under the term "targeting" the complete target management of the marketing department (Kahn, 2000), Closely connected with the overall target management are two fundamental questions:

1. Do we realize the full potential of our target groups or is there room for improvement?
2. How can we identify new target groups based on our segment analysis and then access their attractive potential accordingly?

Especially the second question is one of the main drivers behind the activities of a provider in the mobile payment market since the top management wants to access this attractive market.

Most researchers from empirical research in Chapter 5 (e.g., Mallat, 2006; Mallat and Tuunainen, 2008; Schierz et al. 2009; Yang et al., 2012) suggested that companies should specifically target and position their MPS solutions to well suit individual behavioral patterns and prior experience.

The positive influence of *innovators* and *early adopters* on the intention to use MPS has already been stated. Targeting MPS to them and using them as multiplier for more skeptical consumer groups would probably enable to conquer the mass-market faster.

However, management implication is to expand the user base and their sources of revenue by leveraging the effect of social influences. Eighty-eight percentage use the mobile phone to gather information about products and services and their benefits or to search for certain brands and offers. Men (78%) and consumers younger than 30 years (89%) are the heavy users of mobile web services. A refined segmentation revealed that the *digital natives* and the *digital immigrants* could become the most relevant target groups for strategic marketing activities in MPS as they account for 88% of the users (MCR, 2013), as described in Section 1.6.1. The reason for the high attractiveness of these consumers is that they go online on a daily basis and rather have trust in mobile payments than those who go online less frequently. But also among the frequent Internet users, 30% of mobile shoppers think that payments in mobile commerce are not secure enough. For 19%, the purchase is still too complicated, and for 13%, the payment process is not convenient enough. Fifty-six percentage that have downloaded an App from the App Store pay attention to which features the App has possibly access on (MCR, 2013).

One possibility to considerably increase the efficiency of targeting activities is the implementation and deployment of Customer Relationship Software tools since one of the central values these systems have for companies is that all customer data are centralized and that new target groups can be much more easily identified when customer behavior patterns are at the basis of the analysis (Sheth and Sisodia, 1999).

18.6.3 Managing competition

Ideally, the (competitive) positioning of a company is the result of successful STP activities.

If companies want to be successful, the top management has to firstly make sure that all relevant resources are in place and that there is no shortage. Second, it has to identify the key resources (i.e., technical features, etc.), which are very likely to be crucial for the market success. Last but not the least, the management has to make sure that know-how will be shared and embedded within the mobile payment unit and that the key resources (if it is possible to identify them) will be locked-in accordingly.

But as for companies that would like to participate in the emerging mobile payment market, it is vital to constantly track market facts such as the technology development of the existing products, platforms, and services of competitors such as *Google Wallet* or *PayPass* and all competitive activities of major players such as PayPal in this dynamic market environment. A solid theoretical understanding of possible strategic paths and learning from competitive activities, including online banking and e-commerce, would support protagonists who aim at gaining or maintaining competitive advantages in this field.

With regard to the launch of a new payment product, the CBV gives fruitful theoretical insights and normative recommendations on all three levels. It is key, for instance, not to overestimate the industry level; just because an industry like mobile payment is attractive, it does not mean that a company in the industry will be able to reap its potential profits. On the strategic group level, the management has to focus on raising mobility barriers in order to keep out potential competitors. In the light of enormous technology progress over the last years, this will be very difficult to realize.

According to leading researchers (Karnani and Wernerfelt, 1985; Barnett, 1993; Barnett et al., 1994; Grimm and Smith, 1997), strategies taken by one company have knock-on effects on the strategies and the actions taken by its rivals (Gimeno and Woo, 1999). The most promising level for big players in the mobile payment market is to derive their strategic actions from the action–reaction level of its main competitors. Two factors are responsible for the level of amplitude concerning the action–reaction pattern: interfirm rivalry and the assessment of the action–reaction. This means that the higher the interfirm rivalry is and the stronger the reaction of the main rivals is assessed, the more prudent and thoughtful the management will be to launch strategic attacks. Thus the action–reaction pattern on the firm level is the core element of the concept of competitive dynamics. A consequence of it can be to follow the strategy of coopetition (Brandenburger and Nalebuff, 1996). Benefits of these cooperations between competitors are cost-sharing and taking advantage of complementarities between different players. Banks have gained a great deal of experience in operating mass-market payment systems, which might be critical for a wide adoption of a mobile payment solution. They also have experience in risk and fraud management that other players, like providers, do not have. In contrast, providers have strong partnerships with mobile phone manufacturers, which might help to develop payment-enabled mobile handsets. Costs of cooperation are twofold. First, partners will have to share not only costs but also revenues. For instance, in case of cooperation between banks and providers, if mobile payment adds little value relative to existing payment solutions (e.g., payments by card), this may impede the constitution of a partnership. Second, like in any joint venture, there are costs of coordination. As they have different objectives and interests for the development of mobile payments, one could assume that the costs of cooperation between banks and providers are probably high. Finally, the incentives of different players to develop mobile payments differ. Finally, the cooperation models for the development of mobile payments could be impacted by the regulation of nonbank players. The provision of mobile payment solutions could increase the presence of nonbanks in retail payment systems and, thus, could require a regulatory intervention to limit the potential risks involved in the economy. So far, in developed countries, regulators have often tolerated the use of mobile phones for small transactions without requiring any banking license. However, this situation could be called into question if the volume of mobile payments were to become important. Vodafone has, for instance, teamed up with Visa in order to successfully develop the German mobile payment market only recently.

18.6.4 Positioning of MPS

What does successful positioning in the MPS market imply?

"Those providers will succeed that offer customers the highest degree of 'convenience and value-perception.'" (M-Payment Expert A.M. Bajora, citated in Hecking, 2012).

Thus, for successful mass-marketing of MPS, Yang et al. (2012) suggest that service providers increase the individual's perceived convenience and decrease their perceived risk and cost about the services. Therefore, service providers need to extend their services and offer apart from convenient and fast payment, saving tickets, price comparisons, or search functions to find the closest shop with the lowest price, for instance (Fingerhut, 2014). They would need to develop advertising to emphasize the value of MPS service vs. alternative payment methods such as convenience in terms of speed and at the same time show how compatible the method is with daily life situations as services are accessible everywhere the consumer is.

Schierz et al. (2009) constated, from this result, that for the adoption of MPS, people must find them to be reconcilable with their existing behavioral patterns. Firms that opt to engage in MPS should fulfill their promises, exhibit an ethical behavior, and inform the consumers, taking into account individual user needs and preferences of their target groups, in order to achieve trust (San-Martin and Lopez-Catalan, 2012).

PayPal demonstrates that strategies built on trust are successful in the long run as researchers showed when discovering the influence of trust on consumer satisfaction.

Confidentiality, data integrity, authentication, and nonrepudiation are the most important requirements for e-commerce and mobile payment transactions to build trust among risk-averse consumers (Mallat, 2006). Other aspects Mallat's (2006) research found to be significant for trust *building* are anonymity and privacy, which consumers rather associate with banks than with telecom providers (Kreimer and Rodenkirchen, 2010). Practical implications would be

- Security aspects in software and hardware
- Technical security standards at technical construction of the service
- Reliability as brand benefit in the communication strategy

Influences from friends and social networks have to be considered as critical, especially in the preadoption stage, but "this is especially the case among highly collective-culture countries as China, where individuals are more easily influenced by other's opinions than those, living in the low collective-culture countries as USA, Great Britain, Australia, etc" (Yang et al. 2012, p. 137). Recent attempts of mobile service providers to promote special fares and flat rates targeted at *digital natives* by asking their friends to participate in and installing a bonus system for new contracts ("Friends gain Friends" by Base/Eplus) are going in this direction.

MPS providers can positively influence the consumer perception by

- Improving the conditions for retailers to perceive MPS as attractive enough to adopt it
- Showing in their communication executions different situations for MPS usage to influence the value perception of MPS being more convenient and adding value than alternative payments
- Reducing the payment to maximum €50, spending per buy

Ideally, MPS providers adapt to consumers' needs and wants and offer NFC technology in places where time is a critical factor, for example, airports, train stations or bus/tram stops, taxicabs, staff canteens, public libraries, and university cafeterias.

But time saving also implies the expectation that the payment process is as short as possible. Consumers are not willing to queue up longer in supermarkets because the mobile payment procedure takes longer than card payment would take (Hecking, 2013).

To overcome negative perceptions, service providers would need to

- Design various assurance procedures
- Create *easy-to-understand* design of the verification process during an NFC transaction,
- Offer a long-term user-preferential plan
- Provide a packaging cost scheme

San-Martin and Lopez-Catalan (2012) questioned whether it would prove beneficial for mobile service providers to encourage impulse buying via promotions, etc., since they found out that

- It may reduce satisfaction
- It may lead to a greater number of complaints, refunds, and returns
- It probably sparks negative word of mouth concerning the provider and regarding m-commerce as a whole

In the postadoption stage, the factor "satisfaction" with MPS might play a more dominant role and, on the provider's side, the ability of the provider to cope with system errors (Yang et al., 2012).

18.6.5 Summary

The following table (Table 18.1) provides an overview of the authors' management recommendations for segmenting, targeting, and positioning of a new payment product.

Table 18.1 Overview of STP Options for MPS

Segmenting	Targeting	Positioning
Segmenting the potential users according to their personal traits, especially to their degree of pre-disposition to adopt innovations and to their degree of individual mobility.	Digital natives and people with high personal mobility first	Emotional benefit and convenience communication Show situations of MPS usage: Time saving, queue avoidance/access from allover)
Identify the early adopters of MPS as they are likely to adopt innovations first.	Early adopters and innovators first	Focus communication on high tech future trends and image factors
Segmentation according to the individual risk disposition of consumers.	Customers with low risk disposition and need for control separately	Communicate Transparency of procedures, Security issues, and data safety to build trust
Gender segmentation.	Men with high mobility first Women	Men: Rational benefit communication and execution Women: Emotional benefit communication to build trust
Segmentation according to their habitual banking/payment/m-commerce behavior (card/cash/online banking etc.)	Credit card Customers first Online banking Users first M-commerce Users first	Offer the highest degree of convenience ("MPS is more Convenient than..."), individually related to specific habitual preferences of the respective target group

18.7 Conclusions

With regard to the overall objective to derive management recommendations from empirical research on MPS acceptance and adoption, theoretical research streams and empirical research were able to give fruitful insights.

If one analyzes the success of disruptive technologies and applications, the following important parameters have to be seamlessly working together:

- Constant innovation capability, a reliable technology behind the product, progress in the technological infrastructure, high acceptance rate of retailers and consumers.

If MPS providers and retailers start to create the preconditions for usage intention, that is, wide access and compatibility—most preferably in offering the most convenient and secure solution—the mobile payment method could receive more attention. But it needs a concrete information policy from banks and credit card companies and more intense sales activities by the respective technology providers to convince consumers and retailers that this technology is worth investing as it will help them increase the retailer's sales significantly and build long-term consumer relations.

However, it is almost a fact that only a smaller group of dominant platforms and technologies will emerge from the array of confusing standards competing in the market today; there will remain an ecosystem of players who are all building trust and can plausibly explain that their services offer more convenience than conventional payment methods will demonstrate enough value to consumers to replace credit cards and cash in the long term.

References

Abell, D. F. 1978. Strategic windows. *Journal of Marketing*, 42(3), 21–26.

Abell, D. F. 1980. *Defining the Business*. New York: Prentice-Hall.

Abell, D. F. 1999. Competing today while preparing for tomorrow. *Sloan Management Review*, 40(3), 73–81.

Ahamad, S. S., Udgata, S. K., and Sastry, V. N. 2012. A new mobile payment system with formal verification. *International Journal of Internet Technology and Secured Transactions*, 4(1), 71–103.

Alimi, V., Rosenberger, C., and Vernois, S. 2013. A mobile contactless point of sale enhanced by the NFC and biometric technologies. *International Journal of Internet Technology and Secured Transactions*, 5(1), 1–17.

American Marketing Association. 2014. https://www.ama.org/resources/Pages/Dictionary.aspx?dLetter=M (accessed August 21, 2014).

Bain, J. 1956. *Barriers to New Competition*. Cambridge, MA: Harvard University Press.

Bamasak, O. 2011. Exploring consumers acceptance of mobile payments—An empirical study. *International Journal of Information Technology, Communications and Convergence*, 1(2), 173–185.

Barnett, W. P. 1993. Strategic deterrence among multipoint competitors. *Industrial and Corporate Change*, 2(1), 249–278.

Barnett, W. P., Greve, H. R., and Park, D. Y. 1994. An evolutionary model of organizational performance. *Strategic Management Journal*, 15(1), 11–28.

Barney, J. B. 1991. Firm resources and sustained competitive advantage. *Journal of Management*, 17(1), 99–120.

Bauer, A. 2012. *Vertrauen in Mobile Payment Dienste—Über die Rolle von Vertrauen in der Konstruktion und Kommunikation mit Mobile Payment Diensten*. Hamburg, Germany: Tredition Verlags GmbH.

Bourreau, M. and Verdier, M. 2014. Cooperation models for the development of mobile payment solutions. http://www.pymnts.com/journal-bak/lydian-payments-journal-2010/cooperation-models-for-the-development-of-mobile-payment-solutions (accessed on January 28, 2014).

Brandenburger, A. M. and Nalebuff, B. J. 1996. *Co-opetition*. Cambridge, MA/New Haven, CT: Harvard University Press/Yale University Press.

Brush, L. D., Bromley, D. A., and Hendricks, P. 1999. The relative influence of industry and corporation on business segment performance: An alternative estimate. *Strategic Management Journal*, 20(6), 519–547.

Castanias, R. P. and Helfat, C. 1991. Managerial resources and rents. *Journal of Management*, 17(1), 155–171.

Cockburn, I. M., Henderson, R. M., and Stern, S. 2000. Untangling the origins of competitive advantage. *Strategic Management Journal*, 21(10–11), 1123–1146.

Collis, D. J. 1991a. Organisational capability as a source of profit. Discussion Paper 91-046. Harvard Business School, Boston, MA.

Collis, D. J. 1991b. A resource-based analysis of global competition: The case of the bearings industry. *Strategic Management Journal*, 12(Special Issue Summer), 49–68.

Conner, K. R. 1991. A historical comparison of resource-based theory and five school of thought: Do we have a new theory of the firm? *Journal of Management*, 17(1), 121–154.

Cool, K. and Schendel, D. E. 1987. Strategic group formation and performance: The case of the U.S. pharmaceutical industry, 1963–1982. *Management Science*, 33(9), 1102–1124.

Dahlberg, T., Mallat. N., Ondrus, J., and Zmijewska, A. 2007. Past present and future of mobile payment research: A literature review. *Electronic Commerce Research and Applications*, 7(2), 165–181.

Daskapan, S., Van den Berg, J., and Ali-Eldin, A. 2010. Towards a trustworthy short-range mobile payment system. *International Journal of Information Technology and Management*, 9(3), 317–336.

D'Aveni, R. A. 1994. *Hypercompetition: Managing the Dynamics of Strategic Maneuvering*. New York: The Free Press.

D'Aveni, R. A. 1996. *Strategic Supremacy: How Industry Leaders Create Growth, Wealth and Power through Spheres of Influence*. New York: The Free Press.

Day, G. S. 1990. *Market Driven Strategy: Processes for Creating Value*. New York: The Free Press.

Day, G. S. and Montgomery, D. B. 1999. Charting new directions for marketing. *Journal of Marketing*, 63(Special Issue), 3–13.

Dierickx, I. and Cool, K. 1989. Asset stock accumulation and sustainability of competitive advantage. *Management Science*, 35(12), 1504–1510.

Dos Santos, B. L. and Pfeffers, K. 1995. Rewards to investors in innovative information technology applications: First movers and early followers in ATGMs. *Organisation Science*, 6(3), 241–259.

Dosi, G. 1982. Technological paradigms and technological trajectories—A suggested interpretation of the determinants and directions of technical change. *Research Policy*, 11(3), 147–162.

Edlund, H. E. 2011. Payments made by means of mobile phones. *International Journal of Private Law*, 4(2), 266–273.

Fingerhut, C. 2014. Handy statt Geldbeutel und Schlüssel?—ECC Köln und goetzpartners untersuchen die Anforderungen deutscher Konsumenten an Mobile Wallets. http://www.ecckoeln.de/News/Handy-statt-Geldbeutel-und-Schl%C3%BCssel%3F-%E2%80%93-ECC-K%C3%B6ln-und-goetzpartners-untersuchen-die-Anforderungen-deutscher-Konsumenten-an-Mobile-Wallets (accessed on January 28, 2014).

Francis, L., Hancke, G., Mayes, K., and Markantonakis, K. 2010. On the security issues of NFC enabled mobile phones. *International Journal of Internet Technology and Secured Transactions*, 2(3/4), 336–356.

Fuchs, P., Mifflin, K., Miller, D., and Whitney, J. 2000. Strategic integration: Competing in the age of capabilities. *California Management Review*, 42(3), 118–147.

Gerpott, T. J. and Kornmeier, K. 2009. Determinants of customer acceptance of mobile payment systems. *International Journal of Electronic Finance*, 3(1), 1–30.

GfK. 2011. Mobile payments: The importance of trust and familiarity and the need for co-operation. Global report 2011. http://www.gfk.com/group/press_information/press_releases/007888/index.en.html (accessed on January 31, 2014).

Gimeno, J. and Woo, C. Y. 1999. Multimarket contact, economies of scope, and firm performance. *The Academy of Management Journal*, 43(3), 239–259.

Green, P. E. 1977. A new approach to market segmentation. *Business Horizons*, 20(1), 61.

Grimm, C. M. and Smith, K. G. 1997. *Strategy as Action: Industry Rivalry and Coordination*, 11th ed. Cincinnati, OH: South-Western College Publishing.

Hatten, K. J. and Schendel, D. E. 1977. Heterogeneity within an industry. *Journal of Industrial Economics*, 26(2), 97–113.

Hecking, M. 2012. http://www.manager-magazin.de/unternehmen/it/0,2828,852945-2,00.html (accessed September 4, 2012).

Hecking, M. 2013. http://www.manager-magazin.de/unternehmen/handel/0,2828,880753-2,00.html (accessed May 5, 2013).

Hoskisson, R. E., Hitt, M. E., Wan, W. P., and Yiu, D. 1999. Theory and research in strategic management: Swings of a pendulum. *Journal of Management*, 25(3), 417–456.

Ilinitch, A. Y., D'Aveni, R. A., and Lewin, A. Y. 1996. New organizational forms and strategies for managing in hypercompetitive environments. *Organization Science*, 7(3), 211–220.

Ilinitch, A. Y., D'Aveni, R. A., and Lewin, A. Y. 1998. Managing in *Times of Disorder: Hypercompetitive Organizational Responses*. London, U.K.: Sage Publications.

Kahn, B. 2000. Turn your customers into advocates. *Financial Times Mastering Management*, Part Two, October 9, 2000, pp. 14–15.

Karnani, A. and Wernerfelt, B. 1985. Multiple point competition. *Strategic Management Journal*, 6(1), 87–96.

Kotler, P. 1989. *Marketing-Management: Analyse, Planung und Kontrolle*, 4th ed. Stuttgart, Germany: Poeschel.

Kotler, P. 1997. *Marketing Management, Analysis, Planning, Implementation and Control*, 9th ed. New York: Prentice Hall.

Kotler, P., Kartajaya, H., and Setiawan, J. 2010. *Die neue Dimension des Marketing. Vom Kunden zum Menschen*. Frankfurt am Main, Germany: Campus-Verlag.

Kreimer, T. and Rodenkirchen, S. 2010. Mobile Payment—Umfrage des E-Commerce Center im Auftrag von KPMG—Anforderungen, Barrieren, Chancen.

Kuratko, D. F., Montagno, J. S., and Hornsby, R. V. 1990. Developing an entrepreneurial assessment instrument for an effective corporate entrepreneurial environment. *Strategic Management Journal*, 11, 49–58.

Langlois, R. N. and Steinmueller, W. E. 2000. Strategy and circumstance: The response of American firms to Japanese competition in semiconductors, 1980–1995. *Strategic Management Journal*, 21(10–11), 1163–1173.

Little, A. D. 2012. Global m-payments report (update of the report from 2009). http://www.adlittle.com/reports.html?view (accessed on January 31, 2014).

Liu, Y., Wang, S., and Wang, X. 2011. A usability-centred perspective on intention to use mobile payment. *International Journal of Mobile Communications*, 9(6), 541–562.

Lu, H., Claret-Tournier, F., Chatwin, C. R., Young, R. C. D., and Liu, Z. 2009. An agent-oriented mobile payment system secured using a biometrics approach. *International Journal of Agent-Oriented Software Engineering*, 3(2/3), 163–187.

MacDonald, M. and Dunbar, I. 2012. *Market Segmentation. How to Do It, How to Profit from It*. Oxford, U.K.: Elsevier Butterworth-Heinemann.

Mallat, N. 2006. Exploring consumer adoption of mobile payments—A qualitative study. *Journal of Strategic Information Systems*, 16(4), 413–432.

Mallat, N. and Tuunainen, V. K. 2008. Exploring merchant adoption of mobile payment systems. *E-Service Journal*, 6(2, winter), 24–57. Indiana University Press, Bloomington, IN.

McBurnie, T. and Clutterbuck, D. 1988. *The Marketing Edge*. London, U.K.: Penguin Books.

MCR. 2013. Mobile communications report Austria 2013. http:///www.mmaaustria.at/news-press/data3571 (accessed on January 27, 2014).

Montgomery, C. A. 1995. Of diamonds and rust: A new look at resources. In: Montgomery, C. A. (Ed.), *Resource-Based and Evolutionary Theories of the Firm: Towards a Synthesis*. Norwell, MA: Kluwer Academic Publishers, pp. 251–268.

Nelson, R. R. and Winter S. G. 1977. Dynamic competition and technical progress. In: Nelson, R. and Balassa, R. (Ed.), *Economic Progress, Private Values and Public Policy: Essays in Honor of William Fellner*. Amsterdam, the Netherlands: North-Holland, pp. 57–101.

Ondrus, J. and Pigneur, Y. 2009. Near field communication: An assessment for future payment systems. *Information Systems and E-Business Management*, 7(3), 347–361.

Pavlou, P. 2003. Consumer acceptance of electronic commerce: Integrate trust and risk with the technology acceptance model. *International Journal of Electronic Commerce*, 7(3), 69–103.

Peteraf, M. 1990. The resource-based model: An emerging paradigm for strategic management. Working Paper Number 90-29. Northwestern University, Evanston, IL.

Peteraf, M. 1993. The cornerstones of competitive advantage: A resource-based view. *Strategic Management Journal*, 14(3), 179–191.

Peteraf, M. and Shanley, M. 1997. Getting to know you: A theory of strategic group identity. *Strategic Management Journal*, 18(Special Issue), 165–186.

Porter, M. E. 1980. *Competitive Strategy: Techniques for Analyzing Industries and Competitors*. New York: The Free Press.

Porter, M. E. 1981. The contributions of industrial organization to strategic management. *The Academy of Management Review*, 6(4), 609–620.

Porter, M. E. 1985. *Competitive Advantage*. New York: The Free Press.

Porter, M. E. 1987. From competitive advantage to corporate strategy. *Harvard Business Review*, 65(3), 43–59.

Proctor, T. 2000. *Strategic Marketing: An Introduction*. London, U.K.: Routledge.

Pousttchi, K. and Hufenbach, Y. 2013. Enabling evidence-based retail marketing with the use of payment data—The mobile payment reference model 2.0. *International Journal of Business Intelligence and Data Mining*, 8(1), 19–44.

Rasche, C. 2000. Der resource based view im lichte des hybriden wettbewerbs. In: *Die Ressourcen-und Kompetenzperspektive des Strategischen Managements*, Gemünden, H. G., Hamann, P., Hinterhuber, P., Specht, G., and Zahn, E. (Eds.). Wiesbaden, Germany: Deutscher Universitäts-Verlag, pp. 69–125.

Reed, R. and DeFillipi, R. J. 1990. Causal ambiguity, barriers to imitation and sustainable competitive advantage. *Academy of Management Review*, 15(1), 88–102.

Rogers, E. M. 1983. *Diffusion of Innovations*, 3rd ed. New York: The Free Press.

Rouse, M. 2005. Coopetition. http://searchcio.techtarget.com/definition/co-opetition (accessed on January 28, 2014).

Rouse, M. J. and Daellenbach, U. S. 1999. Rethinking research methods for the resource-based perspective: Isolating sources of sustainable competitive advantage. *Strategic Management Journal*, 20(5), 487–494.

San-Martin, S. and Lopez-Catalan, B. 2012. How can a mobile vendor get satisfied consumers? *Industrial Management and Data Systems*, 113(2), 156–170.

Schierz, P. G., Schilke, O., and Wiertz, B. W. 2009. Understanding consumer acceptance of mobile payment services: An empirical analysis. *Electronic Commerce Research and Applications*, 9(3), 209–216.

Schmalensee, R. and Willig, R. D. 1989. *Handbook of Industrial Organization*, vol. 1. Amsterdam, the Netherlands: North-Holland.

Sheth, J. N. and Sisodia R. S. 1999. Revisiting marketing's lawlike generalizations. *Journal of the Academy of Marketing Science*, 27(1), 71–87.

Steennot, R. 2011. Allocating liability in the event of fraudulent use of electronic payment instruments and the Belgian mobile payment instrument PingPing. *International Journal of Private Law*, 4(2), 274–289.

TNS Infratest. 2012 (and 2013). Online Banking. Mit Sicherheit! Initiative 21 und Huawel Technologies stellen Studie zur mobile Internetnutzung vor. Neuauflage, April 2013.

Tornatzky, L. and Klein, K. 1982. Innovation characteristics and innovation adoption-implementation: A meta-analysis of findings. *IWW Transactions on Engineering Management*, 29(1), 28–45.

Venkataraman, S. 2008. Mobile payment implementation: A reference framework. *International Journal of Business Information Systems*, 3(3), 252–271.

Vesal, S. N. and Fathian, M. 2012. Efficient and secure credit card payment protocol for mobile devices. *International Journal of Information and Computer Security*, 5(2), 105–114.

Wernerfelt, B. 1984. A resource based view of the firm. *Strategic Management Journal*, 5(2), 171–180.

Wind, Y. 1978. Issues and advance in segmentation research. *Journal of Marketing Research*, 15(3), 317–337.

Wurster, R. A. 2013. A new model for acceptance and adoption of mobile payment services. *Proceedings of the 6th International Conference on Management and Industrial Engineering—ICMIE 2013: Management—Facing New Technology Challenges*, Bucharest, Romania, October 31–November 2, 2013, pp. 136–143.

Yang, S., Lu, Y., Gubta, S., Cao, Y., and Zhang, R. 2012. Mobile payment adoption across time: An empirical study of the effects of behavioral beliefs, social influences and personal traits. *Computers in Human Behavior*, 28, 129–142.

chapter nineteen

Success factors influencing consumers' willingness to purchase brands advertised through the mobile phone

Carla Ruiz-Mafe, Inés Küster Boluda, and Christian Damián García

Contents

19.1 Introduction

The rapid diffusion of mobile phones throughout the world is changing the way consumers live, work, and interact with one another (Leong et al. 2013). In fact, the widespread use of social networking site mobile applications is a significant indicator of mobile addiction (Salehan and Negahban 2013). Technological developments in the mobile telephony industry, together with an integration with other devices like tablets and laptops and increasing use by consumers, are making the medium increasingly significant for the strategic planning of firms' marketing communications.

With a worldwide penetration rate of 96% by 2013 (International Telecommunication Union 2013), mobile phones have become the most ubiquitous of all technologies. This rapid proliferation of mobile phones along with their technological development has created a whole new channel for advertising (Saadeghvaziri and Seyedjavadain 2011).

The use of mobile phones as a communication channel has transformed the way consumers process advertising. Being able to receive mobile advertising at the right time and place can increase consumers' receptivity to promotions that arrive on their mobile phones. The increasing role of mobile phones in consumers' everyday lives has led to continuous growth in mobile advertising budgets as marketers realize that being connected all the time and everywhere through mobile phones represents a great opportunity to advertise their products or services, build and develop customer relationships, and receive direct responses from those customers (Google 2012, Martí et al. 2013).

As a consequence of marketers' interest, it is anticipated that 20.6 billion dollars per annum will be spent on mobile advertising worldwide by 2015 (Gartner 2011). In 2011 alone, mobile advertising expenditure increased by 358% in Europe and by 722% in Spain (Inmobi 2012). The study Mobile media: consumer insights across Europe by IAB Europe (2011) conducted in 19 European countries indicates that two out of three mobile Internet users are interested in advertising formats in this medium. With the proliferation of display campaigns, geolocation marketing, and couponing, mobile phones are becoming consolidated in the European market as an effective channel for brands to relate with their potential customers. Hence, the importance of understanding the factors can help marketing managers to improve the effectiveness of their mobile marketing campaigns (IAB Europe 2011).

Mobile advertising allows consumers to access advertising messages at their convenience. When receiving mobile advertising messages, consumers can read them, eliminate them, or purchase the brands advertised. So factors leading to mobile advertising acceptance can be considered to play a key role in the success of mobile advertising (Martí et al. 2013). Although firms are making increasingly significant investments in mobile marketing, the nature and implications of this channel are still not fully understood, and studies are needed to generate an understanding of how their effectiveness can be improved (Bauer et al. 2005, Wei 2008, Bellman et al. 2011, Yang et al. 2013). The reviews by Varnali and Toker (2010) and Shankar and Balasubramanian (2009) show significant gaps in this area of study. Previous research (Yang et al. 2013) shows that responses to mobile advertising depend on a two-dimensional structure of attitudes: technology-based evaluations (utilitarian considerations) and emotion-based evaluations (hedonic considerations). Mobile advertising is affected both by the characteristics of ad communication and by users' voluntary choices of mobile technology. Therefore, to better understand how mobile advertising works and to make it more effective, it is important to explore both utilitarian and hedonic characteristics of mobile advertising messages together.

This chapter aims to present an in-depth study of key drivers of consumers' positive intention to buy the brands advertised through mobile phones by analyzing five message characteristics: (1) direct discounts (2) informativeness, (3) credibility, (4) attitude toward mobile advertising, and (5) entertainment.

The chapter's specific goals are to

1. Identify consumer segments more likely to purchase brands advertised through the mobile phone
2. Analyze the role of utilitarian (direct discounts, informativeness, and credibility) and hedonic (attitude toward mobile advertising and entertainment) message factors on consumer mobile purchase behavior
3. Provide empirical research on the Spanish market with managerial implications for advertisers on how to maximize the effectiveness of mobile advertising in their marketing communications

The chapter is divided into three parts. In the first part, we include the literature review on key drivers of consumer adoption of mobile advertising, and we present a conceptual model focusing on the rationale of the constructs used. In the second part, methodology design, sample, and measures are presented and validated. Finally, the results are presented and managerial implications are discussed.

19.2 *Factors influencing consumer willingness to purchase brands advertised through the mobile phone*

Mobile advertising has been conceptualized as "the set of actions that enable firms to communicate and relate to their audience in a relevant, interactive way through any mobile device or network" (MMA 2010:7). This set of actions includes the sending of SMS, the use of graphic or display formats, marketing using search engines through mobile Internet, Bluetooth technology, couponing or the use of applications, and entertainment content, including mobile advergaming (MMA 2010).

The special characteristic of mobile technology is that advertising through this medium can be used to deliver advertising messages that are radically different from those in traditional advertising (Salo and Tähtinen 2005). The main benefits of the mobile medium include high penetration, covering all sectors of society, Internet connection, with online and offline formats, screen in the medium itself, low cost (Anckar and D'Incau 2002, Facchetti et al. 2005), immediate response (Jelassi and Enders 2004), measurable response, fast and concrete information on cost per impact, cost per response and conversion cost, and easily transportable, so as mobile advertising messages are ubiquitous, the receiver can be contacted anywhere anytime (Jelassi and Enders 2004).

The success of advertising activities on mobile phones depends largely on utilitarian and hedonic characteristics of the message, and therefore, advertisements need to be developed to suit this context (Haghirian et al. 2005, Yang et al. 2012). Past research has identified a number of factors linked to the advertising message predetermining mobile advertising adoption by consumers. This section shows a description of the impact of utilitarian (direct discounts, informativeness, and credibility) and hedonic (attitude toward mobile advertising and entertainment) factors on consumers' willingness to purchase brands advertised through the mobile phone.

19.2.1 *Utilitarian antecedents*

19.2.1.1 *Direct discounts*

The direct discounts variable refers to economic stimuli or incentives, monetary or otherwise, that advertisers and telephony operators can offer consumers to obtain their permission to send advertising to their mobile terminals. These incentives often take the form of discounts, promotions, or gifts.

Previous research has shown intriguing insights with respect to the effectiveness of different types of incentives. Varnali et al. (2012) suggest that consumer personality traits are stronger determinants of responses to mobile advertising campaigns than permission and incentives. On the other hand, after investigating whether consumers are willing to receive mobile advertising if they are provided with certain reward in exchange, like free telephone time, Tsang et al. (2004) show that giving incentives is positive for consumers and favors the effectiveness of advertising messages (Newell and Meier 2007). In this line, Pavlou and Stewart (2000) claim that in an interactive context, the benefits of advertising for consumers can influence, at least indirectly, the measures of advertising effectiveness. Barutçu (2007) found that price-sensitive customers have more positive attitude toward mobile advertising as a consequence of incentives like electronic discount coupons. In short, the literature indicates that receivers react very positively to incentive transfer advertising (Varshney 2003), and so direct discounts have a positive impact on consumer purchaser intention.

19.2.1.2 Information content

Informativeness is defined as the extent to which advertising media provide users with resourceful and helpful information (Chen and Wells 1999, Ducoffe 1996) and is strongly related to advertising value in traditional media vehicles (Ducoffe 1995). Bauer et al. (2005) found that mobile advertising messages delivering a high information value lead to positive attitude toward mobile advertising, which in turn leads to the behavioral intention to use mobile advertising services. Tsang et al. (2004) also found informativeness positively correlated to the overall attitude toward mobile advertising, along with entertainment and credibility.

Informativeness is closely related to advertising value (Ducoffe 1995). It has been shown that the quality of the information on a firm's website directly influences customer perceptions of the firm and its products (Siau and Shen 2003). Therefore, information delivered to customers through mobile devices must also show qualitative characteristics like accuracy, timeliness, and usefulness for the consumer (Siau and Shen 2003). In addition, users demand rapid access to the information they are seeking when using the content. Information is regarded as a very valuable incentive in mobile advertising (Haghirian et al. 2005). Several authors show a positive effect of good information in mobile advertising (e.g., Ducoffe 1996, Barwise and Strong 2002, Bauer et al. 2005, Haghirian et al. 2005, Haghirian and Inoue, 2007).

19.2.1.3 Credibility

Mckenzie and Lutz (1989, p. 53) define advertising credibility as "consumer perception of the truthfulness and credibility of advertising in general," whereas Pavlou and Stewart (2000, p. 70) refer to it as "predictability of compliance with the implicit and explicit requirements of an agreement."

According to Tsang et al. (2004), credibility is one of the factors that shape attitudes, and the significance of its impact can be demonstrated. Haghirian and Inoue (2007) investigated the antecedents of Japanese consumer attitude to mobile advertising and found that informative content and credibility of the advertising message have a significant impact on consumer attitude to mobile advertising. Research by Haghirian et al. (2005) shows a positive impact of credibility on the perceived value of mobile advertising.

Credibility is also influenced by the advertising medium. For example, Marshall and Woonbong (2003) found that a message on the Internet achieves less credibility than a printed message unless the message is advertised by a strong brand. Therefore, we can assume that a high level of advertising message credibility is positively associated with purchase intention.

19.2.2 Hedonic antecedents

19.2.2.1 Attitude toward mobile advertising

Attitude is an individual's positive or negative evaluation of a given object or behavior (Fishben and Ajzen 1975, Ajzen 1991) and includes feelings and affective responses. It refers to an individual's general willingness to engage in a given behavior. Attitude stems from individual beliefs about the behavior and its results and the importance attached to those beliefs. The social psychology literature clearly suggests that attitude has affective and cognitive components (Bagozzi and Burnkrant 1985, Chaiken and Stangor 1987, Weiss and Cropanzano 1996). The affective component of attitude refers to how much a person likes the object of their thoughts (i.e. a product, service, etc.) (Mcguire 1985) and measures the degree of emotional attraction to the object. The cognitive component refers to an individual's concrete beliefs concerning an object (Bagozzi and Burnkrant 1985) and consists in value-based evaluation, judgment, reception, and perception of the object (Chaiken and Stangor 1987). In particular, the cognitive dimension of attitude directly influences the use of an individual's information systems, whereas the affective dimension needs to be treated as a result variable.

Attitude toward advertising is a combination of cognitive, emotional, and behavioral processes (Neal et al. 2004); According to Batra and Ray's (1986) model, exposure to an advertising message from a specific brand first affects consumers' attitude toward the advertisement and finally their purchase intention (Mackenzie and Lutz 1989). In the context of mobile advertising, Batra and Ray's (1986) model may mean that advertising messages sent to mobile phones generate a given attitude toward the message, which in turn will affect the purchase intention.

Mitchell and Olson (1981) study positive and negative effects of advertising and report that consumer attitude to advertising affect brand attitude and purchase intention through their feelings and emotions over the advert itself. This may lead to changes in the conative component, such as greater purchase intention or purchase itself (Lutz et al. 1983, Lutz 1985, Batra and Ray 1986, Mckencie et al. 1986).

19.2.2.2 Entertainment

Entertainment denotes full capacity to satisfy consumer needs of "escapism, fun, aesthetic enjoyment or emotional liberation" (Mcquail 1983). Mobile entertainment services are undoubtedly one of the most important drivers of mobile marketing (Barutçu 2007) because fast and easy access to entertainment is always attractive for many customer segments as it gives the opportunity to kill time and have fun in situations where it is impossible to access home entertainment devices (Varshney et al. 2000). Furthermore, entertainment services can increase customer loyalty as it provides added-value services for them (Haghirian et al. 2005),

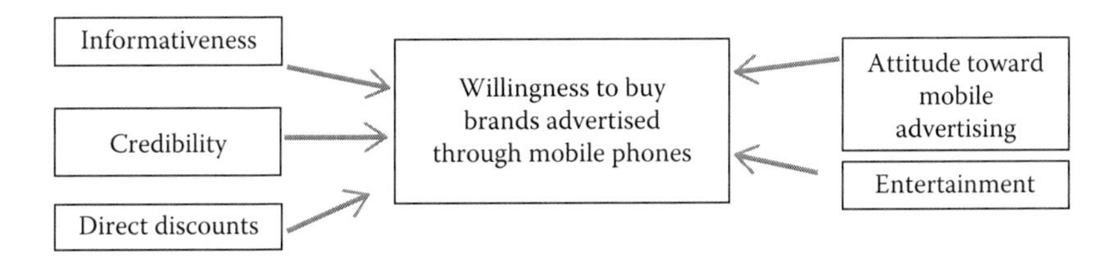

Figure 19.1 Conceptual model.

involving them deeply and familiarizing them with advertised services and products. A high degree of pleasure and involvement during the interaction leads to subjective perceptions of affection and a positive state of mind in consumers (Hoffman and Novak 1996). The feeling of enjoyment associated with advertising plays the most important role in creating general attitude toward them (Shavitt et al. 1998).

Entertainment has been identified as an emotional factor contributing to the formation of consumers' attitude toward advertising (Wang et al. 2002). Pleasurable contexts—like those supported by entertainment—and mood created by these contexts have a positive influence on the attitude toward the ad (Moorman et al. 2002) and on the effects of the ad (Srull 1983, Goldberg and Gorn 1987). It also has been found that consumers' feelings of enjoyment associated with advertisements play the greatest role in accounting for their overall attitude toward the ads (Shavitt et al. 1998). Entertainment has been found to be an influencing factor affecting the intention to use mobile services (Leong et al. 2013) and consumers' attitude toward mobile advertising (Tsang et al. 2004, Saadeghvaziri and Seyedjavadain 2011, Martí et al. 2013).

Therefore, we can state that perceived entertainment will have a favorable impact on purchase intention for the brands advertised in mobile ads.

The conceptual model of mobile electronic commerce adoption which will be contrasted in the Spanish market (see Figure 19.1) is an outcome of the literature review presented earlier.

19.3 Case study: Mobile advertising adoption by Spanish mobile users

19.3.1 Methodology

After utilitarian and hedonic antecedents of mobile advertising adoption have been discussed, in this section, an empirical study of the Spanish market is presented.

A field study was conducted with mobile telephony users who received mobile advertising. The target population was young people, mainly university students toward the end of their degree courses.

The research instrument employed to obtain information was a survey with close-ended questions. Studyby questionnaire has the advantages of convenience and rapid collection of data (Soroa-Koury and Yang 2010) and the ability to allow inferences on consumer behavior for a given population based on a sample (Babbie 1990). The questionnaire was developed and tested with five focus groups—PhD students (3), professors of marketing at the University of Valencia (1), and professionals of marketing and information system activities (1)—to examine the dynamics of mobile users with different levels of previous experience and familiarity with mobile advertising formats. The pretest instructed respondents to fill out the questionnaire and report any feedback. As a result of the pretest, some redundant questions were eliminated, and some of the scales were adapted in order to facilitate understanding and avoid erroneous interpretations.

Sample selection procedure was convenience nonprobability sampling. This type of sampling had been used in many prior studies in the area of mobile advertising with a large proportion of young university students (e.g., Yang 2007, Jayawardhena et al. 2009, Soroa-Koury and Yang 2010). Young adults were chosen because consumers under 30 years old are the first adopters of mobile services in general (Bigné et al. 2007), and they are familiar with the services and use them in a greater proportion than the general population (Wilska 2003, Karjaluoto et al. 2005).

Questionnaires were delivered to and collected from volunteer participants over the age of 15. A total of 250 consumers were contacted during the survey; 219 agreed to participate in this study. Among the questionnaires received, 208 were completed and analyzed. Only 34% of the sample does not have a smartphone.

Demographics are shown in Figures 19.2 through 19.4 (see the following text).

The average age of the sample is 32 years old, with 43% men and 57% women. Of the sample considered, 78% have completed university studies, and 24.4% have completed secondary school. Also, 27% of the participants

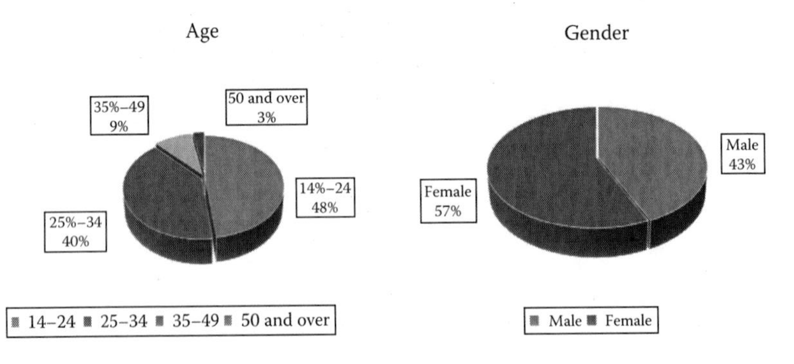

Figure 19.2 Sample demographics: age and gender.

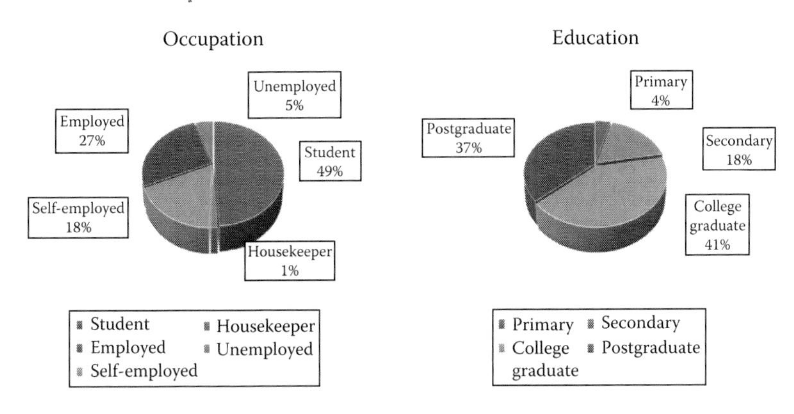

Figure 19.3 Sample demographics: occupation and education.

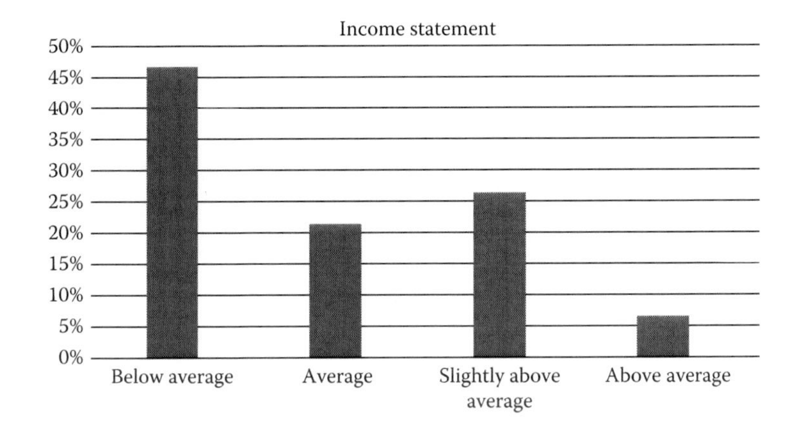

Figure 19.4 Sample demographics: income.

were employed, and 49% were students. Less than 50% of the sample has an income below the average.

All the variables were measured using Likert scales developed in previous research, thereby ensuring content validity (see Table 19.1).

19.3.2 Results

The conceptual model was tested with a structural equations model using EQS 6.1 for Windows. However, before contrasting the model, the scales for the six constructs were assessed using confirmatory factor analysis to determine scale reliability and validity (Anderson and Gerbing 1988). First of all, the scale items were analyzed to ensure they measured the latent variables and so the scores must correlate highly with each other and be internally consistent. Model fit proved satisfactory in the various indexes used: NFI = 0.912, NNFI = 0.959, CFI = 0.963, IFI = 0.964, SRMR = 0.079,

Table 19.1 Measurement of Variables

Variable	Sources
Entertainment	Tsang et al. (2004), Yang (2007)
EN1. I enjoy receiving mobile advertising messages	
EN2. Receiving mobile advertising messages is entertaining	
EN3. Receiving mobile advertising messages is enjoyable	
Credibility	Xu (2006)
CR1. I trust the content of mobile advertising	
CR2. Mobile advertising content is credible	
Informative content	Xu et al. (2009)
CI1. Mobile advertising is a good source of product information	
CI2. Mobile advertising provides me with information on products that are relevant for me	
CI3. Mobile advertising is a good source of up-to-date product information	
CI4. Mobile advertising makes product information accessible	
CI5. Mobile advertising is a convenient source of product information	
Direct discounts	Unni and Harmon (2010)
IN1. This service will offer me exclusive promotions	
IN2. This service will save me time when making purchase decisions	
IN3. Using this service will lead to better purchase decisions	
Attitude to mobile advertising	Gardner (1983), Cox and Cox (1988), Mckenzie and Lutz (1989), Mckenzie and Spreng (1992), Wansink et al. (1994), Zhang (1996), Jin and An (1998), Kempf and Smith (1998)
PM1. It's good	
PM2. I like it	
PM3. It's interesting	
PM4. I have a favorable opinion	
PM5. It's pleasant	
Purchase intention	Taylor and Todd (1995)
IC1. I intend to buy it in the near future	
IC2. I may buy it in the near future	
IC3. I will certainly buy it in the near future	
IC4. I hope to buy it in the near future	

Table 19.2 Model Reliability and Convergent Validity

Concept	Items	Load	Standardized load	T-Value	Cronbach	IFC	AVE
F1. Direct discounts	IN1	1.520	0.833	14.520	0.911	0.913	0.779
	IN2	1.697	0.917	16.990			
	IN3	1.681	0.895	16.312			
F2. Mobile advertising attitude	PM1	1.717	0.947	18.281	0.934	0.986	0.895
	PM2	1.768	0.963	18.857			
	PM3	1.678	0.917	17.239			
	PM4	1.666	0.915	17.180			
	PM5	1.736	0.927	17.583			
F3. Informativeness	CI1	1.579	0.844	14.936	0.943	0.943	0.767
	CI2	1.629	0.874	15.796			
	CI3	1.654	0.892	16.339			
	CI4	1.609	0.883	16.064			
	CI5	1.649	0.886	16.152			
F4. Credibility	CR1	1.762	0.963	18.443	0.926	0.928	0.865
	CR2	1.623	0.896	16.341			
F5. Entertainment	EN1	1.707	0.963	18.823	0.968	0.968	0.911
	EN2	1.675	0.964	18.856			
	EN3	1.595	0.936	17.867			
F6. Purchase intention	IC1	1.943	0.980	19.485	0.974	0.974	0.903
	IC2	1.862	0.937	17.955			
	IC3	1.815	0.942	18.114			
	IC4	1.924	0.941	18.075			

and RMSEA = 0.056 (Browne and Cudeck 1993, Hu and Bentler 1999). Cronbach's alpha value is good for all constructs as it exceeds the minimum threshold of 0.7 (Fornell and Larcker 1981), and the average variance extracted is above the cutoff value of 0.5 (Fornell and Larcker 1981) for all the variables. Thus, the proposed scales have appropriate reliability.

A model is regarded as having convergent validity if the scale items are significantly correlated with the latent variables they are supposed to measure. Hair et al. (1998) recommend that in addition to being significant, average loading on each factor should be above 0.7. Table 19.2 shows that all the factors were significant with average loadings above 0.7.

Discriminant validity was evaluated using (1) the confidence interval test (Anderson and Gerbing 1988), which determines validity if the interval does not include the value 1 and (2) the variance extracted test (Fornell and Larcker 1981) confirming that the squared covariance for each pair of factors is lower than the variance extracted for each factor. Both these methods confirm validity (see Table 19.3).

Table 19.3 Discriminant Validity

	F1	F2	F3	F4	F5	F6
F1. Incentives	0.883	(0.837; 0.709)	(0.878; 0.766)	(0.861; 0.737)	(0.877; 0.769)	(0.779; 0.623)
F2. Mobile advertising attitude	0.773	0.91	(0.831; 0.707)	(0.806; 0.662)	(0.857; 0.749)	(0.751; 0.591)
F3. Informativeness	0.822	0.769	0.876	(0.905; 0.809)	(0.856; 0.740)	(0.703; 0.519)
F4. Credibility	0.799	0.734	0.857	0.93	(0.860; 0.748)	(0.730; 0.642)
F5. Entertainment	0.823	0.803	0.798	0.804	0.954	(0.773; 0.625)
F6. Purchase intention	0.701	0.671	0.611	0.642	0.699	0.949

Notes: On the diagonal: confidence interval for correlation between factors, Diagonal: square root of the variance extracted, Below the diagonal: estimated correlation between factors.

Table 19.4 Results

Relationships	Load	T-Value
Direct discounts → Willingness to purchase	0.337	2.799
Informativeness → Willingness to purchase	n.s.	−1.356
Credibility → Willingness to purchase	n.s.	1.189
Entertainment → Willingness to purchase	0.263	2.343
Attitude toward mobile advertising → Willingness to purchase	0.228	2.43

The structural model was evaluated using the maximum robust likelihood method with EQS 6.1 software for Windows. Goodness of fit statistics suggests that the structural model offers a good fit: NFI = 0.912, NNFI = 0.959, CFI = 0.963, IFI = 0.964, SRMR = 0.079, and RMSEA = 0.056.

Results show (Table 19.4) that direct discounts and attitude toward the mobile advertising ad and entertainment significantly affect consumers' brand purchase intention whereas information content and credibility do not have a significant influence. Direct discounts is the variable with greater influence on the willingness to purchase brands advertised through mobile phones.

19.4 Conclusions and implications

The recent development of mobile advertising has brought significant changes in the communication process, which in turn is changing digital consumer behavior. Nowadays, new terminals provide consumers with more entertainment, information content, and incentives for mobile communications, which requires further investigation into the effectiveness of mobile advertising (Shankar et al. 2010).

This chapter not only contributes to the body of academic knowledge of mobile commerce but also offers recommendations for effective management of mobile advertising campaigns. First, the study confirms the significant influence of attitude toward mobile advertising, incentives, and entertainment on purchase intention, indicating the importance of the mobile medium not only for communicating particular promotions but also for creating and managing brand image mainly by creating emotional ties with consumers through the use of entertainment. Given the importance of attitude toward mobile advertising, firms should systematically analyze the factors that can improve that attitude. Second, mobile advertising campaign managers should issue messages that offer entertainment, making use of the increasing number of multimedia tools on mobile terminals because they are more appreciated than the information content. Users want entertainment on their mobile phones and prefer to use other more traditional means when finding out about a product, service, or brand. Another conclusion is the mobile phone's potential for couponing as the value of economic incentives through this medium has been demonstrated. Advertisers and advertising professionals should exploit this, making use of the mobile phone's ubiquity to do the promotion for their products at the appropriate time and place and persuade consumers to choose the advertised brand at the moment of purchase. Advertisers and advertising professionals should take good note of this. A successful information campaign must be conducted through entertainment, for example, with a pedagogical advergame or a video that is fun as well as informative, or by including a direct discount along with information.

These conclusions should not be generalized unreasonably, given the limitations of this study. The first limitation stems from the small convenience sample. Most of the participants are university students. Bamba and Barnes (2007) suggest that the student population is a core target market for mobile operators and advertisers. Another possible limitation is that the study has focused on measuring attitudes, which do not always become behaviors.

In view of the things discussed so far, we recommend future lines of research: contrasting the proposed model with a representative population sample and an evaluation of the results by establishing comparisons according to demographic classifications and conducting experiments to evaluate real purchase behavior.

References

Ajzen, I. 1991. The theory of planned behaviour. *Organizational Behaviour and Human Decision Processes* 50(2): 179–211.

Anckar, B. and D. D'incau. 2002. Value creation in mobile commerce: Findings from a consumer survey. *Journal of Information Technology Theory and Application* 4(1): 43–64.

Anderson, J. and D. Gerbing. 1988. Structural equation modelling in practice: A review and recommended two-step approach. *Psychological Bulletin*, 103: 411–423.

Babbie, E.R. 1990. *Survey Research Methods*. Belmont, CA: Wadsworth.

Bagozzi, R.P. and R.E. Burnkrant. 1985. Attitude organization and the attitude–behaviour relationship: A reply to Dillon and Kumar. *Journal of Personality and Social Psychology* 49(1): 16.

Bamba, F. and S.J. Barnes. 2007. SMS advertising, permission and the consumer: A study. *Business Process Management Journal* 13(6): 815–829.

Barutçu, S. 2007. Attitudes towards mobile marketing tools: A study of Turkish consumers. *Journal of Targeting, Measurement and Analysis for Marketing* 16: 26–38.

Barwise, P. and C. Strong. 2002. Permission-based mobile advertising. *Journal of Interactive Marketing* 16(1): 14–24.

Batra, R. and M.L. Ray. 1986. Affective responses mediating acceptance of advertising. *Journal of Consumer Research* 13(2): 234–249.

Bauer, H., S. Barnes, T. Reinhardt, and M. Neumann. 2005. Driving consumer acceptance of mobile marketing: A theoretical framework and empirical study. *Journal of Electronic Commerce and Research* 6(3): 181–192.

Bellman, S., R. Potter, S. Treleaven-Hasard, J. Robinson, and D. Varan. 2011. The effectiveness of brand mobile phone apps. *Journal of Interactive Marketing* 25: 191–200.

Bigné, E., C. Ruiz, and S. Sanz. 2007. Key drivers of mobile commerce adoption. An exploratory study of Spanish mobile users. *Journal of Theoretical and Applied Electronic Commerce Research* 2(2): 48–61.

Browne, M. and R. Cudeck. 1989. Single sample cross-validation indices for covariance structures. *Multivariate Behavioral Research* 24: 445–455.

Chaiken, S. and C. Stangor. 1987. Attitudes and attitude change. *Annual Review of Psychology* 38: 575–630.

Chen, Q. and W.D. Wells. 1999. Attitude toward the site. *Journal of Advertising Research* 39(September/October): 27–37.

Cox, D.S. and A.D. Cox. 1988. What does familiarity breed? Complexity as a moderator of repetition effects in advertising evaluation. *Journal of Consumer Research* 15(Junio): 111–116.

Ducoffe, R.H. 1995. How consumers assess the value of advertising. *Journal of Current Issues and Research in Advertising* 17(1): 1–18.

Ducoffe, R.H. 1996. Advertising value and advertising on the web. *Journal of Advertising Research* 36(5): 21–36.

Facchetti, A., A. Rangone, F.M. Renga, and A. Savoldelli. 2005. Mobile marketing: An analysis of key success factors and the European value chain. *International Journal of Management and Decision Making* 6(1): 65–80.

Fishbein, M. and I. Ajzen. 1975. *Belief, Attitude, Intention and Behavior: An Introduction to Theory and Research*. New York: Addison-Wesley.

Fornell, C. and D. Larcker. 1981. Evaluating structural equation models with unobservable variables and measurement error. *Journal of Marketing Research* 18: 39–50.

Gardner, M.P. 1983. Advertising effects on attributes recalled and criteria used for brand evaluations. *Journal of Consumer Research* 10(December): 310–318.

Gartner. 2011. Gartner says worldwide mobile advertising revenue forecast to reach $3.3 billion in 2011, Press release, June 16. www.gartner.com/it/page.jsp?id¼1726614 (accessed June 12, 2012).

Goldberg, M.E. and G.J. Gorn. 1987. Happy and sad TV programs: How they affect reactions to commercials. *Journal of Consumer Research* 14: 387–403.

Google. 2012. Our mobile planet. http://services.google.com/fh/files/blogs/our_mobile_planet_spain_es.pdf (accessed June 10, 2013).

Haghirian, P. and A. Inoue. 2007. An advanced model of consumer attitudes towards advertising on the mobile internet. *International Journal of Mobile Communications* 5(1): 48–67.

Haghirian, P., M. Madlberger, and A. Tanuskova. 2005. Increasing advertising value of mobile marketing—An empirical study of antecedents. *Proceedings of the 38th Hawaii International Conference on System Sciences*, Big Island, HI.

Hair, J., R. Anderson, R. Tatham, and W. Black. 1988. *Multivariate Data Analysis*, 4th edn. Englewood Cliffs, NJ: Prentice Hall.

Hoffman, D.L. and T.P. Novak. 1996. Marketing in hypermedia computer-mediated environments: Conceptual foundations. *Journal of Marketing* 60: 50–68.

Hu, L. and P. Bentler. 1999. Cutoff criteria for fit indexes in covariance structure analysis: Conventional criteria vs new alternatives. *Structural Equation Modelling* 6(1): 1–55.

Iab Europe. 2011. Mobile media: Consumer insights across Europe. www.iabspain.net (accessed October 12, 2012).

InMobi. 2012. Perspectivas moviles. www.inmobi.com/insights/consumerresearch (accessed October 22, 2012).

International Telecommunication Union. 2013. The World in 2013: ICT Facts and Figures. www.itu.int/en/itud/statistics/documents/facts/ictfactsfigures2013.pdf (accessed April 10, 2013).

Jayawardhena, C., A. Kuckertz, H. Karjaluoto, and T. Kautonen. 2009. Antecedents to permission based mobile marketing: An initial examination. *European Journal of Marketing* 43(3): 473–499.

Jelassi, T. and A. Enders. 2004. Leveraging wireless technology for mobile advertising. *Proceedings of the 12th European Conference on Information Systems*, Turku, Finland.

Jin, H.S. and S. An. 1998. Effects of celebrity endorsement on advertising liking and purchase intention. *Proceedings Conference of the American Academy of Advertising*, Tokyo, Japan.

Karjaluoto, H., I. Karvonen, M. Kesti, T. Koivumaki, M. Manninen, J. Pakola, A. Ristola, and J. Salo. 2005. Factors affecting consumer choice of mobile phones: Two studies from Finland. *Journal of Euro-Marketing* 14(3): 59–82.

Karjaluoto, H., H. Lehto, M. Leppäniemi, and C. Jayawardhena. 2008. Exploring gender influence on customer's intention to engage permission-based mobile marketing. *Electronic Markets* 18(3): 242–259.

Kempf, D.S. and R.E. Smith. 1998. Consumer processing of product trial and the influence of prior advertising: A structural modeling approach. *Journal of Marketing Research* 35(August): 325–338.

Kesti, M., A. Ristola, H. Karjaluoto, and T. Koivumäki. 2004. Tracking consumer intentions to use mobile services: Empirical evidence from a field trial in Finland. *E-Business Review* 4: 76–80.

Leong, L.Y., K.B. Ooi, A.Y.L. Chong, and B. Lin. 2013. Modeling the stimulators of the behavioral intention to use mobile entertainment: Does gender really matter? *Computers in Human Behavior*, 29(5): 2109–2121.

Lutz, R.J. 1985. Affective and cognitive antecedents of attitude toward the ad: A conceptual framework. In Alwitt, L. and A.A. Mitchell (eds.), *Psychological Processes and Advertising Effects: Theory, Research and Application*. Hillsdale, NJ: L. Erlbaum.

Lutz, R.J., S.B. Mackenzie, and G.E. Belch. 1983. Attitude towards the Ad as a mediator of advertising effectiveness: Determinants and consequences. In Bagozzi, R.P. and A.M. Tybout (eds.), *Advances in Consumer Research*, Vol. 10, pp. 532–539. Ann Arbor, MI: Association for Consumer Research.

Mackenzie, S.B. and R.J. Lutz. 1989. An empirical examination of the structural antecedents of attitude toward the ad in an advertising pretesting context. *Journal of Marketing* 53(2): 48–65.

Mackenzie, S.B., R.J. Lutz, and G.E. Belch. 1986. The role of attitude toward the ad as a mediator of advertising effectiveness: A test of competing explanations. *Journal of Marketing Research* 23(2): 130–143.

Mackenzie, S.B. and R.A. Spreng. 1992. How does motivation moderate the impact of central and peripheral processing on brand attitudes and intentions? *Journal of Consumer Research* 18(Marzo): 519–529.

Marshall, R. and N. Woonbong. 2003. An experimental study of the role of brand strength in the relationship between the medium of communication and perceived credibility of the message. *Journal of Interactive Marketing* 17(3): 75–79.

Marti, J., S. Sanz, C. Ruiz, and J. Aldás. 2013. Key factors of teenagers' mobile advertising acceptance. *Industrial Management & Data Systems* 113(5): 732–749.

Mcguire, W.J. 1985. Attitudes and attitude change. In Lindzey, G. and E. Aronson (eds.), *Handbook of Social Psychology*, Vol. 19, pp. 233–346. New York: Random House.

Mcquail, D. 1983. *Mass Communication Theory: An Introduction*. London, U.K.: Sage Publication.

Mitchell, A.A. and J.C. Olson. 1981. Are product attribute beliefs the only mediator of advertising effects on brand attitude? *Journal of Marketing Research* 18(3): 318–332.

Mobile Marketing Association. 2010. 3rd Study on Marketing and Mobile Advertising investment in Spain. http://recursos.anuncios.com/files/387/64.pdf (accessed March 12, 2011).

Moorman, M., P.C. Neijens, and E.G. Smit. 2002. The effects of magazine-induced psychological responses and thematic congruence on memory and attitude toward the ad in a real-life setting. *Journal of Advertising* 31(4): 27–40.

Neal, C., P. Quester, and D. Hawkins. 2004. *Consumer Behaviour: Implications for Marketing Strategy*, 4th edn. Boston, MA: McGraw-Hill.

Newell, J. and M. Meier. 2007. Desperately seeking opt-in: A field report from a student-led mobile marketing initiative. *International Journal of Mobile Marketing* 2(2): 53–57.

Pavlou, P.A. and D.W. Stewart. 2000. Measuring the effects and effectiveness of interactive advertising: A research agenda. *Journal of Interactive Advertising* 1(1): 61–77.

Quah, J.T.S. and G.L. Lim. 2002. Push selling–multicast messages to wireless devices based on the publish/subscribe model. *Electronic Commerce Research and Applications* 1(3/4): 235–246.

Saadeghvaziri, F. and S. Seyedjavadain. 2011. Attitude toward advertising: Mobile advertising vs advertising-in-general. *European Journal of Economics, Finance and Administrative Sciences* 28: 104–114.

Salehan, M. and A. Negahban. 2013. Social Networking on smartphones: When mobile phones become addictive. *Computers in Human Behavior* 29(6): 2632–2639.

Salo, J. and J. Tähtinen. 2005. Retailer use of permission-based mobile advertising. In Clarke III, I. and T.B. Flaherty, *Advances in Electronic Marketing*. Hershey, PA: Idea Publishing Group.

Shankar, V. and S. Balasubramanian. 2009. Mobile marketing: A synthesis and prognosis. *Journal of Interactive Marketing* 23: 118–129.

Shankar, V., A. Venkatesh, C.H. Hofacker, and P. Naik. 2010. Mobile marketing in the retailing environment: Current insights and future research avenues. *Journal of Interactive Marketing* 24: 111–120.

Shavitt, S., P. Lowrey, and J. Haefner. 1998. Public attitudes towards advertising: more favourable than you might think. *Journal of Advertising Research* 38(4): 7–22.

Siau, K. and Z. Shen. 2003. Building customer trust in mobile commerce. *Communications of the ACM* 46(4): 91–94.

Soroa-Koury, S. and K.C.C. Yang. 2010. Factors affecting consumers' responses to mobile advertising from a social norm theoretical perspective. *Telematics and Informatics* 27: 103–113.

Srull, T.K. 1983. The impact of affective reactions in advertising on the representation of product information in memory. In Bagozzi, R. and A. Tybout (eds.), *Advances in Consumer Research*, Vol. 10, pp. 520–525. Ann Arbor, MI: Association for Consumer Research.

Taylor, S. and P. Todd. 1995. Understanding information technology usage: A test of competing models. *Information Systems Research* 6(2): 144–176.

Tsang, M.M., S.C. Ho, and T.P. Liang. 2004. Consumer attitudes toward mobile advertising: An empirical study. *International Journal of Electronic Commerce* 8(3) 65–78.

Unni, R. and R. Harmon. 2010. Perceived effectiveness of push vs. pull mobile location-based advertising. *Journal of Interactive Advertising* 7(2): 28–40.

Varnali, K. and V.A. Toker. 2010. Mobile marketing research: The-state-of-the-art. *International Journal of Information Management* 30: 144–151.

Varnali, K., C. Yilmaz, and A. Toker. 2012. Predictors of attitudinal and behavioral outcomes in mobile advertising: A field experiment. *Electronic Commerce Research and Applications* 11(6): 570–581.

Varshney, U., R.J. Vetter, and R.K. Kalakota. 2000. Mobile commerce: A new frontier. *Computer* 33(October): 1032–38.

Wang, C., P. Zhang, R. Choi, and M. Dieredita. 2002. Understanding consumers attitudes towards advertising. Paper presented at *the 8th Americas Conference on Information Systems*, Dallas, TX, August 9–11, 2002. http://melody.syr.edu/hci/amcis02_minitrack/rip/wang.pdf.

Wansink, B., M.L. Ray, and R. Batra. 1994. Increasing cognitive response sensitivity. *Journal of Advertising* 23(Summer): 65–75.

Wei, R. 2008. Motivations for using mobile phone for mass communications and entertainment. *Telematics and Informatics* 25(1): 36–46.

Weiss, H.M. and R. Cropanzano.1996. Affective events theory: A theoretical discussion of the structure, causes and consequences of affective experiences at work. In Staw, B.M. and L.L. Cummings (eds.), *Research in Organizational Behaviour*, pp. 1–74. Greenwich, U.K.: Jai Press.

Wilska, T.A. 2003. Mobile phone use as part of young people's consumption styles. *Journal of Consumer Policy* 26(4): 441–463.

Xu, H., L. Oh and H. Teo. 2009. Perceived effectiveness of text vs. multimedia location-based advertising messaging. *International Journal of Mobile Communications* 7(2): 154–177.

Xu, J.D. 2006. The influence of personalization in affecting consumer attitudes toward mobile advertising in China. *Journal of Computer Information Systems* 47(2): 9–19.

Yang, B., Y. Kim, and C. Yoo. 2013. The integrated mobile advertising model: The effects of technology-and emotion-based evaluations. *Journal of Business Research* 66(9): 1345–1352.

Yang, K.C.C. 2007. Exploring factors affecting consumer intention to use mobile advertising in Taiwan. *Journal of International Consumer Marketing* 20(1): 33–49.

Zhang, Y. 1996. Responses to humor advertising: The moderating effect of need for cognition. *Journal of Advertising* 25(Spring): 15–32.

section seven

Emerging frontiers

chapter twenty

Smart city as a service platform

Identification and validation of city platform roles in mobile service provision

Nils Walravens

Contents

20.1 Introduction

In a period of less than a decade, the mobile telecommunications industry has undergone some dramatic changes. New players like Apple and Google have entered the market, launching a wave of innovation through the industry. After several failed attempts, for example, WAP and i-Mode in Europe, a market for attractive mobile services was created and is continually growing. This changing playing field has diverse implications for all actors involved, and many are taking up new places in the value network: New players enter the field, roles and relationships change, and the interests of companies and stakeholders—including their business models—are redrawn (see e.g., Basole, 2009). These shifts appear to be going hand in hand with a trend of *platformization*, that is, diverse companies adopting platform strategies as they vie for dominance in mobile service provision. The different types of platforms we can identify in this sector were only recently made explicit and validated (see e.g., Ballon et al., 2009).

These strategic business model changes are occurring while the context of the main target audience of mobile service providers—the consumer—changes as well. Although it is a much more gradual process, it is now accepted that an increased urbanization will be one of the main societal trends in coming years (Brand, 2006). Since 2007, over half of the world's population lives in cities and the UN predicts this number will only grow, to a predicted 70% by 2050 (UN HABITAT, 2010). As more citizens (and consumers) move to urban areas, actors from ICT and mobile telecommunications naturally become increasingly interested in offering services that are tailored to life in the urban environment. Cities and local governments are at the same time exploring the role that new ICT services and products can play in increasing the quality of life of their citizens. In recent years, this quest is often captured in the *Smart City* concept, which will be briefly touched upon later on.

Currently, cities are struggling to see which roles they can take up in this quickly changing landscape of mobile services. The aim of this chapter is to apply a general mobile services typology to the specific context of the city, creating a framework cities can use to consider their place in the value network. We will explore the different platform roles a city government can take up in supporting, facilitating, and/or providing mobile services to citizens and travelers by applying and adapting the general mobile service platform typology to this context, based on the thorough analysis of several international cases.

20.2 Framework

As a first step in this chapter, we will briefly summarize the concept of platformization. While there have been some attempts at conceptualizing ICT platforms (Schiff, 2003; Evans et al., 2005; Eisenmann, 2007) and the seminal work by Gawer and Cusumano (2002) identifying platform leadership strategies, these operationalizations do not entirely cover the platform types currently active in the mobile services industry. Evans et al. (2005) provide an overview of different platforms in what they call computer-based industries, discussing the particulars of the Symbian mobile platform, but do not offer an analysis of different platform types *within* a specific sector, in this case, mobile communications. Furthermore, they distinguish between four platform functions (matchmakers, audience makers, transaction-based businesses, and shared-input platforms), which are all roles a mobile platform could take up and thus are not mutually exclusive in this more specific context. Eisenmann (2007) distinguishes between proprietary platforms that consist of a single provider that individually controls its technology and shared platforms in which multiple firms collaborate in developing the platform's technology and then compete in offering users different but compatible versions of the platform. However, most, if not all, platforms operating in mobile services today externalize mostly proprietary technologies but at the same time usually incorporate also some standardized and shared functionalities. Schiff (2003) distinguishes between a platform that delivers an active *matching service* (e.g., an Internet search engine) and a platform that passively mediates (e.g., Adobe Acrobat). However, the various mobile service platform models at the same time provide active matching services (e.g., through personalization features) as passive mediation (e.g., through offering an SDK). Although these authors provide very valuable contributions, the main differentiators between different types of mobile service platforms remain illusive.

What does appear to distinguish the various types is the question whether control over assets is linked to control over customers (Ballon, 2009a). As some examples from the mobile services industry demonstrate, to control some crucial value adding roles does not necessarily mean that the platform owner also has control over the customer (i.e., end user) relationship, or has control over all, or even most, assets needed to *create* the value proposition. By reinterpreting the business models currently employed in the mobile communications industry in this light, four basic platform types can be distinguished. This typology is the result of extensive conceptual and empirical work, thoroughly described in Ballon (2009b). The four general platform types for mobile service delivery, including examples, are represented in Table 20.1.

It is important to note here that when using the term "platform," we do this referring to the concept of two- or multisided markets (Rochet and Tirole, 2003; Ballon, 2009a), rather than a technological platform per se, approaching the issue more from the perspective of business modeling

Table 20.1 Mobile Service Platform Typology

	No control over customers	Control over customers
Control over assets	Enabler platform The platform owner controls many of the necessary assets to ensure the value proposition, but does not control the customer relationship. *Examples: Google Android (pre-Google Play)*	Integrator platform The platform owner controls many of the assets to ensure the value proposition, and establishes a relationship with end users. Entry of *third-party* service providers is encouraged. *Examples: iPhone with iOS and iTunes App Store, Nokia Ovi*
No control over assets	Neutral platform The platform owner is strongly reliant on the assets of other actors to create the value proposition, and does not control the customer relationship. *Examples: LiMo, Bondi, WAC*	Broker platform The platform owner is strongly reliant on the assets of other actors to create the value proposition, but does control the customer relationship. *Examples: GetJar, Appia, Handango*

Source: Ballon, P., Platform types and gatekeeper roles: The case of the mobile communications industry, Proceedings of Druid Conference, June 17–19, 2009a, available online: http://www2.druid.dk/conferences/viewpaper.php?id=5952&cf=32.

literature, rather than that of Information System studies. A platform in this context should be seen as an entity that mediates between at least two sides of a market (e.g., developers, advertisers, and end users), creating new value within the value network through its existence. In practice and in the context of city governments, this *may* crystallize into a technical solution, but could equally turn out to be a more *virtual* approach, for example, several smaller initiatives taken by diverging city government agencies combined. As such, it is not so much the entire geographical region of a city that we regard as a platform in this case, but rather the combined efforts government bodies undertake in bringing mobile services to citizens. This notion is also referred to by O'Reilly (2010) in "Government as a Platform," saying that new forms of government or "government 2.0" should *"use (...) technology—especially the collaborative technologies at the heart of web 2.0—to better solve collective problems at a city, state, national and international level."*

This chapter will apply the preceding typology to city initiatives in order to assess whether and which platform dynamics are at play in the city. In order to achieve this, we will first need to sketch the context of the city more clearly. We will briefly look at attempts to define the Smart City concept and explore how mobile services and platform strategies can fit within it.

20.3 City context

The term *Smart Cities* has been used in different ways: to describe a cluster of innovative organizations within a region, the presence of industry branches that have a strong focus on ICT, business parks, the actual educational level of the inhabitants of a certain city, the use of modern technologies in an urban context, and technological means that increase government efficiency and efficacy and so on (see e.g., Komninos, 2009). Giffinger et al. (2007) describe medium-sized European Smart Cities and define a Smart City using six characteristics in which such a city *performs in a forward-looking way*: Smart Economy, Smart People, Smart Governance, Smart Mobility, Smart Environment, and Smart Living. Caragliu et al. (2009) propose a definition based on the aforementioned study: "We believe a city to be smart when investments in human and social capital and traditional (transport) and modern (ICT) communication infrastructure fuel sustainable economic growth and a high quality of life, with a wise management of natural resources, through participatory governance."

This view is supported by the somewhat technologically deterministic idea of a *control room* or dashboard for the city, providing an architecture and ICT-based overview of all activity in the city as well as the tools to (automatically) interact with infrastructures or adjust parameters to predefined optima (IBM, 2009). These architectural or infrastructural viewpoints are contrasted by a more experimental, bottom-up view on the Smart City. In this perspective, innovation comes from the people *using* the city or is at least cocreated with citizens, a process that can be stimulated by government (Camponeschi, 2011). Examples of this can be found in the growing trend of open data initiatives and *hackathons*, stimulating developers to create applications based on cities' databases, or the use of social media to organize local and ad hoc or more structural events and even protests, placing the *ownership of the city* decidedly with its users (de Lange and de Waal, 2012). Interventions in the public space by small groups of citizens could also be framed in this context and have been referred to as small change or tactical urbanism (Hamdi, 2004). This entails citizens taking the initiative and becoming active agents of change in their city or neighborhood. In such a perspective, what defines the Smart City is not the infrastructures or architectures it offers but the ways in which its citizens interact with these systems as well as each other. In between the top-down or bottom-up views on the Smart City, one could envision the role the city plays as a platform, as sitting somewhere in between, bringing together city governments, private interest, and citizens. While any definitions remain broad, the Smart Cities concept can entail many diverging elements, which are all in some way captured by them. In the urban context, in which mobility, information, and accessibility are all important to inhabitants, mobile services can be one of many tools to make advances in many of the areas listed earlier (Greenfield, 2013; Townsend, 2013). Obviously much is dependent on the policy foci laid out by city governments, and it is likely the emphasis on certain topics will be different among cities.

The Smart City concept has also been criticized, amongst others for its potentially self-congratulatory tone, as well as its focus on I(C)T and the potential consequences toward reinforcing a digital divide (Hollands, 2008). If insufficient attention is paid to this topic, the fear exists that the strong focus on information technologies in the Smart Cities discourse can dramatically impact the digital divide in the negative sense, creating even larger inequalities and social divisions in the city (Graham, 2002), a far cry from what would be labeled smart. This is thus a consideration every policy maker should keep in mind when working on Smart City–type initiatives. With regard to the topic of this chapter, it is clear that the concept of Smart Cities carries relevance, particularly when considering the implications of viewing the city as a platform, but it should not be overstated and seen as a contextual element in the context of mobile services, from the domain of policy making: Mobile services can be a tool, like many others, in increasing the quality of life of citizens, but their application is still very much a matter of local policy decisions.

In what follows, we reexamine the platform typology and clarify its constituting elements in the context of the city. Next, we apply it to various international cases, exposing the different platform strategies cities can undertake.

20.4 Toward a platform typology for the city

Given the more specific nature of the application domain and the elements detailed in the preceding section, we can delineate the constituting components of the platform typology in a more fine-grained way. An important aspect related to the generalizability of the platform typology lies in the definition of the two control parameters: *assets* and *customers*. These parameters need to be well defined and relevant in order to allow the typology to be applied to the somewhat different area of mobile city services. This section will elaborate on that issue.

In the context of the city, we can start by defining the first control parameter, *control over assets*, more clearly. For our purposes, having control over the value-adding, tangible and intangible assets that form the value proposition means that a city is in control of how services are created. For example, this can mean setting up support mechanisms for developers that want to build city services and performing some kind of quality check on these services. Or a city could take a more active role and provide a development and distribution environment, for example, with APIs or an SDK, where developers can create and disseminate services that carry a specific city brand, building a more technology-centric solution. Another example could be opening up data and statistical information about the urban environment. In the latter case, the city is not in direct control over the services that are developed but still leverages a form of control over the assets, namely, the data it chooses to make public, and perhaps more importantly, which data it does not. Additionally, the city could provide the required infrastructure that makes certain city services possible (e.g., sensors, QR-codes, fixed and wireless networks, and so on). Therefore, in the context of the city, we prefer to refer to this parameter as control over data and/or infrastructure. This parameter is guided by questions like:

- To what extent does ownership of the data/infrastructure remain with the city?
- Does the license provided by the city (if any) prohibit certain uses of the data/infrastructure?
- Does the city provide APIs to access their data?
- Are initiatives, programs, or actions related to mobile services mentioned or considered in policy plans?

As far as control over customers is concerned, this parameter refers to the entity the end user (or citizen in this case) has the direct interaction with. In our case, the mobile service that is offered to the citizen is the medium via which local governments reach the public. Hence, this parameter is redefined as whether or not the local city government has *control over the service offering* (or value proposition). By this we mean that the city is in one capacity or another involved in or has control over the actual creation of the service and the resulting product or value proposition that is being offered to the end user. The way the service is branded can be an indicator in this case as well as the guiding questions:

- Is the service developed in-house by the city (or at least by directly contracted developers, executing a clear assignment) or are the developers completely external to the city's organization?
- Does the city have control over or impact on the contents of the service during the creation process?
- Does the city have access to profile/identity information of the end users of the services, or does it gather data or statistics on end users in any way (e.g., related to usage or popularity)?
- Is a form of quality control performed by the city before or after apps are released?

With these parameters more clearly defined, we can explore what the different city platform types would look like. The following sections will describe the platform types in the specific framework of the city, as well as provide at least one case study for each platform. To structure the case studies, we use the framework detailed in Walravens (2012) that describes business model parameters that come into play when a public entity (in this case a city government) becomes an actor in the value network. The framework describes parameters connected to *governance and control* on the one hand and *value and public value* on the other. In this sense, these parameters are directly related to the question of *assets* (or data and infrastructure) and *customers* (or the service offering) in the platform typology. The parameters of this framework are represented in Figure 20.1.

The two columns on the left side of the matrix represent the public and business model parameters related to control and governance and say something about the level of control the city keeps over data and infrastructure. The right two columns discuss parameters related to the value that is being created within the value network and provide more insight into the service offering that is presented to the citizen in our case. This framework helps to take a more structured approach to the case studies and provides a guideline, allowing us to describe the cases in a comparative way, building up more information on the level of control over data and infrastructure on the one hand, and the service offering on

	Value network	Technical architecture	Financial architecture	Value proposition
Business design parameters	Control parameters		Value parameters	
	Control over assets	Modularity	Investment structure	User involvement
	Concentrated vs. distributed	Modular vs. integrated	Concentrated vs. distributed	Enabled, encouraged, dissuaded, or blocked
	Vertical integration	Distribution of intelligence	Revenue model	Intended value
	Integrated vs. disintegrated	Centralized vs. distributed	Direct vs. indirect	Price/quality lock-in effects
	Control over customers	Interoperability	Revenue sharing	Positioning
	Direct vs. mediated profile and identity management	Enabled, encouraged, dissuaded or blocked	Yes or no	Complements vs. substitutes branding
Public design parameters	Public governance parameters		Public value parameters	
Policy goals	Good governance	Technology governance	ROPI	Public value creation
	Harmonizing existing policy goals and regulation Accountability and trust	Inclusive vs. exclusive Open vs. closed data	Expectations on financial returns multiplier effects	Public value justification Market failure motivation
Organizational	Stakeholder management	Public data ownership	Public partnership model	Public value evaluation
	Choices in (public) stakeholder involvement	Definition of conditions under which and with whom data is shared	PPP, PFI, PC...	Yes or no Public value testing

Figure 20.1 Business and public parameters matrix. (From Walravens, N., The public enemy? A business model framework for mobile city services, *Proceedings of the 11th International Conference on Mobile Business*, Delft, the Netherlands, June 21–22, 2012.)

the other. All the information presented here has been gathered through desk research and comes from official policy documents, as well as academic case studies and analyses in specialized media. The first type of platform that is investigated in the context of the city is the Enabler Platform.

20.5 Enabler city platform

Considering the definitions of the control parameters mentioned earlier, we see a few potential models that can be labeled as Enabler City Platforms. One is the case of a city providing open data and statistical information to interested developers. The city does not control the service offering as services are developed by third parties but does control assets in the sense that it decides which data is made available and under which conditions. There are several possible approaches to opening up data: a city can create an online platform where anyone is free to download raw information on the city and create services on top of it. A slightly more closed model requires interested developers to apply with the city based on the service they have in mind, before being allowed access to the data. Whereas some cities take a cautious approach and do not go very far in supporting the developer communities that might be interested by placing particularly large quantities or interesting types of data online, others go further by attempting to stimulate and often create a developer community.

Both in the United States and Europe, the number of cities opening up datasets to developers is steadily growing and the number of cases is *legio*. Depending on the implementation and resources devoted to the initiative, some cities have been more successful than others, inspiring developers to create innovative applications or visualization. In Europe, cities have been obligated to begin opening up datasets under impulse of the European Commission. The Re-use of Public Sector Information Directive, created by the European Commission, was a first step to open up data from the public sector to citizens. The 2003 Directive was implemented into national law of the 27 Member States by May 2008 and "encourages the EU Member States to make public sector information available for private and public organizations as well as for citizens" (EC, 2011). However, the Directive only yielded limited results, as the text remained vague on what types of datasets and in which formats the Member States needed to open up. In December 2011, the European Commission again stressed the importance of open data, even describing it as the new gold with the launch of an "Open Data Strategy," building on the work in the 2003 PSI Directive (EC, 2011). In this perspective, cities truly act as *enablers*, offering relevant information to developers and stimulating them to create new value. An additional way to stimulate a developer community to create quality applications is organizing a competition linked to open data usage, often referred to as Apps for X. These hackathons are being organized all over the world, including in cities like New York, Washington DC (one of the first cities to set up such a competition), Amsterdam, Berlin, Paris, London, and so on. These types of initiatives can encourage the development and deployment of city services in a way that is relatively easy and cost-effective to implement for a city government.

20.5.1 Case: London Data Portal and the London Bike App

The case we selected is the city of London taking up an Enabler Platform role by creating the London Data Portal on which open data about the city was made available to developers. As an example, we include the London Bike App that was created using the data distributed by the city, to illustrate the potential services coming out of these initiatives. London Bike App allows end users to see the live availability of bikes in the sharing system exploited by Transport for London (TfL), the city's official public transportation organization. The application is available for iOS only and costs €0.79 or £0.69. It was developed by an independent developer called Big Ted, Ltd., who made use of the City of London's open data initiative to create the service and keep its information up to date in near real time.

20.5.1.1 Control and governance issues

One of the most important assets in the creation of the service offering of this application is the actual real-time data, which is controlled and offered by TfL and the City of London. In January 2009, TfL launched a so-called Developer Area on its website that gives access to datasets and real-time traffic and public transport information. There are currently 30 feeds available in TfL's Developer Area that can be accessed after information on how and where they will be used as well as the target audience and the estimated audience is provided (Transport for London, 2012a). The value network is somewhat integrated as the data is controlled and choices for availability are made by TfL, but it is relatively easily available. Control over the customer does not lie with TfL, nor the City of London, but rather with the application and its developer, as TfL explicitly prohibits app developers that make use of the real-time data to use TfL branding, templates, or house styles (see the following text). The bike system is heavily sponsored by international finances firm Barclays and is, as a result, mostly known as the Barclays Cycle Hire system in London. Developers are allowed to use this branding (under certain conditions) and so some of the customer relationship may be shared with Barclays. In this case, the developer does not require a registration from end users, so there is no profile or identity management occurring on this side. Although the city is not directly involved in the creation or distribution of this bike application and has no official link to it, the fact that it was possible to develop it is a direct consequence of the city's open data policy, which aligns with certain good governance principles. At the launch of the London Data Store in January 2010, Mayor Boris Johnson emphasized the importance of transparency of government, saying "sunlight is the best disinfectant" (London Data Store, 2010). In a first step, the mayor made all expenditures of the Greater London Authority (GLA) more than £100,000 transparent and available online in an effort to give citizens more insight into government spending and reduce it at the same time. This led to the launch of the London Data Store, an initiative by the GLA to open up as much data it holds as possible. In a debate at the launch of the London Data Store, the late Sir Simon Milton (London's Deputy Mayor for Policy and Planning at the time) highlighted other elements related to good governance, saying the open data initiative was aimed at *bringing significant public benefits* and *improving quality of life* (London Data Store, 2010). He spoke about a "post-bureaucratic age in which decision-making is put back into the hands of the public, that can base itself on correct and open information on government" (London Data Store, 2010). Additionally, the importance of stimulating the local economy was also highlighted as a potential benefit of opening up data, by encouraging local developers and creative people to explore commercial opportunities. These principles are framed in the Freedom of Information Act 2000, enforced by the Information Commissioner and aim to ensure an open government. This legislation addresses public authorities and places two main duties on them: "to adopt and maintain a publication scheme setting out the classes of information which the authority publishes and how it intends to publish the information, to respond to individual requests for information under the general right of access to information, from January 2005" (London Data Store, 2012a). This act gives individuals the right to request any government data, unless it is privacy sensitive or a matter of national security. The GLA aims to involve as many public stakeholders as possible in the Data Store initiative and clearly states it "is committed to influencing and cajoling other public sector organisations into releasing their data here too" (London Data Store, 2012b). Since the TfL started publishing some of their data feeds up to a year before the London Data Store was opened, it was a logical partner and links to the TfL Developer Area are available in the London Data Store as well.

Although the TfL makes an effort to release data that is at least easily machine-readable, the use of interoperable standards varies depending on the type of data. In the case of the Barclays Bicycle Hire, the information is available in an XML-feed since June 2011. This feed allows developers to relatively easily access real-time availability data from the bike-sharing locations and includes the name, location, coordinates, and maximum number of docking points for all operational docking locations, as well as the number of available bikes and available free docking points. Before the feed was made available, the developer of the London Bike App scraped it off the TfL website, using a Google App Engine application (Wise, 2010). Today's feed is updated every 3 min and allows for a maximum of 30 min between capturing and displaying a feed (Transport for London, 2012b). The information is free to use, scalable and robust, as well as the source of data for TfL's own website. It comes with a data dictionary that explains the terminology used in the XML-feed, to make the process easier for developers (Transport for London, 2011). By making use of the XML-standard and providing documentation to developers, TfL makes an effort toward technology governance by being relatively inclusive in its approach to opening up data: Developers with some minor experience should be able to make use of the data. This of course does not naturally imply that the available data is easily accessible or understandable to the general public. It is up to the developer or web designer to take the data and make it easily interpretable to the public, and this is where some issues of inclusivity may arise: Although the open data feed is publicly and freely available, developers are free to create commercial applications with them. TfL clearly states, "Can I charge people to use the app I develop using your feeds?—Yes, you are welcome to charge the public for applications you produce using TfL syndicated data" (Transport for London, 2012c). Given this (understandable) policy, developers may be inclined to develop for the largest platforms only, because there may be the most revenue to be made, or simply because they are more skilled in creating software for these platforms. This is the case for the London Bike App, which is only available on iOS, due to the developer's technical experience with the platform (Wise, 2012). Although the GLA does not make recommendations or suggestions toward developing for multiple platforms, it does have an internal Accessible Communications Policy that aims at combatting *the effects of racism, social exclusion and the removal of other discriminatory barriers to full participation in society* (GLA, 2012) by providing translations of documents, help with interpretation of documents and ensuring all GLA websites are available in all browsers and are compatible with all systems. The developer of the London Bike App does encourage creating applications for other platforms or cities and has made the source code for the service available to anyone interested (Wise, 2011).

20.5.1.2 Value and public value issues

The investment in the London Data Portal is carried by the GLA and the platform does not have a direct revenue model, but it is rather aimed at stimulating a developer economy in the city. Initially, the GLA also partnered with Channel 4's 4iP investment fund that made £100,000 available for innovative applications that make use of public data. The program was however stopped after a Channel 4 restructuring at the end of 2010 (Channel 4, 2010). The return on public investment expected by the GLA for its open data initiative was already highlighted earlier and is mainly focused on indirect financial return in the form of an active London-based developer community that attempts to capitalize on the commercial opportunities created by opening up datasets. Next to this multiplier effect, there is also a strong focus in the rhetoric on improving the quality of life of the public and increasing transparency in government (see preceding text). The London Bike App could be seen as a successful product of both these goals. The data is licensed and provided to anyone interested and no particular public–private cooperation model is set up.

There is potential for user involvement, dependent on their level of app development skills or education in understanding and interpreting the data made available by the government. Otherwise, this aspect is less present both in the open data initiative or the specific app in question. The intended value of the cheap application (apart from offering an app for free, €0.79 is the lowest amount developers can charge in the iTunes App Store) is providing better access to the Barclays Bicycle Hire system that consists of 6,000 bikes in 400 docking stations across central London (Transport for London, 2012b). The app can provide a list of bike availability at the 20 docking stations closest to the user's location as well as a map view that offers a real-time overview. Since the Barclays system uses a variable pricing scheme based on the time of day, the app also provides a timer that shows the user when prices will increase, together with a price estimate of the user's current trip. TfL is very clear in its materials that developers cannot position the services and applications they develop as being officially endorsed by or affiliated with the transport company. To this end, TfL have composed clear and detailed guidelines for developers on the use of TfL logos, brands, or house styles (e.g., Transport for London, 2012d,e). These guidelines deal with the use of logos, fonts, lettering, colors, and so on not only in developers' promotional materials but also in the applications' user interface (Transport for London, 2012b).

The preceding paragraphs have already provided an overview of the types of public value the city wants to create as a result of this initiative. The London Bike App is an example of how both goals of stimulating developers to create innovative apps and improving ease-of-use of public services can be achieved by open data initiatives and the Enabler Platform role played by the City of London. While no particular market justification motivation is used by the city and commercial organizations are also welcome to share data in the London Data Store, given the already public nature of the data, no such motivation is really required or expected. The developer does not provide adequate data to determine whether the app can be evaluated as a success, nor is it clear in which ways the GLA evaluates the London Data Store or has developed metrics or KPIs that identify whether the initiative has successfully or sufficiently reached its goals.

20.5.1.3 Discussion

It is clear in this case the city plays a platform role, since it mediates between at least two sides of the ecosystem: the transport company and the data it holds on the one hand and developers looking to create innovative services on the other. The city remains in control of the data or infrastructure required to create the value proposition, as shown by answering the guiding questions introduced earlier: It remains the owner of the data that is opened up; it provides clear guidelines to developers, stating what use of the data is allowed and goes as far as detailing what elements of the GUI should look like; and the initiative and opening up of data is clearly part of a wider policy approach and is supported by the GLA. The city, however, does not control the service offering that is created as a result of opening up data: The developers involved are totally external to the organization; the city only has a very limited amount of control over the final contents of the services, apart from what is mentioned in the developer guidelines; the city does not have access to users' profile or identity information; and no specific quality control is performed by the city before an app is released. Having control over the data and infrastructure, but not over the service offering, leads us to label this case an example of an Enabler City Platform. The case demonstrates how a city can play a facilitating role in an attempt to make sure developers create applications that have use for citizens and can improve their quality of life, that is, by increasing accessibility to public transport information.

Still, an important question that remains with this platform role is what the return is for the city itself. Other than intangible effects, the financial return of simply opening up data is likely to be low, unless close cooperation with private partners is possible. Thus, another approach an Enabler City Platform could take is an outsourced model in which the city does control and provide the relevant assets, but not the service offering, through close cooperation with a private partner. An example of such collaboration can be found in the Google Transit service, which is available on fixed and mobile devices in several major cities around the world. The service provides the user with a number of public transportation options, updated in real-time and placed on the Google Maps interface. Such a service is only possible when a city's public transportation organization works closely together with Google in integrating the, often different, technical backends and offering them to the end user in an attractive way. This type of more closed, direct outsourcing of service creation and provision, while providing the essential raw data, may also constitute an Enabler City Platform.

20.6 Integrator city platform

In an Integrator City Platform, the city remains in control of both the data and infrastructure and provides the service offering. To some extent this is a more closed approach where the city has the highest form of control on the service development and distribution. In the general description of this platform type, it is stated that attracting third parties such as developers is encouraged, and this should not be different in the case of city services. However, when personal data is concerned, it is to be expected city governments will be more careful in sharing this with third parties. This means an Integrator City Platform is likely to crystallize in a more closed form than in the general typology. It also means the Integrator City Platform takes up more roles than was the case in the Enabler model: It has to manage the relationship with the end user and will, depending on the specific case, as a result also create, host, and/or provide services to the citizen. This brings with it an increased cost and the use of resources, which will need to be intercepted by the city government. In return, however, the city gains insight into life in the city and the concerns of its inhabitants and can perform aggregated statistical analysis on the data it collects through the integrated service to deal with policy issues it might want to tackle.

20.6.1 Case: NYC 311

One interesting example, with a very low threshold for citizens, can be found in New York City's "311" service. NYC 311 was launched by the Bloomberg administration in 2003 as a centralized call center, tasked with unifying the nearly 4,000 services offered by over 120 city agencies and organizations (Accenture, 2011). The service is a citizen's first point of contact with the city government for all questions and issues that are not an emergency (for which one would call 911 services) and quickly became successful. This led to the launch of multiple channels to reach 311, that is, an online portal, a text message service, Skype account, Twitter account, blog, and iPhone application (Chaudhry, 2011), all developed as a direct assignment by the city. Today, the service receives around 50,000 calls a day, serves 8 million citizens and reached the milestone of 100 million treated calls in May 2010 (Johnson, 2010). While not solely a mobile offering, NYC 311 offers several functionalities and services related to the citizen's location or living environment. The investments in a texting service, the development of mobile applications, and integration with social media and data-based communication services (such as Skype, Facebook, and Twitter) underline this.

20.6.1.1 Control and governance issues

The mayor's office decided on a hierarchical approach and integrated value network in the organizational structure of the 311 service in which the control over assets of the service is centralized. The service is a business unit of the Department of Information Technology and Communications (DoITT), which is headed by a Commissioner, appointed by the mayor. Day-to-day operations are managed by the 311 Call Center Director, who oversees various staff sublevels. A major contribution to the success of the service appears to lie with the role the mayor played in unifying the over 45 different call centers of the existing city agencies: He imposed a short, 1-year deadline for the launch of the service and mandated that all Commissioners of existing agencies participate in the initiative. Individual agencies were not able to *opt-out* of having its services or information handled by 311 (DiGiulio, 2008). It is likely to assume this complete and obliged public stakeholder selection is one of the main factors contributing to the success of the service. The Department of Housing Preservation and Development's services were the first to go live through the 311 system in 2003, with the service quickly growing to bundle information and services from over 30 city agencies. A second important characteristic of how 311 works together with public stakeholders is in its direct interaction with the 59 local representative bodies known as community boards that represent specific geographical areas of the city. These boards play an important role in land use and zoning, local services, and the general welfare of their community. New York City's 311 liaises with these community boards on a frequent basis, to ensure that particular local issues and complaints reach the correct city agencies (DiGiulio, 2008). Furthermore, 311 works closely with New York City's HHS-Connect initiative, which integrates the information systems of the city's various health and human services agencies and has begun to introduce links and navigation to state and federal services (e.g., when a citizen wants to apply for a driver's license, he is referred to the DMV, a department that falls under the state's competences) (DiGiulio, 2008).

Aspects related to good governance appear to be an integral part of New York City's 311 service as they are in the mission statement of the call center and online service. The administration of mayor Bloomberg lists accessibility, accountability, and transparency of city government and its services as core principles of an open government (DiGiulio, 2008), concepts often used in public governance literature. The city describes these principles as follows:

- "Accessibility—The 311 Customer Service Center provides residents, visitors, and inhabitants of the City with one number to call to access all New York City government information and services while, at the same time, providing a superior level of customer service. Open 24 h per day, every day of the year, and available in any one of 179 languages, the 311 service connects constituents with the appropriate city services and information they have requested—they do not need to know what agency handles their request—they just need to know what issue or question they have, and 311 will direct their inquiry or request to the appropriate party for a response.
- Accountability—The 311 Customer Service Center helps City agencies improve their delivery of services by handling the customer service and call center functions of the service delivery process. In this way, each Agency is able to focus on its core mission and area of responsibility and manage its workload efficiently.
- Transparency—Through accurate and consistent measurement and analysis of service delivery, the 311 Customer Service Center provides insight into ways in which City government can be improved and made more efficient. The city uses data from the 311 Customer Service Center along with Business Intelligence tools and technologies to provide increased visibility into its operations. Whether it's a scorecard indicating an agency's performance, or easily obtained information on a service request made through the 311 center, this information is conveniently available to all constituents" (DiGiulio, 2008).

Not only through the constantly available phone number but also through the variety of other media the 311 service uses, such as social media and online tools, the city government has made a point of being as accessible as possible. Of course, such availability takes time and comes at a cost, resulting in a gradual expansion of the number of ways to reach 311 (e.g., some mobile platforms like Android do not have an official mobile application available yet). The bullet on accountability highlights how other government agencies may benefit from the 311 service by being able to focus more on their core responsibilities, but this works in two directions: The data gathered from the calls and requests coming into 311 may also bring to light how certain city agencies or organizations have been less than successful in their core assignments and may resist the notion of such a service. Additionally, the customer information, which is gathered by the service (such as people's phone number and location), places the relationship with the citizen with 311 as well.

One of the focal points for the service is the high level of customer service it has aimed to offer since its inception. Added to the three principles listed earlier, it becomes clear the city set out to achieve a concrete policy goal (bundling and improving scattered existing city information organizations and services) while adhering to some of the principles essential in what is understood as good governance and following an integrated approach when it comes to selecting and managing the stakeholders involved.

As far as technology governance is concerned, it appears the city of New York makes an effort to be as inclusive as possible toward its citizens. The 311 service originated as a phone number, which can be called free of charge, 24/7, and today offers a text messaging service, various social media account, an iPhone application, and is reachable via TTY or textphone (a device for the hearing-impaired) (Chaudhry, 2011). As far as the official applications developed by the city of New York are concerned, there seems to be a focus on iOS development at the time of writing, with only 2 out of 10 official apps available for both Android and iOS (NYC Digital, 2012). There also appears to be some criticism on the stability and use of the 311 iOS application, even by the mayor himself, as only around 4,000 requests have been made through the app, which was downloaded 23,000 times since its launch in 2009 (Tiku, 2012; Weber, 2012).

One of the most important criticisms on the three core principles discussed earlier is related to transparency and the fact that only a limited set of data gathered through 311 is openly available to developers and the public. While the city of New York does operate an open data portal, NYC DataMine, for a long time only a limited amount of data available there came from 311. While the DoITT is obliged to distribute monthly reports to the city council and local community boards, as well as the public, the data was aggregated and, for example, not machine-readable, allowing little further analysis or development of new services or visualizations. This critique was repeated when the city launched a map of 311 requests in February of 2011, but did not provide access to the raw data, nor the time series of the data giving insight into the evolution of requests and complaints (Judd, 2011). Although a data-driven approach was being pursued in the internal organization of 311 (e.g., sharing data and feeding results into NYCSTAT, the city's tool to track and measure data (Beck, 2011) focusing at least on internal interoperability within the city's services, or the development of an internal tool able to compile data on calls received, selected services, open tickets, and the status of requests, this raw information did not trickle through to the public (DiGiulio, 2008). At the end of 2011, the city began offering an NYC 311 application programming interface (API) to give developers quicker and easy access to the data generated by 311 in JSON and XML format (DuVander, 2011). This illustrates that after a period of hesitation on opening up specific datasets and a certain need becoming increasingly apparent with the public, as well as an international move toward open data (see e.g., DiGiulio, 2008), the city of New York has slowly begun adopting an open approach to the data its services generate.

20.6.1.2 Value and public value issues

As far as financial aspects and the expected return on public investment is concerned, the main focus of the 311 project seems to lie in customer satisfaction and offering a high quality of service, rather than expecting a particular financial return. By consolidating the previously over 40 call centers, the city was able to reduce the distributed costs of these centers and was able to operate more cost-effectively, as it had a better overview (DiGiulio, 2008). In another effort to reduce costs, the city opted for the multichannel approach, which required some initial investment, but is likely to save a lot of money in the longer term. For example, by offering the online 311 system where citizens can submit their requests themselves, some pressure is taken away from the call center, which is constantly seeing an increased call volume. By letting citizens submit requests themselves, the call center does not need to hire more staff as requests increase, keeping costs balanced (DiGiulio, 2008). As far as the investment structure is concerned, the service's operating costs are carried by DoITT, which—as a mayoral agency—is funded directly by the city budget (DiGiulio, 2008).

The main (somewhat intangible) return for the city is the large set of data gathered from the calls and requests coming into 311. By logging, mapping, and tagging all requests coming into the service via its various channels, the city builds up rich information from which it can distil trends or structural issues in particular neighborhoods (Johnson, 2010). When such trends are identified, policy can quickly be altered or tailored to the specific needs of that area. The city does not list particular expected multiplier effects in official policy documents, but given the breadth of the service, it is likely to assume some secondary benefits arise as a consequence of a more efficient handling of citizen requests and concerns by government. For example, it is estimated that because of 311, the burden on the 911 emergency call centers has been reduced by 4 million calls between 2003 and 2009, allowing those centers to operate more efficiently and to focus more on their core tasks (Accenture, 2010). Some duplication of existing services has also been resolved by 311 (Accenture, 2010). Any financial gains from the service (its revenue model) are thus expected to be indirect and the result of increases in efficiency.

In order to develop and deliver the 311 service, the city has set up partnerships with various private actors. One of the largest partners in first developing the call center, followed by the online portal in a later stage, is the management consulting and technology service provider Accenture. The company was contracted by DoITT to provide technical architectures, testing and deployment capabilities, the design, building, and implementation of the online portal, and a large-scale integration of services. Together with the DoITT, the company built a searchable knowledge base and taxonomy of city, state, and federal services; built a new, integrated call center operation; and introduced several operational processes using Oracle Systems technology such as Siebel Customer Relationship Management applications and content management tools from Interwoven (Accenture, 2010). Additional partnerships include an outsource vendor who handles overflow calls when the 311 service sees a peek in calls (e.g., on busy days with extreme weather conditions like snow or heat), a contract with Language Line, which offers translation and interpretation services to be able to offer 311 in 179 languages and a close collaboration with the City University of New York to provide part-time jobs and internships to over 130 students (DiGiulio, 2008).

In the framework of the ideas on open data (briefly touched upon earlier) and the open source movement, there is some criticism on this form of contracting between the government and private companies, particularly when it comes to the reuse of data. When private companies control the data generated by a particular service, they will be less inclined to freely open it, but would rather look to sell it (Farley, 2012). Another specific concern is revenue-sharing contracts that are quite popular in city technology services and in which the city pays the supplier a fixed amount for a product or service, plus a percentage of what the city has as income over a longer period of time. When the city of Chicago privatized its parking meters in 2008 as a quick means of dealing with the recession, it got $1 billion from Morgan Stanley under a revenue-sharing contract. However, the company estimates making around $11 billion over a longer period of time, based on the revenue share (Farley, 2012). As far as the city of New York is concerned, the spending on private contracts has almost doubled between 1996 and now, from $5.7 million to $10.5 million, with an increased focus on technology companies and a limited amount of transparency as to how these contracts come about (Farley, 2012).

Since 2008, 311 organizes a public evaluation of its services. The first survey showed that citizens were pleased with the service as it scored on par with the highest ranked private call centers and much higher than other government call centers (Chaudhry, 2011). A survey at the end of 2011 showed that customer satisfaction had only increased over the years, with the remark that more could be done to ensure accessibility for all (Chaudhry, 2011). 311 also organizes an internal evaluation in the form of key performance indicators (KPIs) that need to be met. These mainly relate to the speed with which calls are answered: 90% of calls should be answered within 30 s, with a maximum answer delay of 3 min (Beck, 2011; DiGiulio, 2008). Apart from the general satisfaction survey and the internal KPIs, there is no broader evaluation of the public value generated by the service and there is apparently no tradition of organizing a public consultation or test of potential new services by the government.

20.6.1.3 Discussion

The City of New York plays the role of a platform in the sense that it mediates between citizens, its various internal organizations and services and other actors in the city. When looking at the city as a platform in the case of 311, it becomes clear the city controls both the data and infrastructure required to create the value proposition as well as the service offering, meaning we categorize it as an Integrator City Platform. The city contracts certain technological aspects with large companies (like Accenture and IBM) and mostly keeps control of the data that is gathered via the service, allowing the structural analysis of the information that comes in, to adjust policies. Although the merging of the different call centers and several efficiency-increasing exercises has cut costs, running and supporting such a widely available service is quite cost-intensive for the city government. Furthermore, the city has created an API to open up some of the data gathered via 311, but this means the city remains in control of how data is opened, under which circumstances and with which limitations. Setting up 311 was part of a longer-term policy road map to increase QoS to citizens, meaning it is not an ad hoc initiative.

The service offering is also controlled by the city as 311 is clearly branded and marketed as a city government service and there is no charging or billing relationship as the service is free. General identity information is used and controlled by the city as well, as a means to improve the service. Many of the services offered under the 311 moniker are developed either in-house by the city or as a result of direct contracting. The city controls the contents of the application and when it is not satisfactory, the developers are called upon to augment the services, as was illustrated earlier (Tiku, 2012). NYC311 takes an integrated approach, resulting in a service of relatively high quality, that is—according to the surveys cited earlier—appreciated by citizens.

20.7 Neutral city platform

In this platform type, the city does not control data or infrastructure assets, nor the service offering. In real life this would crystallize in a city not taking any particular initiative toward deploying mobile services, leaving either private projects to provide the tools available to citizens, with the commercial logic such initiatives in many cases assume, or to more bottom-up projects where enthusiastic individuals or nonprofit groups organize themselves to create city-related services. It is of course a distinct possibility that there is no interest in building out a mobile service offering at the policy-making level of a metropolitan area or that city governments feel paralyzed by the fast pace and technical challenges in the sector. One example of such a neutral platform initiative is the Open 311 set of tools that are being developed since 2010 and supported by Vivek Kundra, the Chief Information Officer of US Government. Since no cities are directly involved in the creation of Open 311, it is less relevant to discuss all the parameters from our framework and so this analysis will be somewhat limited. However, we will provide a description of the various aspects of the standard and the stakeholders involved.

20.7.1 Case: Open 311

Open 311 aims to provide a common and open specification to build services on that allows citizens to report nonemergency issues. The idea behind Open 311 is not developing a single API or technology, but rather providing an open platform (Ashlock, 2009) that allows issue-reporting services in different cities to use the same middleware layer to handle requests, before they end up in the city's own issue-tracking backend system. By using the Open 311 specification, cities can allow third-party developers to create 311 services that perhaps use a different approach, graphical user interface, or visual style than the city's official application. It also allows developers to create services for mobile platforms the city is not present on (in many cases, e.g., Blackberry or Windows Phone), which can be useful for cities when the cost of developing official applications for many different mobile operating systems becomes too great. Since Open 311 is being used by more and more US and European cities with issue-reporting services, it also means these applications created by third-party developers do not have to be linked to one specific city: When Open 311 is used, these services work for all cities that also use the set of standards (mySociety, 2013).

The idea for the specification first came from the OpenPlans organization, a nonprofit consisting of several developers and strategic planners, with the goal of supporting cities in their plans to improve communications and relationships with citizens through technology (OpenPlans, 2013). The idea was quickly picked up and further development of the service is also supported by Code for America, a nonprofit organization funded by several companies, cities, foundations, and donors. Code for America brings together developers, city experts, and enthusiasts to support cities to think about innovative services for their citizens (Code for America, 2012). The organization was started in 2010 and offers a fellowship program for enthusiastic individuals (developers, designers, planners, strategists, etc.) that gives them the chance to work closely together with a city for 1 year, on a project they develop that can help cities to innovate, be more open, increase efficiency, and have impact on society. Ashlock (2009) was directly involved with the development of Open 311 as a presidential fellow, and continues to push for the implementation of the specification in as many cities and issue-reporting services as possible. At the time of writing over 30 cities—including San Francisco, Boston, Baltimore, Washington D.C., and Chicago—supported the Open 311 standard (Open 311, 2013) with many of them encouraging developers to create new applications for the cities' 311 services. In Europe, Open 311 is supported by mySociety, the charitable group behind FixMyStreet, an issue-reporting service that is available in an increasing number of cities on and off the continent (mySociety, 2013).

20.7.1.1 Discussion

The main parties involved with creating, supporting, and developing the Open 311 platform are neutral and nonprofit in nature, with no cities directly involved in their operation. However, Open 311 has a direct impact on individual cities in making the 311 services they offer more accessible to citizens through new channels. Open 311 does not control the service offering to individual citizens, as many people will be unaware of the platform's existence and the tools and resources are made available by the various nonprofits involved. Since Open 311 does not have a direct relationship with the citizen and gathers different APIs and tools to make issue-reporting services interoperable, rather than directly manage any data or infrastructure, it can be categorized as a Neutral City Platform.

20.8 Broker city platform

The Broker City Platform would entail a city has control over the service offering, but is not in control of the assets required to assemble the value proposition, that is, the data and infrastructure. In a real-life setting, this could, for example, mean that the city is involved in some limited service or application development, but does not have control over the required data or infrastructure to put together that service.

20.8.1 Case: I Amsterdam QR spots and Amsterdam partners

City marketing efforts that include the creation of mobile services can serve as illustrations of a brokering role being played by the city. This strategy is clearly present in the city of Amsterdam where a group of organizations have come together under the "I Amsterdam" label to promote the city. The group calls themselves Amsterdam Partners (2011) and consists of governmental, regional, sectorial, and business organizations responsible for the promotion and city marketing of Amsterdam. These organizations provide the data and tools needed to create websites, mobile applications, organize events, advertise, and so on, under the "I Amsterdam" moniker, with the goals of promoting the city both locally and on an international scale. Amsterdam Partners, for example, created the I Amsterdam QR Spots application for iOS. It was launched in January 2012 and offers a quite complete touristic and historical guide, the location of WiFi-hotspots and a QR scanner that allows end users to scan QR codes posted at diverse important locations in the city. However, the city itself did not have direct access to or control over the various datasets and pieces of information required to create the actual service, as it was distributed among the network of Amsterdam Partners.

20.8.1.1 Control and governance issues

The control over the data required to create the value proposition lies with Amsterdam Partners, a large group of stakeholders that requires a relatively strong integration of the day-to-day activities, organized by a management board. A supervisory board, represented by the mayor of Amsterdam, oversees these activities. Although the approach is integrated, it does not place all decision power with the municipality, but rather focuses on clear rules and cooperation. This is described in the policy document detailing the city marketing approach:

> Our starting point has been that the city does not need to manage the entire programme, but should clearly guide a limited number of key issues. The city also has the task of assisting in the clear division of tasks and responsibilities to provide a boost to new issues and in those areas that are forgotten by others. Organisations involved should have the encouragement and opportunities to take their own initiatives but should also know clearly where the city is heading and what the game rules are. These rules will be set up cooperatively with companies, intermediates, the region and partners.
>
> **City of Amsterdam (2004, p. 28)**

Related to the good governance parameter, the main focus in the rhetoric of Amsterdam Partners seems to be on city marketing and attracting new businesses and visitors to the city, as well as promoting certain activities or events more to residents. The application is one tool in reaching the goals that are set out in the consortium's annual report (Amsterdam Partners, 2012):

- Optimize the potential of the Amsterdam brand and the Amsterdam Partners network
- Ensure that the "I Amsterdam" motto acquires optimal visibility
- Ensure that the "I Amsterdam" motto acquires optimal significance

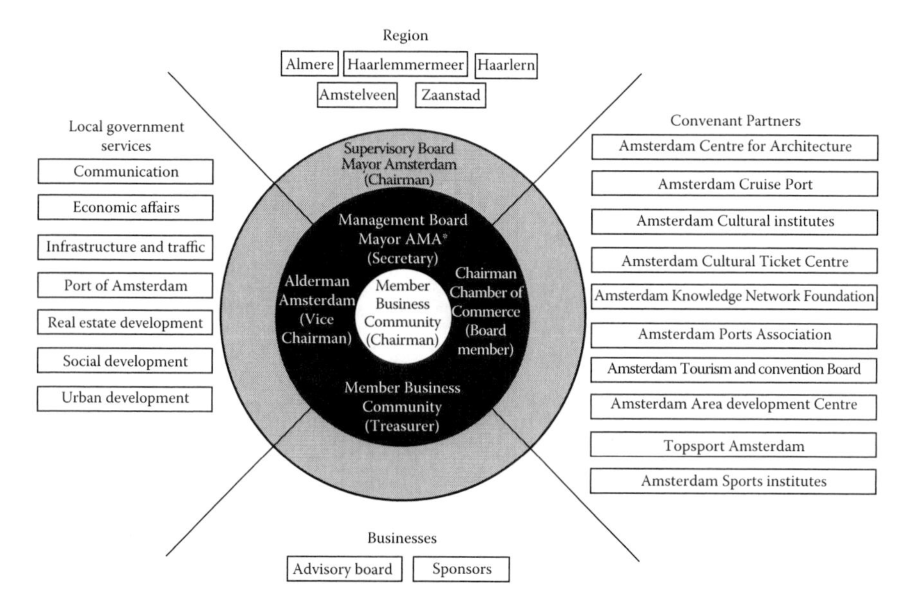

Figure 20.2 Overview of the Amsterdam Partners organization. (From Iamsterdam, Organisation, *Iamsterdam*, available online: http://www.iamsterdam.com/en-GB/Amsterdam-Partners/About-us/Organisation, 2012.)

These three aims are interesting as they clearly illustrate the intention of the city to establish the direct relationship and interaction with the citizen, through the concerted city marketing effort. The stakeholders involved are organized in an interesting construction that brings together diverse actors in the city that could benefit from an optimized city brand. The main organization involved in the development of this application is Amsterdam Partners. This PPP-construction (see following text) brings together local government entities, businesses active in the city, local organizations, and representations of the cities surrounding Amsterdam. The organizations involved are represented in the Figure 20.2 (Iamsterdam, 2012).

These organizations are brought together under a management board and supervisory board, chaired by the mayor of Amsterdam. This even visually illustrates the brokering role the city of Amsterdam and the office of the mayor take up in the city's marketing plans, by mediating between many different types of organizations. Additionally, Amsterdam Partners works closely together with other Amsterdam organizations such as the Amsterdam Tourism and Convention Bureau (ATCB), the Amsterdam Uitburo (AUB) and Amsterdam in Business (AIB). These organizations gear all their activities to one another and present it to the public under the "I Amsterdam" moniker, which was launched on September 23, 2004. To further increase cooperation, these organizations have been merged during 2012 and started activities as a single legal entity from

January 1, 2013, onward (Amsterdam Partners, 2012). The main stakeholders involved in the development of this specific mobile application were the City of Amsterdam Economic Affairs, the Amsterdam Bureau of Monuments & Archeology, the City Archive and the ATCB. One hundred and thirty-nine signs made out of traditional Amsterdam ceramics, containing a short explanation in Dutch and English as well as the QR code, have been placed along important monuments or areas of the city and its surrounding neighborhoods. A Dutch advertising company called Beedesign was responsible for the fixtures and placement of the signs, which were designed and developed by (van Dijk, 2011).

The data and information that is presented to end users in the service offering is not open and provided by the partnering organizations in the I Amsterdam consortium: In this specific case, mainly the Amsterdam Bureau of Monuments & Archeology and the City Archive. Although the information may be publicly available from other sources, it is not the result of an open data initiative (although the city of Amsterdam does have one) and does not foresee interoperability with other services. The application itself is developed and supported by XS2TheWorld, a Dutch internet company with international branches that has international companies such as McDonald's, Coca-Cola, Sanoma, and Porsche as its clients. The developer was contracted by Amsterdam Partners to create and provide support for the I Amsterdam QR Spots service.

20.8.1.2 Value and public value issues

The main investment for the operations of Amsterdam Partners comes from the city of Amsterdam, its surrounding municipalities and contributions from industry partners. In 2011, Amsterdam Partners received almost €1.4 million in grants and contributions, combined with just under €150,000 in revenues from other sources such as retail of merchandise and interest, bringing the total budget of the organization to €1,528,286 in 2011, a slight decrease compared with €1,707,742 in 2010 (Amsterdam Partners, 2012). No further detail is available and it is unclear how much was needed for the development of the application and the placement of the QR codes around the city. The application itself does not have a revenue model as it is offered for free to end users by the city of Amsterdam. Although the expected returns are not directly financial, they are certainly defined as indirect commercial or economic indicators, mainly related to an increased use of city products and services by visitors and residents. This is expressed in the goals Amsterdam Partners sets: *Investments in the Amsterdam brand effectively translate to positive results in numerous fields including revenue, visitor numbers, investments, image and global market position. In turn, this will result in an economically, socially and culturally stronger Amsterdam* (Amsterdam Partners, 2012). Amsterdam Partners describes itself as a public–private foundation, but its legal status, details of the cooperation and how finances exactly flow between the partners remains unclear at the time of writing and would need to be detailed in further research. The service is positioned according to the philosophy of the I Amsterdam project and its target audience is summarized in the annual report as:

- International business world (decision makers)
- Visitors (tourists and business people)
- Residents (the bearers of the motto and ambassadors of the Amsterdam Metropolitan Area) (Amsterdam Partners, 2012)

There is a very limited aspect of user involvement present in the application as end users can scan the QR codes to receive additional information on a certain location. However, the application is also usable and provides the same information without making use of this functionality. There is thus a *gimmick* aspect to this part of the service, although the presence of the QR codes in the city infrastructure may inspire visitors and citizens to make use of the application more. The intended value of the service is explicitly focused on disclosing the historical background of certain areas of the city and specific landmarks, as well as tips on restaurants, bars, or other experiences in the end user's area:

> Visitors with a smartphone can scan the QR code to reveal additional information about each location, including background stories, little-known secrets, historical photos or pictures of the interiors. Based on your current location, you can also get tips about other points of interest in the neighbourhood. There are no fixed routes—instead, you can explore sections of the city at your own pace, based on your interests.
>
> **I Amsterdam (2012)**

The popularity of the I Amsterdam QR Spots applications is hard to measure, given its relatively young nature and the fact that the iTunes App Store does not provide measurements to the public. The two previous apps developed for I Amsterdam (a full touristic guide including similar information and a light version that consumes less data) were already included in the 2011 Annual Report of Amsterdam Partners and were downloaded 120,000 and 60,000, respectively (Amsterdam Partners, 2012). Although the services are briefly evaluated in the annual report of Amsterdam Partners, the expected returns and generated public value appear to be quite vague and are referred to as increasing the visibility of the city and its promotion. Hence, it is hard to judge whether the mobile applications contribute to this and to what extent.

20.8.1.3 Discussion

Given the strong focus on city marketing, the Amsterdam Partners (and thus the city) puts forward, the platform role played by the city is that of a broker. The assets required to compose the value proposition are distributed among the many partners of the city and brought together in the Amsterdam Partners consortium, whose long-term interests are commercial in nature. The data required to put together the service is not directly under the control of the city's central administration, as it is owned and provided by the Amsterdam Bureau of Monuments & Archeology and the City Archive. The data is not openly available and so no particular license is available, nor an API. Control over infrastructure is also distributed as in some cases the monuments involved are owned by the city and private property in others. The I Amsterdam initiative is a large part of the city's policy plans related to city marketing, and the mobile application is a very small part of that.

There is a strong emphasis on the branding and promotion of the city, meaning that the city uses different types of service offerings to create a strong relationship with citizens, of which the mobile app under discussion here is one. The city has control over the service offering as it is developed under a direct assignment by an external developer and does impact the content of the service to some extent since it is determined and provided by partners in the Amsterdam Partners network. A form of quality control is present to make sure the I Amsterdam brand is represented in a positive and consistent manner.

20.9 City platform typology

The following table provides an updated view of the platform typology to address the elements particular to the context of the city that were identified and illustrated by the case studies provided earlier. The definitions used in the general platform typology remain valid, but are not repeated here. This typology can be a first step in thinking about the position a city government wants to take in mobile service creation and provision. Briefly returning to the definition of Smart Cities that was proposed earlier on, and the different domains the concept may refer to, we see that the services created based on the city platform described earlier indeed address many of the issues important for the advancement of the Smart Cities concept. Innovative environmental services, new approaches to mobility and public transportation, increased government efficiency, etc., are domains in which mobile city services have already been effective. Nurturing and fostering these initiatives into fully operational platforms can thus be one way for a city government to build out its Smart City strategy, if such policies are within the scope of a city's priorities. Important to note here, and in line with the criticism formulated on the Smart City concept, such services

Table 20.2 City Platform Typology

	No control over service offering	Control over service offering
Control over data and/or infrastructure	Enabler city platform Facilitating city services. Can be open data initiatives or outsourcing of service creation based on provided datasets. Stimulating development is key. *Example: London Data Store, Google Transit*	Integrator city platform Governmental city services. Somewhat more closed approach, can be high-cost depending on implementation. *Example: NYC311*
No control over data and/or infrastructure	Neutral city platform Commercial, nonprofit, or crowdsourced city services. City government does not take initiative and relies on privately funded or bottom-up projects. *Example: Open 311*	Broker city platform City-branded services. More of a distributed effort, for example, aimed at city marketing, rather than direct service provision. *Example: Amsterdam Partners*

should make all efforts to be as inclusive as possible and thus not be limited to certain technological platforms (e.g., more expensive and complicated smartphone devices), but attempting to cater to the diverse population inhabiting a region (e.g., by also offering services via voice or SMS).

These four different platform types clearly run along the same principles as the initial mobile services typology. However, these city platform types operate on a different level and scale and are more related to how public entities can organize themselves to make new and appealing services offerings for citizens possible. In this sense, it should be interesting to see to what extent these platform types can also be applied to other types of services than mobile only. Initially, it would appear this framework could also be used to assess other digital services offered by cities (e.g., in the case of open data initiatives or that of city marketing, as was illustrated in two cases earlier), but this would have to be the subject of further work (Table 20.2).

20.10 Conclusion

The goal of this chapter was to verify whether a general typology for mobile service platforms would hold up to scrutiny in the more applied and specific context of the city and to what extent city agencies could take up platform roles within the value network. After sketching the context of the city and the business model implications it brings to the creation and delivery of mobile services, the chapter described in more detail how the control parameters can be interpreted in this application of the typology. Next, several cases were presented for the four city platform types. We can clearly identify platform roles that are being taken up by cities internationally. These different roles have their own merits and consequences that depend on the relationship the city has with its citizens (via the service offering) and the data or infrastructure it controls. The thorough case studies showed the different possible approaches cities can take: play a facilitating role by providing developers with data, tools, and resources in the case of the enabler platform, keep tighter control over the service in an effort to increase QoS in the integrator platform, rely on the initiative of commercial or nonprofit entities that can play a supporting role for cities in the neutral platform, or focus solely on communication with the citizen by focusing on city marketing in the broker platform type. This typology can be an initial tool for city governments to consider their own role within the value network and expose the potential platform dynamics at play when they are involved in mobile service provision.

Future research will build on the concepts presented in this chapter and establish whether there are crucial gatekeeping platform roles at play in the creation and provision of mobile services in the context of the city. An even clearer definition of what control over assets and customers means in this context is required, while developing more applied enabling and constraining factors, as well as policy recommendations city governments can go to work with, toward providing compelling mobile city services.

Acknowledgments

This work was performed in the framework of a Prospective Research for Brussels grant, funded by Innoviris and the Brussels Capital Region, and carried out at iMinds-SMIT, Vrije Universiteit Brussel.

References

Accenture. 2011. Transforming customer services to support high performance in New York City Government, Accenture, project report, retrieved on September 15, 2013, available online: http://www.accenture.com/SiteCollectionDocuments/PDF/Accenture_Health_and_Public_Service_NYC_311_Customer_Service_Solution.pdf. Accessed September, 2013.

Amsterdam Partners. 2012. Annual report 2011, Amsterdam Partners, available online: http://www.iamsterdam.com/~/media/DB6E1F7AFF1F4C0EB5B55E3E9937A94B.pdf. Accessed September, 2013.

Ashlock, P. 2009. Open 311 is a specification for an open platform, Open 311 Blog, September 25, 2009, available online: http://open311.org/2009/09/open311-is-a-specification-for-an-open-platform/. Accessed September, 2013.

Ballon, P. 2009a. Platform types and gatekeeper roles: The case of the mobile communications industry, *Proceedings of Druid Conference*, June 17–19, 2009, Copenhagen, Denmark, available online: http://www2.druid.dk/conferences/viewpaper.php?id=5952&cf=32. Accessed September, 2013.

Ballon, P. 2009b. Control and value in mobile communications: A political economy of the reconfiguration of business models in the European mobile industry. PhD thesis, Vrije Universiteit Brussel, Brussels, Belgium, available online: http://papers.ssrn.com/paper=1331439. Accessed September, 2013.

Ballon, P., Walravens, N., and Delaere, S. 2009. A typology of business models for mobile service platforms, *Global Mobility Roundtable*, Cairo, Egypt, November 1–3, 2009.

Basole, R. 2009. Visualization of interfirm relations in a converging mobile ecosystem, *Journal of Information Technology*, 24(2), 144–159, available online: http://www.ti.gatech.edu/basole/docs/Basole.VisualizationConvergingEcosystem.JIT.2009.pdf. Accessed September, 2013.

Beck, K. 2011. 311: The agency that never sleeps, *CRM Magazine*, available online: http://www.destinationcrm.com/articles/Editorial/Magazine-Features/311-The-Agency-That-Never-Sleeps-72864.aspx. Accessed September, 2013.

Brand, S. 2006. City planet, *Strategy+Business*, Spring, available online: https://gbn.com/articles/pdfs/City-Planet_StewartBrand.pdf. Accessed September, 2013.

Camponeschi, C. 2011. *The Enabling City*. Toronto, Ontario, Canada, available online: http://enablingcity.com/. Accessed September, 2013.

Caragliu, A., Del Bo, C., and Nijkamp, P. 2009. Smart cities in Europe, *Proceedings of the Third Central European Conference on Regional Science (CERS)*, Kosice, Slovak Republic, available online: http://www.cers.tuke.sk/cers2009/PDF/01_03_Nijkamp.pdf. Accessed September, 2013.

Channel 4. 2010. 4iP, Channel 4, available online: http://www.channel4.com/programmes/4ip. Accessed September, 2013.

Chaudhry, S. 2011. Customer services in changing times, presentation by the Director of the 311 Call Center, September 27, 2011, available online: http://www.hthts.com/Teleseminars/Saadia_Chaudhry_NYC_311_call_center.pdf. Accessed September, 2013.

City of Amsterdam. 2004. The making of the city marketing of Amsterdam, City of Amsterdam, available online: http://www.iamsterdam.com/~/media/PDF/the-making-of-the-city-marketing-definitief.pdf. Accessed September, 2013.

Code for America. 2012. Fellows, Code for America, available online: http://codeforamerica.org/fellows/. Accessed September, 2013.

De Lange, M. and De Waal, M. 2012. Ownership in the hybrid city, Virtueel platform, available online: http://virtueelplatform.nl/english/news/ownership-in-the-hybrid-city/. Accessed September, 2013.

DiGiulio, T. 2008. Integrating service delivery: Case study—New York City 311, Institute for Citizen-Centered Service (ICCS), Toronto, Ontario, Canada, available online: http://www.iccs-isac.org/en/isd/cs_new_york_311.htm. Accessed September, 2013.

DuVander, A. 2011. 35 New APIs: Medicare, NYC 311 and mobile contact syncing, programmable web, October 2, 2011, available online: http://blog.programmableweb.com/2011/10/02/35-new-apis-medicare-nyc-311-and-mobile-contact-syncing/. Accessed September, 2013.

EC. 2011. Press release: Digital agenda. Turning government data into gold, European Commission, December 12, 2011, available online: http://europa.eu/rapid/pressReleasesAction.do?aged=0&format=HTML&guiLanguage=en&language=EN&reference=IP%2F11%2F1524. Accessed September, 2013.

Eisenmann, T. 2007. Managing proprietary and shared platforms: A life-cycle view, Harvard Business School Technology & Operations Management Unit, Research Paper No. 07-105, June 27, 2007, Harvard Business School, Boston, MA.

Evans, D., Hagiu, A., and Schmalensee, R. 2005. A survey of the economic role of software platforms in computer-based industries, *CESifo Economic Studies*, 51(2–3/2005), 189–224.

Farley, J. 2012. Opening up the inner workings of New York City, *MetroFocus*, April 16, 2012, available online: http://www.thirteen.org/metrofocus/news/2012/04/opening-up-the-inner-workings-of-new-york-city/. Accessed September, 2013.

Gawer, A. and Cusumano, M. 2002. *Platform Leadership*. Boston, MA: Harvard Business School Press.

Giffinger, R., Fertner, C., Kramar, H., Kalasek, R., Pichler-Milanovic, N., and Meijers, E. 2007. Smart cities—Ranking of European medium-sized cities, Research Report, Vienna University of Technology, Wien, Austria, available online: http://www.smart-cities.eu/download/smart_cities_final_report.pdf. Accessed September, 2013.

GLA. 2012. Accessible communications policy, Greater London Authority, available online: http://legacy.london.gov.uk/gla/accessible_coms_policy/docs/access_comms_policy.pdf. Accessed September, 2013.

Graham, S. 2002. Bridging urban digital divides: Urban polarisation and information and communication technologie(s), *Urban Studies*, 39(1), 33–56.

Greenfield, A. 2013. The city is here for you to use, available online: http://www.wired.com/beyond_the_beyond/2013/02/adam-greenfield-the-city-is-here-for-you-to-use-one-hundred-easy-pieces/. Accessed September, 2013.

Hamdi, N. 2004. *Small Change: About the Art of Practice and the Limits of Planning in Cities*. London, U.K.: Routledge, 184pp.

Hollands, R. 2008. Will the real smart city please stand up? *City*, 12(3), 303–320.

Iamsterdam. 2012. Organisation, *Iamsterdam*, available online: http://www.iamsterdam.com/en-GB/Amsterdam-Partners/About-us/Organisation. Accessed September, 2013.

IBM. 2009. How smart is your city? IBM Institute for Business Value, executive report, available online: http://public.dhe.ibm.com/common/ssi/ecm/en/gbe03248usen/GBE03248USEN.PDF. Accessed September, 2013.

Johnson, S. 2010. What a hundred million calls to 311 reveal about New York, *Wired*, November, available online: http://www.wired.com/magazine/2010/11/ff_311_new_york/all/1. Accessed September, 2013.

Judd, N. 2011. New York City releases 311 map, TechPresident, February 16, available online: http://techpresident.com/blog-entry/new-york-city-releases-311-map-says-raw-datas-way. Accessed September, 2013.

Komninos, N. 2009. Intelligent cities: Towards interactive and global innovation environments, *International Journal of Innovation and Regional Development*, 1(4), 337–355.

London Data Store. 2010. Boris Johnson launches London Data Store, London Data Store, available online: http://www.youtube.com/watch?feature=player_embedded&v=NjcZOefdmXE. Accessed September, 2013.

London Data Store. 2012a. The freedom of information, London Data Store, available online: http://data.london.gov.uk/datastore/freedom-information. Accessed September, 2013.

London Data Store. 2012b. Welcome to the London Data Store, London Data Store, available online: http://data.london.gov.uk/#. Accessed September, 2013.

mySociety. 2013. Open 311, What is it, and why is it good news for both governments and citizens? mySociety Blog, January 10, 2013, available online: http://www.mysociety.org/tag/open311/. Accessed September, 2013.

NYC Digital. 2012. NYC Apps, available online: http://www.nyc.gov/html/mome/digital/html/apps/apps.shtml. Accessed September, 2013.

OpenPlans. 2013. About OpenPlans, OpenPlans, available online: http://openplans.org/about/. Accessed September, 2013.

Open 311. 2013. GeoReport v2 Servers, Open 311, available online: http://wiki.open311.org/GeoReport_v2/Servers. Accessed September, 2013.

O'Reilly, T. 2010. Government as a platform, in: Lathrop, D. and Ruma, L. (Eds.), *Open Government*, O'Reilly Media, Sebastopol, CA, available online: http://chimera.labs.oreilly.com/books/1234000000774/ch01.html#government_as_a_platform. Accessed September, 2013.

Rochet, J. and Tirole, J. 2003. Platform competition in two-sided markets, *Journal of the European Economic Association*, 1(4), 990–1029.

Schiff, A. 2003. Open and closed systems of two-sided networks, *Information Economics and Policy*, 15(December), 425–442.

Townsend, A. 2013. *Smart Cities*. New York: Norton & Company.

Tiku, N. 2012. El Bloombito Thinks New York City's 311 App Really Bites, Betabeat, April 17, 2012, available online: http://www.betabeat.com/2012/04/17/gov-2-0-fail-el-bloombito-thinks-new-york-citys-new-311-app-bites/. Accessed September, 2013.

Transport for London. 2011. BCH data dictionary, Transport for London, available online: http://www.tfl.gov.uk/assets/downloads/businessandpartners/BCH_feed_data_dictionary_-_may_2011.pdf. Accessed September, 2013.

Transport for London. 2012a. Developer's area: Get data, Transport for London, available online: http://www.tfl.gov.uk/businessandpartners/syndication/16492.aspx. Accessed September, 2013.

Transport for London. 2012b. Developer's area: Guidelines and support, Transport for London, available online: http://www.tfl.gov.uk/businessandpartners/syndication/16493.aspx. Accessed September, 2013.

Transport for London. 2012c. Developer's area: Common questions, Transport for London, available online: http://www.tfl.gov.uk/businessandpartners/syndication/17252.aspx. Accessed September, 2013.

Transport for London. 2012d. Barclays cycle hire: Basic elements standard, Transport for London, available online: http://www.tfl.gov.uk/assets/downloads/corporate/barclays-cycle-hire-basic-elements-standard-issue05.pdf. Accessed September, 2013.

Transport for London. 2012e. Barclays cycle hire: User terminal display standard, Transport for London, available online: http://www.tfl.gov.uk/assets/downloads/corporate/barclays-cycle-hire-display-screen-standard.pdf. Accessed September, 2013.

UN HABITAT. 2010. State of the world's cities 2010/2011, UN HABITAT, available online: http://www.unhabitat.org/pmss/listItemDetails.aspx?publicationID=2917. Accessed September, 2013.

Van Dijk, E. 2011. Explore a different Amsterdam, Edenspiekermann Blog, December 2, available online: http://edenspiekermann.com/en/blog/explore-a-different-amsterdam. Accessed September, 2013.

Walravens, N. 2012. The public enemy? A business model framework for mobile city services, *Proceedings of the 11th International Conference on Mobile Business*, Delft, the Netherlands, June 21–22, 2010.

Weber, H. 2012. NY mayor fumes over the city's failed 311 App, The Next Web, April 17, 2012, available online: http://thenextweb.com/us/2012/04/17/new-york-mayor-fumes-over-the-governments-latest-failure. Accessed September, 2013.

Wise, N. 2010. XKCD and London Bike App, Fast Chicken, available online: http://www.fastchicken.co.nz/2010/09/13/xkcd-and-london-bike-app/. Accessed September, 2013.

Wise, N. 2011. London Bike App source code now available, Fast Chicken, available online: http://www.fastchicken.co.nz/2011/11/02/london-bike-app-source-code-now-available/. Accessed September, 2013.

Wise, N. 2012. About Nic wise, Fast Chicken, available online: http://www.fastchicken.co.nz/about/. Accessed September, 2013.

chapter twenty-one

Strategic and tactical issues with Apple's mobile maps

Mark R. Leipnik, Sanjay S. Mehta, and Vijayaprabha Rajendran

Contents

21.1 Introduction

On September 19, 2012, Apple introduced the iOS 6 and iPhone 5 with great fanfare (*Time Magazine*, 2013). The most publicized feature of the new iPhone and operating system (OS) was Apple Maps (AM). Apple had long been concerned that Google Maps (GM) was used habitually and exclusively by most iPhone and iPad users to generate maps, view imagery, search for points of interest (POI), and generally navigate through their lives. The loss to Apple of data about location-based searches (e.g., where is the nearest café) and control over spatial content (e.g., locations and attributes of gas stations in a particular town) and advertising revenues to rival Google was deemed a strategic danger. Apple's acquisitions of Tom Tom, Intermap, Digital Globe, Urban Mapping, and Waze data and ownership interests in Yelp, Placebase, Locationary, and C3 Technologies paved the way for Apple to challenge GM (Forbes, 2012). To deter iPhone users from using GM, Apple replaced it with AM (for users who upgraded their iPhones or OS). This change was a shock for millions of iPhone users who tried to engage with location-based services or simply view and query locations. Although AM did offer users turn-by-turn directions, these were no better than those available on vehicle navigation systems. The new *3D* feature was a source of derisive comment due to distortions of features, notably many iconic bridges (like the Brooklyn Bridge) and monuments (like the Eiffel Tower). The most visible distortion was that the feature appeared to be melting into the lower elevation surroundings (such as the river). In other cases, cars appeared to be driving up the side of buildings or ships sailing up the side of bridges. The cause of these issues were problems with the draping algorithm developed by C3 Technologies (Schmidt, 2012).

The release of AM generated instant ridicule in cyberspace, negative comments in mainstream media, critical reviews by pundits, and a downgrade of Apple prospects by financial analysts (CNN, 2012a,b). There were a vast number of errors in the names and locations of features, but the most prominently reported errors involved public facilities, such as train stations, bridges, airports, subway entrances, parks, and zoos. These were either placed in the wrong location or misnamed. This was largely due to Yelp data being focused on rating commercial POI data (i.e., restaurants, dentists) and Tom Tom data being originally intended for a vehicle navigation system and hence mostly oriented toward linear features (i.e., streets) that vehicles could drive on. When Apple users compared AM with GM and Microsoft's Bing Maps (BM), AM generally came in last in accuracy. The navigation feature, which was basically Tom Tom data and algorithms, was in many places inferior to competing vehicle navigation systems from Garmin using Navteq data (Navteq, 2013). Of course it did allow users to use a smartphone to replace a navigation system which was convenient.

Apple's unique *3D* feature, which was designed to give AM a competitive advantage, generated bizarre images of melting bridges, spike islands, black holes, and low-resolution imagery that circulated actively on social and broadcast media. Apple intended AM to be a GM *killer* application. Within months, much of the user base had rejected and ridiculed AM. The *Huffington Post*'s Wall Street 24/7 site ranked AM as the worst new product *flop* of 2012 (Huffington Post Business, 2013). Within 10 days of AM launch, Apple's CEO Tim Cook made a formal apology in an open letter to the public stating that "While we're improving Maps, you can try alternatives by downloading map apps from the App Store like Bing, MapQuest and Waze, or use Google or Nokia maps…". This last alternative, at the time, required users to download Google Chrome and use Safari as a search engine, which a few users did, but the process was complicated. The AM debacle was followed within 3 months by the firing of the three senior Apple executives most visibly and directly involved. Many analysts, Apple users and staff inside Apple stated, "Steve Jobs would never have allowed this to happen" because he always wanted to "grock" (his term for thoroughly understanding) an issue. However, corporate decisions made during Steve Jobs tenure as CEO (i.e., not to pursue geospatial technologies in contrast to competitors Google and Microsoft) meant that Apple had positioned itself for a troubled AM introduction (Isaacson, 2011). The question asked by many loyal users and followers of Apple is "How could Apple get it so wrong?"

21.2 Issues with Apple maps

Some of the issues bedeviling the AM rollout were inevitable, and would have been present with any effort to create worldwide high-quality geospatial data for a wide range of applications important to m-commerce (e.g., POI and turn-by-turn navigation of a network of street centerlines). However, other problems were largely self-inflicted. The most prominent of the self-inflicted issues was the *3D* feature. The *3D* view was designed to be more impressive than anything competitors, such as GM (with their "Google Earth," "Bird's Eye View," and "Streetview") or BM could offer.

While there are undoubtedly many tens of thousands of erroneous features and incorrectly associated attributes that are relatively trivial (such as "Lowry Lane" being called "Lowry Road") among the billions of accurate features in AM's worldwide geospatial data set, some of the errors are more significant and attracted multiple repostings from users in cyberspace and were aired in online and broadcast media. The single most publicized type of error was the *melting bridge*. This was an error associated with draping imagery over a *3D* model of terrain. Actually, the algorithm was developed and patented by C3 Technologies of Sweden, now acquired by Apple. It is a mesh *fishnet* surface used in the draping algorithm, so technically it is not a true *3D* model but a *2.5-dimensional* one. According to Ryan Schmidt, a leading expert on draping imagery over 3D surfaces and terrain who is employed as a senior scientist by Autodesk, Inc., "the underlying algorithms produce a set of 3D points, and the surface you see is a smooth interpolated surface through those 3D points." "The texture is then projected onto this smooth 3D surface." "If not enough points are produced, then the result is going to be oversmoothed…the algorithms work better with a large number of high-resolution photos taken from many viewpoints around the 3D feature." "Some things are just inherently hard to reconstruct this way, because the small repetitive details in the images confuse the image-feature matching process." Bridges in particular have that characteristic. To avoid generating melting bridges and other anomalies, the *3D maps* must be carefully edited. Google has done a lot of this tedious work but it also has a few issues of the same sort as seen in AM, for example, with bridge approach ramps (Schmidt, 2013).

Melting bridges rapidly became the signature of AM's problems. Although Brooklyn Bridge in New York is the most notable, the Lions

Gate Bridge in Vancouver, British Colombia suffered a more dramatic distortion. It not only appeared to be melting, but a barge was shown sailing up one side of a triangular face while cars drove on the other face at an impossible angle. This bridge was also mislabeled with an old highway designation changed decades previously. Other bridges with issues include the Tacoma Narrows Bridge, the Mackinac Bridge, and the bridge below Niagara Falls on the US–Canadian border as well as bridges in Quebec, England, Germany, and Australia that appear to be melting into the waters below them.

How the melting bridge phenomena could have missed developers beta testing is astounding. Though the Hoover Dam itself is well imaged, the bridge at Hoover Dam was not there when C3 Technologies featured it in a product demo video back in 2009. Thus, C3 Technologies may not have updated the imagery in tests and run out new imagery showing the drooping bridge without adequate quality control checks (C3 Technologies, 2011). This rationale does not explain the melting Brooklyn Bridge, which has been there since 1883. Another potential explanation comes from the corporate history of C3 Technologies. It was a division of Sweden's SAAB and its owners were eager to spin off this subsidiary because the parent company, General Motors, had filed for bankruptcy. Apple reportedly paid the equivalent of $700 million for C3 Technologies (*Wall Street Journal*, 2009). In reality, the *3D* maps did not add any significant value over ordinary aerial imagery in the form of improved user understanding or experience. This is partly because the world is generally rather flat (i.e., seen in 2D). Draping imagery on the sides of buildings in urban areas does not do much to improve navigation in an urban environment once the initial *wow* factor has passed (Paar, 2006). Users do not need to see imagery draped across terrain to make informed decisions about which turn to take on their journey or where to go to purchase products. In fact, the application had a tendency to make users physically sick due to vertigo (*The Guardian*, 2013). Oddly, not all bridges in the *3D* maps melted. Noticeable, the Bay Bridge and Golden Gate Bridge in Apple's home turf were not melted, nor were quite a number of bridges in Sweden. Developers likely made sure the *3D* application worked well in *demos* presented to Apple executives featuring places familiar to them in their daily travel route to and from Cupertino, California (YouTube, 2011). It is probably true that top Apple executives were genuinely unaware of the *melting bridge* problem or that it was widespread. The draping algorithm also had other glitches such as areas where no imagery was available and thus *black holes* appeared or where the underlying fishnet mesh peaked through gaps. Frequently, the digital elevation model was out of sync with the imagery, and this created anomalous *spike islands*.

In reality, imagery can never be entirely up-to-date. The satellite systems that gather imagery can only pass over a given spot on the earth every 16–21 days. Depending on the system, processing of imagery requires *orthorectification* (i.e., aerial photographs that are geometrically corrected and scaled) and *georeferencing* (i.e., establishing location using coordinate systems) (Rees, 2001). Nevertheless, the imagery used to support the rollout of AM fell far short of the high standard that BM and GM had established in the minds of users. Specifically, the AM imagery for Davos, Switzerland (home to the World Economic Forum), was not only outdated and of low resolution, but it also represented three distinct seasons simultaneously.

Some AM imagery was at such a low resolution that only very general building outlines were visible. In contrast, GM accesses Digital Globe imagery at 0.41 m resolution, which has over 5300 times more information content than 30 m imagery used for some areas in AM. The m-commerce implications of low-resolution imagery are also important. With 30 m resolution imagery, individual businesses, even the largest *big box* retailer, are difficult to resolve. While in 0.41 m resolution imagery, each business location is easy to discern. There are several areas with low-resolution imagery in AM including the United Kingdom, Australia, Italy, and Cuba.

AM imagery that was only available in black and white was present in areas of Australia, Albania, Germany, Italy, and Croatia. It appears that color imagery was obtained by Apple, but a processing error was made and only a single band of multispectral imagery was used. This is an error typical of image-processing neophytes and one that is glaringly obvious and easy to rectify (Lemeshewsky, 2004).

A grayscale-only image, due to a different cause, showed up for Colchester, England (population 104,390). A large cumulonimbus cloud covered the entire urban area. The label (which is a separate layer of data in AM) was located in the right spot, but the whole of Colchester was under an off-white cloud. As a practical issue, users could still find Colchester and when zoomed in on the image see labeled streets, but the whole area appeared white (the color of the top of the cloud). While finding a cloud-free day in the Highlands of New Guinea or the Costa Rican cloud forest is a challenge, in most parts of the world it is readily available at some time every year. This image did not meet an acceptable standard for an area as developed as England. Similar issues impacted Red Square in Moscow, a town in Germany, and many parts of the Scottish Highlands (which are indeed rather dreary).

A type of imagery-related error with important consequence for m-commerce was present in the case of the Yummy Foods Store on Santa Monica Avenue in Los Angeles. This natural food store was portrayed in AM in the correct location; its listing had a number of positive Yelp reviews associated with it, the turn-by-turn navigation to it was correct, but the imagery showed not a store but a large hole in the ground at that location. Although the imagery was in color and was high enough resolution to show a wheelbarrow, it was local municipal government aerial photography that was several years out of date. This brings to the forefront another significant issue. That is, the recent financial crisis impacted the ability of many local governments to conduct aerial photography in their regions as often as some users would like (*Huffington Post*, 2013). This sort of error is always possible for construction that involves more than merely reusing an existing building but in its demolition. It may also happen in the aftermath of a fire or natural disaster. Since natural disasters, like super-storm Sandy, often trigger subsidized aerial photography efforts, developers of imagery used in mapping applications have to be judicious in not always replacing existing imagery with newer imagery even if it is higher resolution, since a POI for a business location may be on top of a temporary burnt-out or washed-away building or hole in the ground.

The process for updating POI data relies on not only teams in the field working for geospatial companies but also the firms whose units are to be located themselves (Tom Tom, 2013). For example, Royal Dutch Shell, Best Western Motels, and Ace Hardware all have internal teams building geospatial data, as do smaller firms with fewer units (Leipnik and Mehta, 2004). Unfortunately, not all firms are technological savvy enough to get their location data into all the geospatial data bases that are being constructed. These at a minimum include Apple, Google, Bing, MapQuest, Yahoo Maps, and Multi-map in the United Kingdom and affiliated companies Navteq and Tom Tom.

While existing businesses may have a real interest in updating POI data, the same cannot be said of businesses that have ceased operations. This fact is reflected in AM, which has approximately 800 POI's for all the "Woolworth" units in the United Kingdom, even though the discounter closed all stores in 2002. Even more surprisingly, AM had POI's for approximately 200 units of "Our Price," a chain of deep discounters. This firm ceased operation in 1989, which is 3 years before Tom Tom began operations. From a marketing perspective, it is more damaging to not have data for an existing unit, but keeping defunct businesses on the map is an error nevertheless and one that will not be resolved after the fact by the firm in question.

A much more serious deficiency of AM is illustrated by the case of the Pepper School of Music in San Antonio, Texas. As soon as AM replaced GM, the owner of this small business began receiving calls from clients that could not locate his school on Apple devices. He found that in AM, the school's location was off by about 10 blocks and in the middle of a residential area rather than on the arterial street where he operated. He had to verbally explain to clients how to extricate themselves from a tangle of cul-de-sacs and dead-end streets. To add insult to injury, his location was correctly mapped in GM, BM, and MapQuest. Mr. Pepper made repeated efforts to correct his location, contacting first Apple who referred him to Tom Tom who in turn suggested Yelp. He also tried Locationary (a mapping data company owned by Apple), but to no avail. Of late, Apple has tried to improve its POI data by acquiring HotSpot and also has begun to solicit input from small business owners.

The lack of a mechanism, at the time of initial release of data, to report and correct errors in AM data, was probably the second most profound mistake (after the *3D* mapping errors) made by Apple. This was due to the fact that Apple had acquired the rights to use all Tom Tom data in 2011 but was negotiating to actually purchase the firm as late as September 2013 (Forbes, 2012). Each firm seemed to blame the other and Apple's Vice President for Retail, Jerry McDougal, was initially not responsive to calls for Apple to take responsibility for erroneous data. He stated that AM were "good and would get better." What mechanism would rapidly improve AM, without a mechanism for continuous consumer feedback is unclear. It is noteworthy that Google has long encouraged feedback on errors in its geospatial data and makes a simple GIS program called Mapmaker freely available to users who want to try their hand at editing and updating maps and submitting those results for possible inclusion in revised Google offerings (Google, 2013).

The iPhones use a combination of GPS, cellular tower signal triangulation, and Wi-Fi to help determine location. In some cases, such as when the device is inside a structure, GPS does not work because the signal is blocked by a structural component. There are cases where the IP address of the Wi-Fi router is set someplace else and the physical location of the device is not the same as the past physical location of the router. This has caused some users of iOS 6 and iOS 7 devices, when at home, to have their home locations appear a large distance away from where they actually are. This makes it impossible for them to use the navigation feature from their home. In many cases, when they manually enter their home address, they find this is in the wrong location as well. It may be interpolated based on incorrect address ranges or in the case where the address is a P.O. box or rural delivery route, it will not have any corresponding physical address nearby. This is an issue for all cellular devices, not just iPhones, but the Tom Tom data used in AM is more error prone than GM or BM data, at least in the United States.

AM use of Tom Tom data supplemented with Yelp data to create information on POIs is quite limited. For example, Yelp uses a hierarchical algorithm, so it lists restaurants 1st, public facilities 6th, and schools 12th. A test was done (by the authors) of the Yelp listings in Huntsville, Texas. It indicated that there was only one bar (with one review), when there are over a dozen bars. A search for "Scott Johnson Elementary School" (in operation at the same location since 1962) returned "Specs Liquor Warehouse." "Sam Houston State University (SHSU); enrollment 20,000;

founded in 1879)" was not listed at all, but Yelp suggested that both Lone Star College and Sam's Deli in Houston, Texas (which was 80 miles away) as possible substitutes. Conversely, Tom Tom data places SHSU at the correct location, but its attribute data (when accessed with a GIS program) indicates it is located in Phelps, Texas (a town of only 100 people located more than 20 miles to the South East of Huntsville). This is typical of the deficiencies of AM data with respect to universities and other public not-for-profit entities. It is because they are not typical businesses and so do not have the same set of attributes nor do they contact firms like Tom Tom with their data. Even when a university is mapped correctly in AM, the buildings inside the campus are not mapped in most cases (e.g., Louisiana State University appears as an empty space on AM). This is in contrast to GM, which generally have the buildings inside campus identified. Given that a high proportion of college students are incessant users of smartphones, with a disproportionate share using Apple devices, ignoring the on-campus way finding needs of this segment is hard to justify or understand. The Yelp coverage of other public facilities is also meager, in particular coverage of public transportation routes (e.g., bus stops) is basically lacking.

The ratings in Yelp are problematic as well. Zagat or Michelin ratings are generated by *mystery shoppers* (i.e., real diners and buyers) and part of Navteq's offerings and available on Garmin GPS devices. In contrast, on Apple devices, the ratings from Yelp rely on the *user community* (which is anyone that cares to join the site). Few restaurants receive more than 20 ratings and the immediate family and friends of the owners of any restaurant can join Yelp and serve as shills to submit numerous highly favorable ratings. The arithmetic average of 18 spurious ratings and 2 genuine ones will still be a high rating. In fact, most restaurants in AM have uniformly high ratings. This fact defeats the purpose of the ratings (Main Street, 2011).

Although Yelp data is more complete in California and New York City than for example in Texas, it is out of date and misleading there as well. For example, of the four restaurants closed by the Orange County Health Department in 2012, two were listed in September 2013 on Yelp as open and both had four out of five star ratings. In the case of one other closed restaurant, Yelp found a dentist, of the same name, in a nearby city as a not very helpful alternative. Yelp depends on the business continuing to operate or be superseded by an operational business. New stores, bars, and restaurants depend on active Yelp users to give them ratings. While some segments of the population in some regions of the United States are eager to give ratings, many are not, and few businesses outside the United States are even rated. More fundamentally, many of the patrons with the most disposable income may be too occupied with professional and personal interests to take the time to post a rating.

Apple describes AM on its official site as a *new turn*; unfortunately, they often provide users a *wrong turn* (Apple, 2013). Tom Tom has data for the United States, Canada, Mexico, and the European Union (EU), that it claims is *99% complete*. But it does not include Croatia, which is now a member of the EU. Additionally, Japan is excluded from having complete road data and so are China and India along with the two largest European countries, Russia and Ukraine. All of these nations are at best poorly mapped in Tom Tom data and by extension very likely to be badly mapped in AM. In some countries, AM data is almost absent, such as in Belarus and Grenada. For many nations, all the data in AM is basically data mapped by the Defense Mapping Agency in the 1980s (vector data that is publicly available for free download). This is a highly generalized one-to-one million scale Digital Chart of the World data set (now called VMAP 0) and consists of major roads and points representing inhabited places. In many nations, particularly in Africa, it is all that AM has to offer besides imagery. Effectively, it is the lowest common denominator for international geospatial data (NGA, 2013).

21.3 Types of errors

Tom Tom data available in AM is subject to a wide range of errors. The most common being incorrect definition of city limits, inaccurate speeds associated with roads, wrong road type, not including the correct name for a road with multiple alternative names and/or numbers, failure to show a road at all, and connecting roads to each other that do not connect. Transposing road names is one type of error that is rare but very confusing and hard for users to resolve. A more common error is calling a street an avenue, a drive, or a road. More significant is mapping a section of the Interstate with the incorrect name of a surface street. For example, part of Interstate 45 in Huntsville, Texas is incorrectly named Green Briar Drive (the real "Greenbrier Drive" is 2 miles away).

A surprisingly common error in AM is portraying a road as existing, when it does not exist. This error can have three main sources: one is showing a road that once existed but is now abandoned, another is portraying a road that was once planned and laid out on maps but never constructed, and another is due to misidentification of utility right of ways, bridal paths, or fire breaks as roads. A related issue is the long private driveways on farms and ranches that are misidentified as public roads. They may or may not be closed off by entry gates, but driving on them is *trespassing* and likely unlawful. The most defensible type of error is the case where a road is under construction or reconstruction and shown as through. This type of error is a temporary one, although the alignment shown on the map may not be the ultimate alignment.

One facet of AM data is the option to receive real-time traffic updates. This will help with detours and road construction issues, although it does not resolve the countless other errors and issues with AM street data. Generally, only driving roads actively with numerous data collection vehicles, diligent digitizing, constant editing with frequent reference to aerial imagery, and being responsive to user comments can help resolve these problems and update the street data to reflect constant changes.

There is a large category of errors where the street name is incorrect; this can include misspellings such as "Warner Road" instead of "Werner Road" or "Lowry Road" instead of "Lowry Lane" or "Palaside Circle" instead of "Palisade Circle" (all occur in Walker County, TX). Or the rarer, but humorous examples, where "Carse Road" in Vermont became "Curse Road" or where "Satinwood Drive" in Maryland was altered to "Satanwood Drive." The street direction also can be in error and often the address range can be wrong, as can the speed limit. If the speed limit is incorrect, then the algorithm to determine the shortest travel time through several potential routes will often produce errors. Generally, the further back in the name the misspelling has taken place, the less significant it is. Erroneously, naming the midsized town of "Slaithwaite" in the United Kingdom as "Blaithwaite" is a more serious error than misspelling the last letter of the name of a street. Misspelling the first letter in a name makes online searches much more difficult. For example, having "Slaithwaite" named as "Blaithwaite" is a more significant error than spelling it "Slaithwait" or "Slaithwaitee." The fact that there is another nearby town in England named "Braithwaite" just adds to the possible confusion. This mistake was made in 2012 in GM data. Google, to its credit, corrected the error within weeks.

The consequences for a business of having every part of the location correct but having an incorrect street number could be devastating. This error will send potential consumers, relying on smartphones or mobile maps, to another block, perhaps on the other side of town. This incorrect block is probably also on the same *shopping street* and may harbor rival businesses. Therefore, someone searching for a restaurant might easily find a different restaurant and settle for that one instead. Also, misclassification of features in POI data may lead one to the wrong location being visited by consumers (e.g., Voodoo Tattoo in Huntsville, Texas, being classified by Tom Tom as a restaurant).

An interesting miscategorization that has implications for m-commerce and businesses involves data for the town of Lindora, California where a storefront weight loss program was identified as the "Lindora Medical Clinic." This feature in AM covers an entire city block with the pink Hospital color and a Red Cross logo in the center. In fact, this nonexistent hospital occupies all the space in a strip center, which in fact has a McDonald's and a Chevron Gas station, none of which are portrayed. Conversely, the Bethesda Naval Hospital and National Institutes of Health

Complex in Maryland is not portrayed on AM at all. Since this complex has 4 hospitals, 27 research institutes, and 75 buildings on a 300-acre campus, it is a significant omission. The errors involving hospitals could have tragic consequences. One can imagine an iPhone user desperately searching for the nearest hospital and driving to the weight loss program office with a critically injured passenger. The potential liability of AM in this instance is unsettled, but disclaimers that the maps are for reference only might absolve the developers from legal liability (Sovocool, 2008).

Another type of error involves Powell Butte Nature Park in Portland, Oregon, which is portrayed as probably 50 times its true size and occupies most of the Eastern part of that important city. The Main Train station in Helsinki, Finland is also mapped as a park as was Madison Square Garden in New York City. A heuristic for data builders seems to be, "when in doubt, make it a park."

There are numerous search problems that are due to ambiguity in naming. For example, there are at least 11 places named "Riverside" in the United States. One of the great areas of superiority of GM's intelligent search algorithms over rivals like AM and BM are that Google is unsurpassed in giving users options. Thus, a user in Texas who searches for "Riverside" will be given the "City of Riverside" (population 525) as well as options for "Riverside, California" (population over 1,000,000) and the neighborhood of San Antonio called "Riverside" (population approximately 5,000), but not Riverside, New Jersey (population about 8,700). Similarly, a search initiated in New York will omit the two places called "Riverside" in Texas but add the one in New Jersey to the much larger one in CA as well as offer "Riverside Drive" in New York City as an alternative choice. On an iPhone used in Riverside, Texas, a search for "Riverside" will yield only Riverside, California. To find a map showing where the City of Riverside in Texas is, one needs to specify "Riverside, Walker County, Texas." This type of issue affected users in England who searched for "London" and were given maps of London, Ontario as the default option. An Apple user interviewed on a London Street by CNN in a report in October 2012 about issues with AM, indicated he was unperturbed because he "liked Canada and Canadians."

Another iconic AM error was associated with labels for oceans that were misplaced by tens of thousands of miles and appeared both at sea and in the middle of land masses. Thus, the "Arctic Ocean" appears in blue script off the south coast of Tamil Nadu, India. The Indian Ocean appears in Northern Greenland, the North Pacific Ocean is in the middle of Africa. Other labels that were vast distances away from their true location include "Berlin" appearing at the South Pole, and the label "Cayman Islands" appearing on the Isle of Man in the Irish Sea. Issues with labels being in the wrong language and alphabet were associated with AM labels in Chinese characters appearing in Krakow, Poland.

Airports that were misplaced were a major potential source of inconvenience to travelers. A hurried traveler, who searched for "Heathrow Airport" while in London, would find that the drop pin was located on Bath Street near Vicarage Road in the residential Hounslow West neighborhood of London, miles away from the correct location. Conversely, a search for "London Heathrow Airport" placed a drop pin on the terminal building. Airport errors plagued the entire AM product because the underlying Tom Tom data did not include a good set of airports in their POI data set. A travel writer in Australia missed his flight because the drop pin for the international airport in Melbourne (the second largest in the country) was located on a dead-end street on the far side of the runways and inaccessible to the terminal. One would need to know the correct street address for the terminal or spot the terminal on imagery and manually enter a destination point, for AM to give correct turn-by-turn directions to this busy airport. Other airports with issues included San Francisco, Akron, Fairbanks, and Kathmandu.

The turn-by-turn navigation features in AM often led users to the end of dead-end streets portrayed as through roads on the maps and routes. Other routing errors included routing cars across open water where there were no bridges or diversions of the route that should have been along a coastal road to inaccessible off-shore islands. Specific examples of this error include the routes along the coasts in England, Denmark, and Spain. In one case in Denmark, a bridge was avoided in *turn-by-turn* directions and drivers were routed to a ferry service that cost more and took much longer. Surprisingly, this error also occurs with respect to the route between San Francisco and Marin County, California. One would suppose that every Apple employee in Cupertino could choose the route using the Golden Gate Bridge (which was not melted, incidentally) but the routing function chooses to have users take a ferry service, which is nice for a leisurely excursion by a sightseeing pedestrian, but impractical or impossible for a daily commuter.

In a number of other cases, the directions were more than misleading, they were dangerous. For example, the interchange between two interstate highways in Ohio was shown so that drivers would have been directed to drive off an overpass rather than take the correct flyover ramp. In other cases AM users have been left stranded on dead-end roads in isolated places, such as the *outback* of Australia.

21.4 *Wide world of woes*

It is probable that there are errors in AM data for every country on earth even the smallest ones, since errors in the correct boundaries of the Vatican City, Monaco, San Marino, Andorra, and Lichtenstein have all been identified in AM as well as GM data. However, to get a feeling for the vast scope of errors and the diversity of different issues that exist with AM data, a list of selected nations with telling or unusual errors follows. It should help readers comprehend the global magnitude of the spatial woes facing Apple.

In *Albania,* many parts of the country are only in black and white imagery due to a failure to separate the bands on available imagery. Also, the same area is displayed at differing resolutions and in different seasons. The discrepancies in tone from one image to another are dramatic and disconcerting.

In *Argentina,* the Bonin Islands are portrayed as spikes rising almost vertically from the water. The port complex at Puerto Madero not only gets labeled with one drop pin, but six separate alternative drop pins for it are scattered at somewhat random locations within the city.

In *Australia,* Mildura, Victoria (population 35,000) is located 384 km from its true location. Most cities in Victoria can be found only when their name is supplemented by the state name in searches. The Sidney Opera House (one of the world's most iconic buildings) is missing and the area is only identified as "Bennelong Point." Cairns, Australia are portrayed over 100 km from its correct location, with no roads reaching it. In many cases, town names in Australia were replaced with those of businesses located in the same city. Thus, "Tatura," Victoria, was replaced with "Commercial Hotel (Tatura)." The Yeronga Hospital (a major medical center in Brisbane) was identified as a hotel. Another error in Brisbane was identification of a church as a retail business, despite bearing the name "Christian Brethren Church." Some of the errors in Australia probably related to aboriginal place names being unfamiliar and difficult to differentiate or enter for the staff creating the maps. Thus "Mildura" (a city) is conflated with "Mandurah" (a large bay missing from the map) and with "Murray Regional Nature Park." All these places have M, U, and A, and a typo in one can cause a chain reaction of misidentifications. Several travelers to Mildura were left stranded far from any services overnight in the Murray Nature Park, this led the Victoria State Police to call a news conference warning motorists not to rely on AM for navigation (ABC News, 2012).

There is virtually no data for *Belarus* in AM. The paucity of data in AM for some countries, in contrast with the detailed data in BM and GM, indicates the long distance that Apple has to go to have a truly global set of geospatial data. For example, in the capital of Minsk there is only one road portrayed in AM (an outer loop highway). This is one of three concentric loop highways and one of eight highways and hundreds of arterial streets that are portrayed in Minsk alone in GM.

Several Latin American countries have problems. In *Brazil,* several buildings and trees in Rio De Janeiro were melting in the 3D view. A football stadium in Sao Paulo, Brazil was labeled only in Japanese characters.

In *Chile*, portions of Punta Arenas' street network are portrayed as sitting in the Pacific Ocean. Over 50 streets are shown in the sea with the streets being up to 1000 m off the mapped shoreline. *Mexico* had a similar issue with streets in Cancun apparently at sea, but in that case in a different ocean.

China has issues with labels that only appear in poorly transliterated English. BM has far more streets labeled in both Beijing and Shanghai in Chinese characters or in *pinyin* transliterations (perhaps as a result of the alliance of Navteq with the Chinese government). China Nav Info, which is 50% owned by Navteq (which was acquired by Nokia and now part of Microsoft), has been designated by the Chinese Government as the sole developer of vehicle navigation systems data in China. Therefore, Microsoft has a growing advantage over both its rivals in China (Leipnik et al., 2012). Missing major rivers are a common theme in China. For example, the Yangtze River (China's largest river) is missing completely from Chung King, a city of 20 million people and the fourth largest in the country. While in the most populous city, the Pearl River is missing from Shanghai's Pudong district. This district alone has over three million inhabitants. Completely omitting these rivers, which divide and define these cities, is an inexcusable cartographic lapse.

Additional countries have specific issues. For example, large areas of *Croatia*'s coast are only portrayed in black and white imagery. In the *Czech Republic*, official Slavicized names of many places are uniformly replaced by earlier German names not used since 1945 (Similarly, a square in Berlin was labeled "Adolf Hitler Platz"). *The Falkland Islands* are filled with spikes, possibly due to issues of this area being disputed territory between the United Kingdom and Argentina. The same error occurs in *Tahiti*, which is an undisputed possession of France. In *Finland*, Helsinki's main train station and the surrounding area are identified not as a station but as a park with a uniform green fill as well as a tree marker symbol. In *France*, the images of the Eiffel Tower lying apparently on its side in the 3D view is startling but it is not as distorted as the famous Millau Viaduct Bridge that consists of eight separate triangles. Also, the village of Cadeillan (population of 74 people) appears in the middle of an undeveloped open space several kilometers from its true location with no roads leading to it. In *Germany*, the Winninger Bridge is melting. The main cathedral in Cologne is colored and labeled as a train station. The main Lutheran cathedral in Copenhagen, *Denmark* is labeled as a Burger King restaurant. In *Ireland*, the "Airfield" urban farm in Dublin was portrayed with an airport symbol. The potential that misguided pilots could attempt to land at the large farm with the airplane symbol in it led the Irish Attorney General to make a public announcement on Irish Television that he had formally contacted Apple about the error that he deemed a danger to public safety. The Dublin Zoo was also missing. In *Italy*, the central square in Milan looks as if a nuclear blast has taken place (with scores of deformed buildings).

In *India*, the Ganges River is missing in Varanasi. This is the holiest of rivers to the Hindus in the holiest of cities. It is a very large feature not portrayed on AM, although the bridges across it are shown and labeled as bridges. In Kolkata, the Hooghly River is portrayed, but its location has been shifted over about 1000 m to the East, so that it is under a dense street network and the bridges across it though portrayed have no river under them. It is possible that imagery from a period of monsoonal flooding contributed to this type of error.

In *Japan*, the world's busiest train station at Shinjuku is not shown in its correct location but in a nature park of the same name and multiple train and subway stations are shown in the center of water features, such as bays and rivers. These features are mislabeled with Tokyo Bay labeled as the North Pacific Ocean. POI for many features were missing, in particular numerous small eating establishments were omitted. In fact, AM was unable to locate the Apple Store in Ginza district of Tokyo. It is noteworthy that Japan presents a cartographic challenge, because few streets have names and houses are numbered by block in chronological order of date of construction, not spatially. These challenges were overcome by domestic spatial data developers Increment P Corporation, and Mapion over the past decade. Apple initially opted to not purchase their data. There were such problems with AM of Japan that apple released a separate update before waiting for its planned 6.1.3 revision with much better data based on purchase of domestically developed Japanese data (Nine-to-FiveMac, 2013). The problems with AM in Japan were blamed for poor sales of iPhones there in 2012 in a *New York Times* article that highlighted the success of rival domestic Japanese web map maker Mapion, which began experiencing a huge increase in page views on its website reaching 86 million per month in late 2012 (Tabuchi, 2012).

In *New Zealand*, the drop pin and label for the main train station at Auckland was portrayed in the waters of the harbor. In *Norway*, an edge of frame issue with glare from the lens of the aerial imaging platform caused an explosion of bright white grading into bands of purple and green over the town hall of Lenvik. In *Russia*, clouds are covering portions of Red Square in Moscow. In fact, most of Russia has only major connecting roads (i.e., between Moscow and Saint Petersburg) portrayed. In *Serbia*, the Danube River is missing from its capital, Belgrade. As this is largest river in the region and forms the part of the nation's border, so it is a significant lapse.

Large areas of Capetown, *South Africa* appear under cloud cover. Capetown is in the most technologically advanced African nation and one with reasonably good geospatial data and a number of mapping websites. In one of Tom Tom's more audacious claims, the company went from claiming that it had geospatial data in only a handful of African countries in 2011 to claiming the majority of Africa as mapped (at least to some extent) in 2013. The real quality and quantity of this data is highly questionable.

Goteborg, *Sweden*'s second largest city (with a population of over 504,000) is not labeled at any scale in AM. This error is probably due to the city having multiple spellings and the staff working on the maps putting off labeling the area until the name was resolved. The drop pin for the airport in Jonkoping appears over a farm and the runway has multiple hills due to errors in the draping algorithm.

In the *Ukraine*, the base of Russia's Black Sea Fleet in Savastopol is missing entirely from AM, apparently due to it being a high security area. Kiev (spelled Kyiv in *Ukrainian*) is labeled "Klyv" in AM. One can see how the use of the Ukrainian spelling plus misinterpreting an "I" as an "L" can cause this sort of error but one feels that to misspell the name of a city of six million and the most sacred in the Russian orthodoxy religion is a major error, especially since it is such a short name. Only about 10% of the roads in Ukraine are portrayed in AM and many towns are missing as well.

In the *United Kingdom*, a series of convoluted interrelated errors involve the historic city of Stratford-upon-Avon (which is a tourist mecca and the birthplace of Shakespeare). These errors involve simultaneously leaving the city of Stafford-upon-Avon off the map and list of towns and replacing the polygon defining the urban area with a pink fill and a Red Cross symbol and labeling the whole place "Stratford-upon-Avon Hospital." A search for Stratford-upon-Avon does not find the town but instead zooms to Birmingham and locates Stratford-upon-Avon TV LTD (a television repair service). The United Kingdom has a lot of complicated hyphenated place names that make mapping a challenge. An iconic error involved the Manchester United Football Club not being linked to any place, despite having a huge stadium and being the most important sports franchise in Europe. Instead, the Sale United Football Club in Manchester is mapped in response to a search for "Manchester United Football Club." Sale United is an amateur boys club with a few teenage and preteen squads. This error was wildly reposted on websites since it confused clubs at opposite ends of the spectrum of possible significance. Other mistakes include moving Blackpool to the Cornish coast and spelling Doncaster as "Duncaster," Westminster as "Westminister," and Torquay as "Torbay." Paddington Station in London was identified as a park, while Aldewych subway station, which has been closed for many years, was portrayed as open. The errors involving London's public transportation network led to the posting of signs in underground stations offering official paper maps *for the benefit of lost iPhone users.*

A large portion of the city of Valencia, *Venezuela* is missing and many of the areas on the west side of this city have incorrect names. As is common, the space where features should be filled with streets is filled with green to indicate it is parkland or open space. Possible reasons for this lack of information include lack of cooperation from local authorities in a chaotic country, heavy tree canopy, and possibly confusion about the correct name for a region that exists with similar names in several places (such as Spain and California).

While many of the errors stated earlier are specific to a particular country, some errors in AM are truly international in scope. Thus, one does not know which country to associate with a search for "Rome" that yields a map of North Korea or a search for "London" that yields a map of Ontario, Canada. AM also identifies *Taiwan* as a province of China, for political reasons. Some errors involve multiple countries, such as a portion of the French border portrayed crossing Lake Geneva and entering what is Swiss territory and swallowing up the town of Céligny in Switzerland before re-crossing the lake to reenter actual French territory. Another territorial error involves the border of Lithuania and Belarus with a small enclave of Lithuania as a seeming island inside the nation of Belarus. The disputed island of Senkaku (also called Diaoyu) appears twice with a small separation on the map of the East China Sea. If this uninhabited place were in fact two identical islands in reality, the ongoing dispute between China and Japan over the sovereignty of island could be easily resolved.

In defense of Apple and Tom Tom, AM has extremely few errors in the *Netherlands* and more generally in the Benelux countries. Tom Tom is headquartered in Amsterdam, Holland and it and predecessor Teleatlas have for over two decades produced high-quality data for Holland, Belgium, and Luxembourg. In fact, GM showed a car parked on the side of a wall in Westenbergstraat, Holland while Apple did not have this or any reported errors in Holland. However, Apple does have imagery depicting cars driving two stories up the side of a building in Madrid, Spain. Tom Tom POI data are quite good for Shell stations worldwide as both firms are Dutch based. California-based Chevron also has high-quality AM data. But AM data is poor for Texas-based Valero and Exxon-Mobile. One might suppose that the *death of distance* would make it equally easy for geospatial data to be dispatched from Texas to Cupertino or Amsterdam as for data from Holland or California would be, but the human contacts that make sharing data more feasible may be present among a group of Dutch or California based GIS analysts based in Holland or California and not among those based in other nations or metropolitan areas (Cainrcross, 2006). While it might be construed as a case of *damming with faint praise*, Apple also has far superior maps compared with Bing or Google for North Korea.

21.5 Response of consumers and competitors

Although the troubles of AM elicited reactions ranging from anger and pity to elation and self-satisfied smirks among consumers in the many videos posted online, most responses of competitors were muted. Nevertheless, there were responses such as the advertising campaign launched in print and billboard media and frequently reposted on the Internet of Samsung in Australia. The Victoria State Police issued a formal warning to motorists, in a televised news conference, that due to the Mildura and other errors, the use of iPhone could lead to more travelers being stranded many kilometers from any services in the *outback*. This gave official imprimatur to the poor quality of AM relative to rival Google's offerings available on Samsung smartphones. A Samsung Galaxy SII phone was featured in advertisements with a Victoria map on it and the caption "Oops...should have got a Samsung...Get navigation you can trust." Microsoft also later chimed in citing the *reality distortion field* surrounding overhyped Apple product releases and touting its *surface* tablet and BM tools and imagery (Keizer, 2013).

The criticism of AM became viral in the form of numerous parody videos on YouTube. It is ironic that in addition to losing access to GM, users of iOS 6 also lost easy access to YouTube where they could view dozens of videos lampooning AM. The most frequent parodies were Hitler first looking at a paper map of Berlin in his bunker, than going into an obscenity-laced rant about AM and their errors, the lack of the ability to get GM, and the lack of mechanisms to report problems. Other popular parodies included the *Batman Dark Knight* being repeatedly lost using an iPhone and Jack Nicholson's character from *The Shining* using an iPhone to try to chase *Danny* through a frozen outdoor maze and hearing the perky then increasingly the testy voice of *Seri* leading him to a frozen death. Parodies of Apple commercials featured lost iPhone users standing in front of a pizzeria and unable to find a pizza place in Yelp. There was also a clever British parody where a supposed Apple executive named Nigel Seep explains, with a clipped British accent, that if the iPhone 5 is held in a certain angle the phone's *quantum* processor will access a parallel universe that features melting bridges, the London Eye Ferris Wheel without connecting spokes, and Basingstoke plopped unceremoniously in the middle of farm fields. This parody also highlights the antenna issues with the iPhone 4S (*New York Times*, 2010). The initial denial of that issue by Apple presaged the more damaging initial denial that AM were seriously flawed. These parodies were popular enough to attract a cumulative 2 million plus views within less than a year of their creation. Of course, like many aspects of the web, their popularity rapidly wore off. But with consumers making decisions about which mobile device to buy on average every 2 years in the United States, these videos were a potentially powerful influence on buying decisions of many consumers (Phone Arena, 2011). In particular, they probably influenced that group of Apple users who opted to download GM as their default mapping choice when it became easy in late 2012. These more *tech savvy* users are a key demographic that Apple has probably lost for good, in terms of mobile mapping and location-based services.

21.6 Corporate response

The explanation for the many issues with AM is not just a failure to perform adequate quality control and beta testing before product release. Nor is it an overly ambitious release target date, or even a failure of leadership on the part of the three senior executives in charge of AM, all of whom were fired within 3 months of the first release of AM. These executives were Scott Forstall, who personally unveiled AM in September 2012 and was manager for all mobile applications at Apple. He was fired in late October 2012. Mr. Forstall had reportedly refused to apologize, something that Apple CEO Timothy Cook did soon after. At the same time John Browett, head of Retail at Apple, was terminated. His connection to AM was less direct than that of Mr. Forstall. The lack of a means for consumers to report problems or rectify errors with AM, which was one of his responsibilities, probably led to his termination (Wingfield and Bilton, 2012). These two terminations were followed a month later by that of Richard Williamson, the manager directly responsible for AM implementation (Chen and Wingfield, 2012). These changes probably are cosmetic since the issue with AM is a deeper strategic flaw at Apple where it appears the company supposes that acquiring another firm or the rights to use its data can solve longstanding issues without the need for developing internal expertise. Microsoft had its own GIS software at one time (MapPoint) and almost two decades of experience with imagery on TerraServer and its own BM (and it now owns Navteq). Apple has never had much experience with geospatial technologies and deferred to rival Google, which with GM, Google Earth, Bird's Eye View, and Streetview, is the worldwide leader in online maps and imagery and has been a geospatial technology leader for over a decade. Apple could not replicate this expertise simply by acquiring Yelp, and Locationary, and proposing to buy Tom Tom and licensing all their maps and other data.

The errors in AM have had a tremendous impact on the valuation and stock price of Apple. While the decline in the stock price cannot be solely attributed to the AM debacle, there is no doubt that it played an important role in influencing stock analysts and it reduced consumer confidence regarding Apple products. If Apple can fix its problems with AM, consumers will forget it in a few years. By 2013 it was clear that the majority of Apple iPhone users had adopted the new OS and phones and about two thirds were using AM rather than taking the time to go the Apple I Tunes store and download GM. Figure 21.1 portrays the decline in Apple stock price from an all-time high of over $700 a share to a low (after the apologies, terminations of executives, and reversion to access GM in early January) of less than $400 per share. The beginning of the steep decline coincides with the release of the iOS 6 and AM on September 19, 2012.

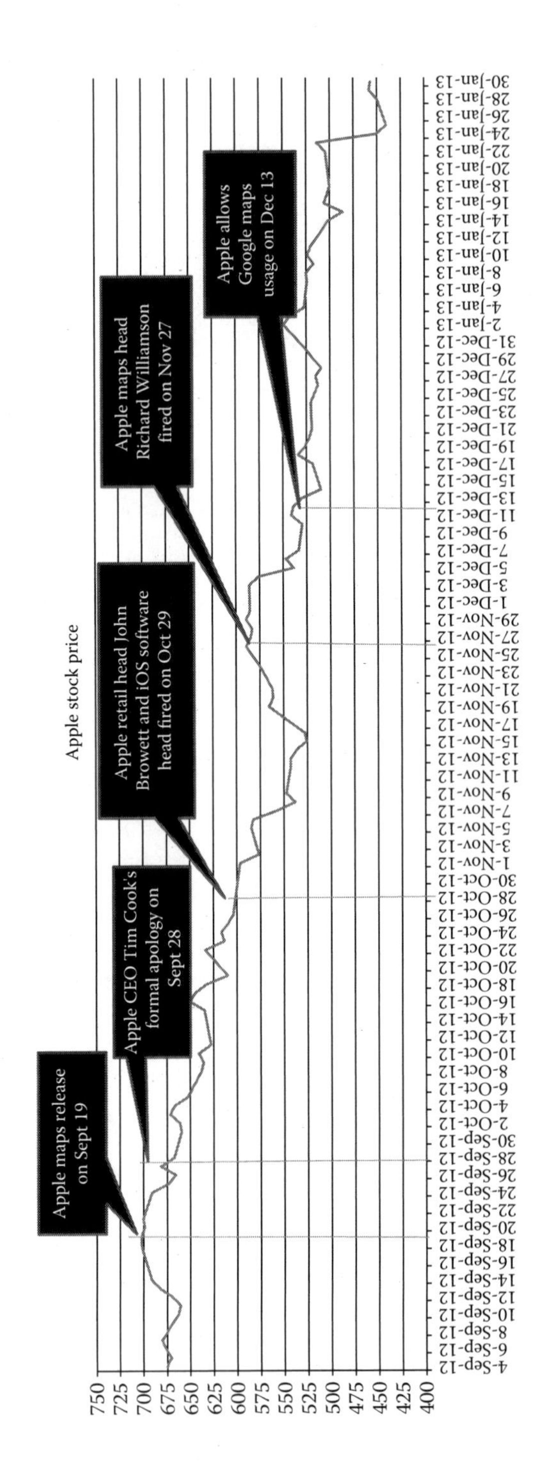
Apple stock price
Apple maps release on Sept 19
Apple CEO Tim Cook's formal apology on Sept 28
Apple retail head John Browett and iOS software head fired on Oct 29
Apple maps head Richard Williamson fired on Nov 27
Apple allows Google maps usage on Dec 13
750
725
700
675
650
625
600
575
550
525
500
475
450
425
400
4-Sep-12
6-Sep-12
8-Sep-12
10-Sep-12
12-Sep-12
14-Sep-12
16-Sep-12
18-Sep-12
20-Sep-12
22-Sep-12
24-Sep-12
26-Sep-12
28-Sep-12
30-Sep-12
2-Oct-12
4-Oct-12
6-Oct-12
8-Oct-12
10-Oct-12
12-Oct-12
14-Oct-12
16-Oct-12
18-Oct-12
20-Oct-12
22-Oct-12
24-Oct-12
26-Oct-12
28-Oct-12
30-Oct-12
1-Nov-12
3-Nov-12
5-Nov-12
7-Nov-12
9-Nov-12
11-Nov-12
13-Nov-12
15-Nov-12
17-Nov-12
19-Nov-12
21-Nov-12
23-Nov-12
25-Nov-12
27-Nov-12
29-Nov-12
1-Dec-12
3-Dec-12
5-Dec-12
7-Dec-12
9-Dec-12
11-Dec-12
13-Dec-12
15-Dec-12
17-Dec-12
19-Dec-12
21-Dec-12
23-Dec-12
25-Dec-12
27-Dec-12
29-Dec-12
31-Dec-12
2-Jan-13
4-Jan-13
6-Jan-13
8-Jan-13
10-Jan-13
12-Jan-13
14-Jan-13
16-Jan-13
18-Jan-13
20-Jan-13
22-Jan-13
24-Jan-13
26-Jan-13
28-Jan-13
30-Jan-13
Apple stock price

(a)

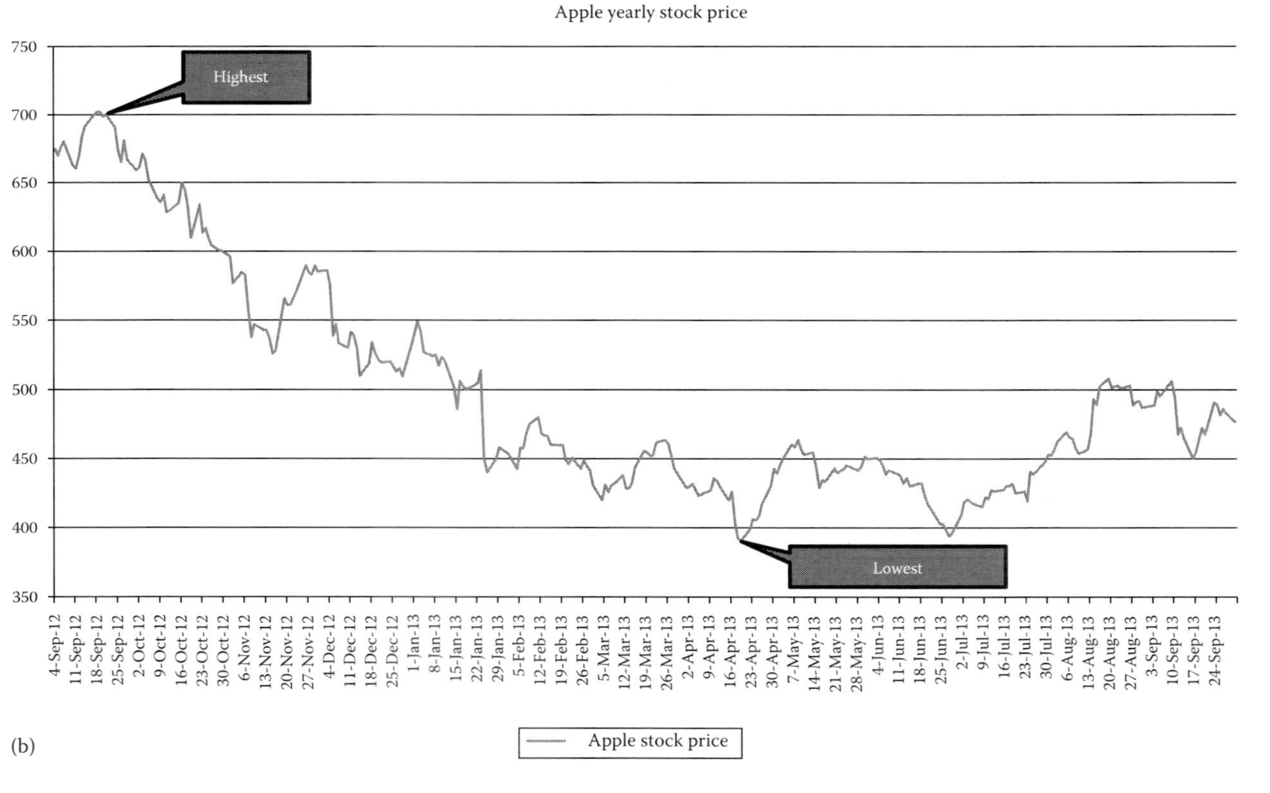

Figure 21.1 Apple stock price. (a) x-axis: Important AM-related dates, 2012 and 2013. (b) y-axis: Price per share of Apple stock. (From http://finance.yahoo.com/. Accessed March 10, 2014.)

21.7 Conclusion

The mistakes that Apple made, with respect to AM, was not in attempting to take on GM and BM directly. If Apple had not launched AM, it would have continued to lose advertising revenue, POI, and search results. Being *third*, not to mention starting to fall behind new mapping *aps* in India and China was not a good strategy. Where Apple made the worst blunder in its corporate history was in trying to do in only a few years what had taken rivals over a decade. AM was also underinvested and oversold. It was not an adequate strategy to acquire Tom Tom data rights and buy Yelp, C3 Technologies, etc. Apple needed to invest tens of billions of dollars and millions of person hours in gathering geospatial data. It needed to buy more rights to high-resolution imagery from SPOT or Digital Globe. It needed to purchase the best available data whether commercial data like Mapion or Increment P Corporation data in Japan or public (though far from free) data like the Ordnance Survey data in the United Kingdom or government data in Singapore and Taiwan. Apple also needed to better understand that geospatial data is error prone and rather than roll out AM with unjustified hubris, it should have anticipated a flood of error reports and created a mechanism for users to report errors and be rewarded for doing so (like Google has). This is particularly true for businesses that were misplaced or miscategorized. Instead, Apple shifted blame to vendors like Tom Tom and Yelp who in return shifted blame back to Apple. Proper beta testing should also have revealed the vast number of iconic errors that permeate AM, many of which could have been easily corrected.

References

ABC News (2012), Apple maps strands motorists looking for Mildura, http://www.abc.net.au/news/2012-12-10/apple-maps-strands-motorists-looking-for-mildura/4418400. Accessed March 14, 2014.

Apple (2013), Apple maps home page, http://www.apple.com/ios/maps/. Accessed March 15, 2014.

Cairncross, F. (2006), *The Death of Distance: How the Communications Revolution Is Changing Our Lives*, Harvard University Business School Press, Boston, MA.

Chen, B. and N. Wingfield (2012), Apple fires a manager over its misfire on maps, *New York Times*, November 29, 2012.

C3 Technology (2011), Background on C3 technologies, http://9to5mac.com/2011/10/29/apple-acquired-mind-blowing-3d-mapping-company-c3-technologies-looking-to-take-ios-maps-to-the-next-level/. Accessed March 12, 2014.

CNN (2012a), Initial reaction to AM, http://www.cnn.com/2012/09/20/tech/mobile/Apple-maps-complaints/index.html. Accessed March 10, 2014.

CNN (2012b), Forestall leaving Apple, http://www.cnn.com/2012/10/29/tech/mobile/forstall-leaving-Apple/index.html. Accessed March 15, 2014.

Forbes (2012), Apple acquisition of Tom Tom, http://www.forbes.com/sites/timworstall/2012/12/12/is-Apple-going-to-buy-tomtom-to-sort-out-the-maps/. Accessed March 14, 2014.

Google (2013), Google mapmaker homepage, http://www.google.com/mapmaker. Accessed March 15, 2014.

Huffington Post (2013), Cuts hit golden state hard, February 25, 2013, http://www.huffingtonpost.com/2013/02/25/california-cuts_n_2762028.html. Accessed March 12, 2014.

Huffington Post Business (2013), Wall street 24/7: The worst new product flops of 2012, http://www.huffingtonpost.com/2012/12/23/worst-product-flops_n_2347424.html. Accessed March 10, 2014.

Isaacson, W. (2011), *Steve Jobs*, Simon & Schuster, New York, pp. 21–26.

Keizer G. (2013), Microsoft exec scoffs at Apple, *Computerworld*, October 23, 2013, pp. 3–5. http://www.computerworld.com.au/article/529912/microsoft_exec_scoffs_talk_apple_free_iwork_threatens_office/. Accessed March 8, 2014.

Leipnik, M. and S. Mehta (2004,) Geographic information systems in electronic marketing, in *Advances in Electronic Marketing*, T.B. Flaherty and I. Clarke III (eds.), Idea Group Publication, Hershey, PA.

Leipnik, M., X. Ye, and G. Gong (2011), Geo-spatial technologies and policy issues in china: status and prospects, *Regional Science Policy and Practice*, 3(4), 339–356.

Lemeshewsky, G.P. (2004), Visual enhancement of multispectral imagery, *Proceedings of the International Society for Optical Engineering*, San Diego, CA http://pubs.er.usgs.gov/publication/70026546. Accessed March 9, 2014.

Main Street (2011), The real value of a yelp review, http://www.mainstreet.com/article/lifestyle/food-drink/real-value-yelp-review.

Navteq (2013), Navteq maps, http://www.navteq.com/company.htm.

NGA (2013), Vmap 0 data, National Geospatial-Intelligence Agency, http://earth-info.nga.mil/publications/vmap0.html. Accessed March 8, 2014.

New York Times (2010), Executive leaves after iPhone trouble, By Miguel Helft, August 7, 2010, http://www.nytimes.com/2010/08/08/technology/08apple.html. Accessed March 13, 2014.

Nine-to-Five Mac (2013) Apple releases updates to maps of Japan without 6.1.3 update, http://9to5mac.com/2013/03/11/apple-releases-enhancements-to-maps-in-japan-without-6-1-3-update/. Accessed March 16, 2014.

Parr, P. (November 1, 2006), Landscape visualizations: Applications and requirements for 3D visualization software for environmental planning, *Computers Environment and Urban Systems*, 30(6), 815–839.

Phone Arena (2011), Americans replace their cell phones every 2 years, Finns—Every six, a study claims, by Victor H, July 11, 2011, http://www.phonearena.com/news/Americans-replace-their-cell-phones-every-2-years-Finns—every-six-a-study-claims_id20255. Accessed March 17, 2014.

Rees, W.G. (2001), *Physical Principles of Remote Sensing*, 2nd ed., Cambridge University Press, London, U.K., p. 343.

Schmidt, R. (2012), Interactive modeling with mesh surfaces, *SIGGRAPH On-Line Conference Proceedings*, Los Angeles, CA, http://s2012.siggraph.org/attendees/sessions/100-152. Accessed February 6, 2014.

Schmidt, R. (2013), Personal communication from Autodesk, Inc. Senior Scientist Ryan Schmidt commenting on cause of draping errors in "3D" maps.

Sovocool, D. (2008), GPS: Charting new terrain: Legal issues related to GPS-based navigation and location systems, Reuters Findlaw, http://corporate.findlaw.com/litigation-disputes/gps-charting-new-terrain-legal-issues-related-to-gps-based.html#sthash.TY6wZCHt.dpuf. Accessed March 8, 2014.

Tabuchi, H. (2012), Apple maps errors send Japanese to homegrown app, *New York Times*, November 24, 2012, p. B1.

The Guardian (2013), Apple map makes users sick, September 27, 2013, p. C1, http://www.theguardian.com/technology/2013/sep/27/ios-7-motion-sickness-nausea. Accessed March 14, 2014.

Time Magazine (2013), Apple earnings may disappoint everybody today, October 28, 2013, pp. 37–38, http://business.time.com/2013/10/28/apple-earnings-may-disappoint-everybody-today/. Accessed March 14, 2014.

Tom Tom (2013), The process of collecting POI data, http://www.tomtom.com/en_us/legal/privacy/.

Wall Street Journal (2009), GM ready to sell bankrupt SAAB unit, October 12, 2009, p. B3, http://www.marketwatch.com/story/gm-reportedly-nears-deal-to-sell-saab. Accessed March 7, 2014.

Wingfield, N. and N. Bilton (2012), In shake-up, Apple's mobile software and retail chiefs to depart, *New York Times*, October 29, 2012, p. B1.

YouTube (2011), Comparison of Apple and Google image rendering capabilities, http://www.youtube.com/watch?v=JyMU0Luu6E8. Accessed March 12, 2014.

Index

L

M

N

W